SECRET

THE SECOND WORLD WAR
1939—1945
ARMY

SPECIAL WEAPONS AND TYPES OF WARFARE

VOLUME I—GAS WARFARE

COMPILED BY
LIEUTENANT-COLONEL D. J. C. WISEMAN

THE WAR OFFICE
1951

The Naval & Military Press Ltd

Published by

The Naval & Military Press Ltd
Unit 10 Ridgewood Industrial Park,
Uckfield, East Sussex,
TN22 5QE England

Tel: +44 (0) 1825 749494
Fax: +44 (0) 1825 765701

www.naval-military-press.com
www.military-genealogy.com

ROYAL ARMOURIES

The Library & Archives Department at the Royal Armouries Museum, Leeds, specialises in the history and development of armour and weapons from earliest times to the present day. Material relating to the development of artillery and modern fortifications is held at the Royal Armouries Museum, Fort Nelson.

For further information contact:
Royal Armouries Museum, Library, Armouries Drive, Leeds, West Yorkshire LS10 1LT
Royal Armouries, Library, Fort Nelson, Down End Road, Fareham PO17 6AN

Or visit the Museum's website at
www.armouries.org.uk

FOREWORD

This book is one of a series of volumes, compiled by authority of the Army Council, the object of which is to preserve the experience gained during the Second World War, 1939–1945, in selected fields of military staff work and administration. The author has been given access to official sources of information, and every endeavour has been made to ensure the accuracy of the work as a historical record. Any views expressed and conclusions drawn are those of the author, and do not necessarily reflect those of the Army Council, which, so far as they relate to current training, are to be found in the official manuals, training memoranda, etc., issued from time to time by the War Office.

For the operational background, the reader is referred to the Official History of the War.

THE WAR OFFICE,
1951.

(C29058) B 2

CONTENTS

PAGE

CONTENTS—*continued*

CHAPTER II.—ORGANIZATION

CONTENTS—*continued*

(C29058) B 4

CONTENTS—*continued*

APPENDICES—*continued* PAGE

ABBREVIATIONS

A.A.	Anti-Aircraft.
A.D.G.B.	Air Defence of Great Britain.
A.F.V.	Armoured Fighting Vehicle.
A.O.D.	Advance Ordnance Depot.
A.R.P.	Air Raid Precautions.
A.T.S.	Auxiliary Territorial Service.
B.C.V.	Bulk Contamination Vehicle.
B.E.F.	British Expeditionary Force.
B.E. shell	Base Ejection Shell
C.C.D.D.	Controller, Chemical Defence Development.
C.D.E.S.	Chemical Defence Experimental Station.
C.D.R.D.	Chemical Defence Research Department.
C.I.G.S.	Chief of the Imperial General Staff.
C.I.O.S.	Combined Intelligence Objectives Sub-committee.
C.W.	Chemical Warfare.
C.W.S.	Chemical Warfare Service (U.S.A.).
D.C.W.	Director of Chemical Warfare.
D.E.-in-C.	Deputy Engineer-in-Chief.
D.G. of A.	Director General of Artillery.
D.G.M.P.	Director General of Munitions Production.
D.M.I.	Director of Military Intelligence.
D.M.O.	Director of Military Operations.
D.M.T.	Director of Military Training.
D.R.A.	Director of Royal Artillery.
D.S.D.	Director of Staff Duties.
D.S.W.V.	Director of Special Weapons and Vehicles.
D.W.V.	Director of Weapons and Vehicles.
E.-in-C.	Engineer-in-Chief.
E.T.O.U.S.A.	European Theatre of Operations United States Army.
F.F.C.	Field Force Committee.
F.I.A.T.	Field Intelligence Army Technical (U.S.A.).
G.H.Q.	General Headquarters.
G.I.O.	Gas Identification Officer (Civil).
G.O.C.-in-C.	General Officer Commanding-in-Chief.
G.S.O.	General Staff Officer.
H.E.	High Explosive.
I.S.C.C.W.	Inter Service Committee on Chemical Warfare.
L. of C.	Lines of Communication.
L.M.G.	Light Machine Gun.
M.A.P.	Ministry of Aircraft Production.
M.G.O.	Master General of Ordnance.
M.I. Room	Medical Inspection Room.
M.T.	Mechanical Transport.
O.C.T.U.	Officer Cadet Training Unit.
O.W.P.C.	Organization and Weapons Policy Committee.
P.A.D.	Passive Air Defence.
Q.M.G.	Quarter-Master-General.
R.A.C.	Royal Armoured Corps.
R.A.E.	Royal Australian Engineers.

ABBREVIATIONS—*continued*

R.A.F.	Royal Air Force.
R.A.O.C.	Royal Army Ordnance Corps.
R.A.S.C.	Royal Army Service Corps.
R.C.E.	Royal Canadian Engineers.
S.A.P.	Semi-Armour Piercing.
S.C.I.	Smoke Curtain Installation.
S.H.A.E.F.	Supreme Headquarters Allied Expeditionary Force.
S.S.	Schutz-Staffel (German special guards).
U.P.	Unrotating Projectile.
" V.E. " day	Termination of hostilities in Europe.
" V.J." day	Termination of hostilities in Japan.
V.Q.M.G.	Vice Quarter-Master-General.
W.E.	War Establishment.

CHAPTER I.—MAJOR POLICY

SECTION 1.—1925 TO THE OUTBREAK OF WAR

The Geneva Protocol, 1925

Since 1925, the policy of the British Government has been governed by the Geneva Gas Protocol which prohibited the use in war of asphyxiating or poisonous or other gases or bacteriological methods of warfare. Most, but not all, of the principal powers subscribed to this, the notable exceptions being the United States of America, and Japan. The British Government reserved the right to retaliate in kind should gas be used against us. It was accepted that the Government would endeavour, at the beginning of a war, to obtain from the enemy government, if it was not a party to the Geneva Protocol, an engagement not to use offensive gas as a weapon of war. In December, 1938, it was decided that it would be advisable, before the outbreak of a war, to obtain such an assurance from the enemy Government, irrespective of whether it was a signatory to the convention or not, since this would serve to accentuate the enemy's responsibility should he refuse to give an undertaking, or having given it, break his word. The last official communication of our Ambassador in Berlin, Sir Neville Henderson, was a request for such an assurance from the German Government. This was later given through the Swiss Minister in London on the understanding that we should similarly observe the provisions of the Protocol. Our own intention to abide by the terms of the Protocol had been reaffirmed in the Anglo-French declaration on the methods of warfare of 2nd September, 1939, with a similar proviso.

Defensive Policy—1925 Onwards

No country could allow its safety to be dependent on the observance by other states of rules of war to which they were pledged. This particularly applied to gas warfare, where no agreement or conventions could stop the discovery of new gases or circumvent the ease with which any nation, bent on doing so, could evade international prohibitions or limitations on the manufacture of war chemicals in peace.

Our policy from 1925 onwards, both in research and training, was primarily directed to the defensive. The ability of every man in the Fighting Services to perform all his duties by day or night while protected against gas has always been a fundamental principle underlying the training of the Army, and defensive equipment has been provided accordingly. Initially this consisted of the respirator only, but as circumstances changed and the air threat from blister gas developed, a correspondingly wider range of equipment was provided (*see* Chapter VI).

Offensive Policy—1925 to 1939

In regard to the offensive use of gas, research continued. In so far as the Services were concerned, Government policy, laid down in 1925, was that Service schools and commands should study the offensive employment of gas, not only because the study of offensive gas was necessary for the study of protection against it, but also in order that we might be in a position to retaliate immediately, should this course be forced upon us by our opponents. Steps were also to be taken to secure the bulk production of gas when required.

For various reasons no action was taken to implement this policy. For a time the study of offensive gas was banned, and it was not resumed until 1936. In that same year, an interdepartmental committee was set up to consider plans for producing, storing and charging the gas required for Army and R.A.F. use.

Pilot plant had already been designed and was in operation at Sutton Oak ; during 1937–1938 a factory to manufacture mustard gas was established at Randle (Merseyside), initially on a small scale, pending final consideration by the Cabinet on the capacity it was desirable to establish in peacetime in the light of the strategic and political considerations involved.

By this time Germany had emerged as our most likely opponent in a future war, and German gas warfare preparations were, therefore, the " yardstick " with which to measure our own preparedness. Intelligence indicated that Germany was making the most thorough preparations for the employment of gas in war and in this respect was ahead of this country. The form of her preparations showed that she had the capability (though not necessarily the intention) of employing gas against the civil population.

It was considered that retaliation in kind would be essential should gas be used against us. Since our major gas effort would have been made from the air, early planning visualized 90 per cent. of gas production being allocated to the R.A.F. and 10 per cent. to the Army. The problems connected with retaliation are, therefore, reviewed below from the angle of the major gas arm.

Gas was visualized not as an alternative to, but as supplementary to, high explosive bombs and incendiaries. It was thought that its use, in conjunction with these, against a civil population would unquestionably hasten the dislocation, disorganization and demoralization of the community. Further, defence against gas would necessitate diversion of a considerable portion of available effort and resources from offensive to defensive activities. It was not part of our strategic policy to direct such attacks against the civil population, unless such a policy was forced on us by way of reprisal. Our policy was that retaliatory gas attacks would be directed only against limited military objectives ; against such objectives gas would augment the effectiveness of our attacks to a degree unobtainable by other means. It had to be assumed that, in a war against a first class power, the maximum effort in the air would be deployed from the outset, and the means could not, therefore, be found further to increase our bombing effort by way of retaliation for the use of gas. On the other hand, the effect of our air effort could be increased by employing gas in substitution for a proportion of high explosive and incendiary bombs.

Although gas in conjunction with high explosive and incendiary bombs increased the effectiveness of these weapons, gas was also regarded as a potent weapon in itself—particularly in the effect of aerial gas attack on land forces.

High altitude spray attacks against all echelons of an army in the field, with the exception of the foremost area, would seriously reduce mobility. Apart from the casualties which might be caused, the necessity for changing and decontaminating clothing and for observing passive air defence measures, such as taking cover and wearing anti-gas clothing, would impose delay and cause severe dislocation and loss of fighting value. Low spray attacks on concentrations of troops at base ports, entraining or embussing centres, or on columns on the march, would produce considerable casualties. In addition contaminated areas would have to be contended with.

It was therefore considered imperative that we should be in a position to use gas from the air against the enemy's ground forces immediately it was employed against our armies, since the advantage which might accrue to the enemy if we were unable to do so might well prove decisive. Quite apart from the desirability, from the point of view of morale, of turning against the enemy the weapon he had chosen to employ, important strategic and tactical aims could be served in so doing.

The capacity for retaliation in relation to the amount of gas required by the R.A.F. was based on the proportion of 25 per cent. gas to 75 per cent. H.E. and incendiary. A gas provisioning policy based on 25 per cent. of the bomb lift would have enabled a comprehensive range of targets to be engaged effectively with gas and would have permitted heavy gas concentrations at frequent intervals against targets particularly vulnerable to gas attack. On this basis R.A.F. requirements alone were estimated at 470 tons a week.

On the other hand, in view of our policy to refrain from using gas unless it was first used against us, it was necessary that any gas provided should be additional to, and not in substitution for, high explosive and incendiary bombs. It would have been illogical to reduce our offensive or defensive capacity in more important directions in order to include an ideal scale of provision for a weapon which it was hoped would not be used. It was considered that gas provision was a direction in which some risk might legitimately be taken ; an arbitrary figure of 50 per cent. of the ideal requirement was, therefore, advocated.

On the political side it was not felt that the public would be likely to be perturbed by the knowledge that we had the means for effective retaliation. If anything, this knowledge would have a reassuring effect at home. On the whole public reaction to A.R.P. measures had been apathetic rather than alarmist. The provision of gas with which to reply to enemy attacks was a natural corollary to the provision of gas masks for our own protection.

Reaction abroad was thought likely to do nothing but good. The knowledge that we were in a position to retaliate was felt to be a strong deterrent to potential enemies initiating gas warfare. On the other hand, positive knowledge that we did not possess gas was likely to be dangerous and to encourage its use against us, since the enemy might be tempted to use it at the outbreak of war in the hope of gaining the advantage through surprise and of forcing us into submission before we had time to develop our manufacturing resources. The correctness of such views was proved after the war when it was confirmed from enemy documents that Allied gas potential and preparedness had been a weighty factor in influencing the German High Command against initiating gas warfare, particularly in the closing phases of the war (*see* Chapter VIII).

In November, 1938, in the light of the appreciation outlined above, the Cabinet agreed that provision for the use of gas as a retaliatory measure was necessary, and authorized the completion of the Randle Works so that they could produce a total weekly output of 300 tons of mustard gas. It was realized that it would be 12 to 18 months before this capacity could be established. A war reserve of 2,000 tons of mustard gas was also authorized in order to bridge the period while production was being increased to full war output. A small pilot plant existed for the manufacture of tear gas to meet Army requirements. Our production policy in relation to types of war gases is referred to in Chapter III, Section 5.

The above Cabinet decision only related to the arrangements to be made in the United Kingdom and did not include preparatory measures in other parts of the world where British forces were stationed in peace. In May, 1939, it was decided to extend these measures to provide the necessary equipment for retaliation by British forces in Egypt. From the strictly military standpoint of the possible effectiveness of gas under Middle East conditions of terrain and types of target, there was not a strong case for the provision of equipment and the storage of gas in Egypt, but political considerations outweighed all others. Our Ambassador in Cairo, the G.O.C.-in-C. in Egypt and the Egyptian Prime Minister were unanimous that, in the light of the information that the Italians in Libya were ready to use gas and had the means at their disposal, the absence of the power to retaliate immediately would have a deplorable effect on the morale of the Egyptian population as well as of our own troops. It had already been decided that a knowledge of United Kingdom preparedness might deter the Germans from using gas, and similarly there was no reason to doubt that knowledge of gas preparedness in Egypt would deter the Italians from using gas in the event of war.

Policy in relation to offensive preparedness has been discussed above from the point of view of the major gas arm—the R.A.F. Army policy for the offensive use of gas is discussed in Chapter II, Section III, Chapter IV and Chapter V. Naval policy was not to use gas against enemy ships. The object of Naval warfare is to sink the enemy, and the use of gas shell at the expense of high explosive was, therefore, deprecated. Designs for chemical shell for 6-in. Cruiser and 4·7-in. Destroyer guns were being developed for possible use against shore positions when supporting a landing. Trials were in progress on low priority so that the final designs could embody any features found necessary as a result of Army trials with base ejection shell then being undertaken. If it had been found necessary to use Fleet Air Arm aircraft for gas attack the weapons developed by the R.A.F. would have been used.

Liaison with the French became well established shortly before the war. In the summer of 1939 a French mission visited England and staff conversations took place in which all the cards were put on the table as regards research, production methods, and stocks of offensive and defensive gas stores, and in which the whole tactical and strategical doctrine underlying the gas policy of both nations was reviewed.

SECTION 2.—THE OUTBREAK OF WAR TO DUNKIRK

The position on the outbreak of War

The war came upon us before our plans for production of gas and gas weapons were implemented. The almost complete lack of immediately available gas weapons on 3rd September, 1939, can be seen from the details given in Appendix II.

From the defensive aspect the position was better. Although provision was still in process of being extended to meet the expansion of the Territorial Army and the introduction of militia training which had begun earlier in the year, the first contingent of the Field Force were complete in defensive equipment except for certain items of anti-gas clothing for decontamination tasks in two divisions. A.A. units of Air Defence Great Britain had respirators, capes and eyeshields, the remainder of the Regular Army and the Territorial Army (less A.D.G.B.) had training but no war equipment, and the militia had respirators only. Reserves were incomplete. Overseas garrisons (other than India) were complete in anti-gas equipment except for Malaya, Ceylon, Bermuda and Jamaica. In India reserves for troops earmarked for overseas were complete in essential items such as respirators, but even a small training scale of respirators was far from complete for the remainder of the Army in India.

Had the Germans initiated gas warfare in September, 1939, our immediate ability to retaliate would have been negligible. Our stock of mustard gas was 500 tons and plant capacity had only reached 90 tons a week. A beginning had been made in August, 1939, to charge 250-lb. mustard gas aircraft bombs. The charging of 250-lb. S.C.I.s for aircraft spray began in September, 1939, and of 500-lb. S.C.I.s and of 30-lb. bombs in October, 1939. Army weapons were non-existent except for some 4·5-in. and 6-in. Howitzer empties of the 1914–18 war pattern which were sent for charging and became available between October and December, 1939.

In October, 1939, the Cabinet approved proposals, which had been under consideration by Service Departments immediately before the war in relation to the increased size of the Army and R.A.F. then planned, to increase war gas production as shown in Appendix II. This involved the establishment of new factories and increased storage facilities at dispersed sites.

In November, 1939, the War Cabinet approved the despatch of gas and equipment to France on the understanding that every possible precaution would be taken to prevent observation. The necessary R.A.F. and R.A.O.C. personnel trained in handling gas munitions were despatched to France at the end of that month and the Base Ammunition Depot at Fécamp was ready to receive these munitions early in December.

In October, 1939, there was a scare that the Germans might use Arsine, against which our respirators gave inferior protection ; immediate steps were taken to rectify this. Details of the scare and of the measures taken will be found in Chapters III and VI.

April 1940—Position if Germans had used Gas in Invasion of the Low Countries

In April, 1940, the position was reviewed in relation to the possibility that the Germans might initiate gas warfare in an attempt to break through on the Western Front. The fact that they had up till then not used gas implied little but that they had not considered it expedient to do so. It was felt certain that they would use it if and when they considered that, if gas were used by both sides, the balance of advantage would be in Germany's favour. It is now known that the Germans did not want gas warfare to break out at this stage because they thought it would slow up their anticipated rapid advance.

The War Cabinet had given covering authority to the Chiefs of Staff to put into effect pre-arranged plans for checking, by air action, a German advance into the Low Countries ; it was suggested that similar authority should be given to the commanders concerned to use gas in retaliation (including high spray) without further reference to the War Cabinet. In such circumstances immediate retaliation was considered vital since, apart from considerations of morale, a tactical advantage would have accrued to the Germans had they been allowed to use a weapon of which one of the main effects was the slowing down of all movement without a corresponding brake being applied—particularly at a time when each side would require to hamper the forward movement of the other. There was not, however, an exact analogy between the covering authority given for air action and the proposal for similar authority to include retaliation with gas including high spray. High spray is an indiscriminate weapon and might have done serious harm to the civil population. In an invasion of the Low Countries it would have inevitably entailed the gassing of Belgian and Dutch civilians. On the other hand, local retaliation with gas bombs, low spray and such shells as were available, against strictly military objectives would have meant the abandonment of what was then considered to be our principal air gas weapon.

A possible alternative would have been to use high spray only at night, when the majority of civilians would have been under cover, and low spray by day. In any case it was the prevalent opinion that night was the best opportunity to use high spray, since considerable movement was to be expected then and the factor of surprise could be best exploited.

It was appreciated that the enemy would rapidly take precautions which would reduce the casualty effect of spray, and for this reason the maximum effects were to be expected during initial attacks. It was the view of the Air Officer Commanding-in-Chief the British Air Forces in France, in agreement with Army Commanders, that the maximum effort should be devoted to the use of spray for a short period rather than a sustained effort for a longer period, in order to take full advantage of the factor of surprise. At the same time attacks would have been made with gas bombs on railheads, bridges, etc. where high concentrations of blister gas would have been likely to impede the enemy's operations.

The position regarding supplies of air gas weapons had improved. As far as ground weapons were concerned, although we could have added, if required, a very small quota to operations by the French, no large scale gas operations would have been possible by the Army—nor was there any likelihood of this before the spring of 1941. The stock position of air and ground gas weapons in April, 1940, is shown at Appendix II.

In the air at maximum effort our stocks of S.C.Is. would have lasted one week—after which they would have had to be returned to England for recharging. A period of a month would have elapsed before they could have arrived back, and in this interval a sustained effort would have been maintained with 250-lb. and 30-lb. mustard gas bombs, although there were then some limitations to the types of aircraft on which the latter could have been used.

The fact that at the moment spray weapons were in better supply to the B.E.F. than mustard gas bombs added force to the proposals outlined above. Had experience shown that spray was a less worth-while weapon than was then assumed, as was recognized later in the war (*see* Chapter V), the improving supply of mustard gas bombs would have enabled us to transfer a greater weight to that form of attack, although for some time our total gas effort would have been likely to be below the optimum.

While the above plans were in course of discussion and submission to the War Cabinet for approval the invasion of the Low Countries occurred without gas being used. The pace of events was such that no decision was taken, before the evacuation from Dunkirk, regarding the delegation to commanders of authority to retaliate.

SECTION 3.—THE LATTER PART OF 1940

Our Own Capacity to Wage Gas Warfare

The major part of our stocks of gas weapons were evacuated from France, and these formed the bulk of stocks in the United Kingdom immediately available after Dunkirk. The Livens projector equipment of the chemical warfare companies was lost. A special effort to evacuate our spray apparatuses was made since it was thought that the means of high altitude spray were an Allied secret. All these apparatuses were evacuated with the exception of two S.C.Is. which had been given to the French Government for experimental purposes.

Although large stocks of defensive equipment had been lost in France, production capacity was such that deficiencies were quickly made good (weekly output of the chief items between April and December, 1940 approximately averaged :—Respirators 75,000, Eyeshields (singles) 900,000, Ointment 600,000 tins). Personal equipment was carried to the end and much of it brought to England. The carrying of it had not been found to be harassing, partly because men were in the habit of carrying it and partly owing to the sense of protection afforded by it.

Our immediate concern was with the possibility of invasion and of the Germans initiating the use of gas in this connection, or alternatively with the possibility of the Germans using gas, without invasion, against military objectives and against the civil population in the United Kingdom. The entry of Italy into the war also necessitated taking stock of our gas position in the Mediterranean.

The need was felt for a body to co-ordinate, between the three Services, plans for retaliation should this become necessary, and also to keep the position in relation to our resources and the best means of their employment under constant review. The Inter-Service Committee on Chemical Warfare was, therefore, set up in September, 1940, as a War Cabinet Sub-Committee to advise, and report to, the Chiefs of Staff Committee. The first survey of the position by this Committee revealed that, although our ability to retaliate from the air had increased since April, 1940, the deficiency of Army gas weapons was lamentable, and our production capacity was much below that authorized by the War Cabinet in October, 1939.

The Prime Minister was greatly concerned at the position which he considered to be most dangerous ; for a time weekly progress reports were submitted to him. An enquiry was undertaken by the Lord Privy Seal (Mr. C. R. Attlee) to discover the reasons for the lack of storage capacity which had been authorized in 1938 and also for the lack of empty weapons for charging. This enquiry found that no blame could rightly be attributed to any one individual, but that the fault mainly lay in serious defects in organization. There had been a tendency to set up committees of experts who substituted enquiry for action, and the sum total of their recommendations was a counsel of perfection. There was no one person who insisted on a workable solution being quickly found and driven forward to a conclusion. There had been many changes in the organization of Departments and responsibility had become sub-divided in such a way that no one person below the Minister could exercise any real control. Measures were taken to rectify these organizational defects.

War gas production had lagged owing to lack of storage capacity and lack of empty weapons for charging. These hindrances were particularly felt in gas shell production. The Ministry of Supply had been asked to provide ten 25-pounder gas shell for every hundred shells of other types. Apart from set-backs in developing a satisfactory design, the greatest difficulty had been experienced in meeting the demands for 25-pounder H.E. shell, and, in consequence, supply of empty gas shell had lagged.

In the case of gas storage, exaggerated views had been taken about the dangers of gas and the need for secrecy. The protection to be provided for bulk storage of gas had been decided on, in 1938, as protection against a 500-lb. S.A.P. bomb. This had led to the drawing up of very elaborate requirements. A prolonged search for a suitable tunnelled site ensued and work had not been started on the main storage project until after the outbreak of war. There had been a lack of balance between an absolutely satisfactory scheme and the completion of the work within a reasonable period.

The stock position as at 1st October, 1940, is given at Appendix II. Phosgene had not up till then been a charging for air bombs, but steps were taken so to charge a proportion of 250-lb. bombs and all phosgene which could be made available for gas warfare purposes was diverted to charging these bombs during the winter of 1940–41. Reasonable stocks of Livens drums charged phosgene were available, but the 5-inch rocket had not yet been perfected. Production capacity of phosgene was small and the majority of existing output was required for essential war purposes other than chemical warfare.

Had gas warfare started at this juncture, our offensive effort with mustard gas would have been limited by the rate at which we could have produced the empty shells and bombs and recharged spray containers, rather than by our capacity for producing gas. This production capacity, together with existing stocks, was sufficient to charge all R.A.F. and Army weapons as soon as they were produced, provided that output was not seriously reduced by air bombardment. Owing to the lag in the establishment of the new and dispersed factories and storage facilities for mustard gas, the majority of our production of this gas suitable for aircraft spray, almost all the charging apparatus, and about half the reserve storage were concentrated at one place—Randle. The plant was easily identifiable and had the Germans initiated the use of gas they would almost certainly have tried to knock out the plant, although the lay-out of the gas in store was such that a heavy scale of attack would have been required to have seriously affected the gas stored. A direct hit on the charging plant would, however, have been serious. Had such an attack been made we should have been reduced to the use of mustard gas bombs, for which reserves of mustard gas were stored in the Welsh hills, where plant was also then coming into operation.

Appreciation of Enemy Intentions

Although there were numerous reports from secret sources about the preparation for gas warfare by Germany, the evidence did not justify the conclusion that the use of gas was more imminent then than at any period since the beginning of the war. It was appreciated that the Germans were well prepared offensively and defensively, although it seemed probable that the defence of their civilian population, particularly in relation to the availability of civilian respirators, was efficiently organized only in the more vulnerable areas of the Western Zone.

It was felt that their use of gas would probably depend on whether or not they considered we were in a position to retaliate effectively. If they had had reason to doubt the effectiveness of our power of retaliation they might well have employed gas as the main surprise factor in any attempted invasion of this country. Our Air Staff held the opinion that the Germans were capable of attacking on a heavy scale ; on the basis of 50 per cent. of the total offensive load, it was assessed that they were capable at maximum effort for two or three days of up to 2,500 tons of gas a day, with a sustained effort for a period up to a month of some 880 tons a day. In some quarters this was considered to be an over-estimate.(1)

The view of the Air Staff was that, in support of an invasion, the main weight of attack would be concentrated in very heavy raids on a limited number of military objectives. Although the civil population were likely to be involved in attacks on focal points, gas would not primarily be employed against them or against industrial areas, unless the Germans considered that owing to lack of protective measures the effect on morale would be serious. Military objectives would be best attacked by a combination of spray and gas bombs, and, assuming that German views were likely to coincide with those of our own Air Staff, *viz.*, that spray was primarily a weapon for use against an army in the field, the principal weapon against civilian and industrial targets would be the gas bomb, which would be most effective if combined with high explosive and incendiary. Low spray, however, might be directed at such objectives as docks and ports.

Our Defensive Position

Had the weakness of our offensive potential become known to the Germans, it might have encouraged them to initiate gas warfare, particularly if our defensive preparedness was also assessed as being of a low standard. It was therefore essential that our defensive measures should be tightened up where necessary. On the whole our defensive position was good. On the civil side

---

Note (1).—(*a*) Up to the time of writing a search of the mass of captured documents has not revealed any production or stock figures as far back as 1940. Assuming the proportion of the total gas holdings (including gas in bulk) which was charged into air gas weapons at this date to be the same as the proportion so charged of the total holdings captured on conclusion of the war, and by deduction from the total gas stocks shown in a captured document for September, 1943 and January, 1942, it would seem that the sustained scale of air effort of which the Germans were capable at this date was over-estimated by about 75 per cent.

(*b*) As far as can be ascertained, as a result of interrogation of high ranking German officers and such documentary evidence as has come to light, no proposals for the use of gas were ever included in German plans for an invasion of Britain.

the whole population possessed respirators, although some apathy existed and respirators were not universally carried. The Government essential services were provided with some form of protection against mustard gas. The main defects were that the position in London for cleansing contaminated clothing was unsatisfactory owing to lack of co-operation of the London laundries, but the position in other parts of the country was better. Government Departments in London were ill-prepared to deal with considerable numbers of employees arriving in a contaminated state and steps were taken to implement simple cleansing facilities.

The Army was on the whole better protected against gas than any other Army in the world, although impregnated clothing was not yet available (*see* Chapter VI). Reserves were incomplete and until these were brought up to scale little could be spared for the Home Guard. The Home Guard had been provided with one eyeshield and one pot of anti-gas ointment for each man ; the ultimate scale was to be two of each item. Only a token issue of Service respirators had been made, but all the Home Guard were equipped with civilian respirators, although these would not have stood up to hard wear and tear. Training in the Army in the use of equipment was not of the overall high standard desirable, but the need for this was constantly emphasized.

R.A.F. preparedness corresponded to that in the Army. In the Navy, gas attacks against ships were not to be expected except when in harbour, but seamen, when ashore, were fully provided for. There were some deficiencies in oilskins for mustard gas protection for personnel of shore establishments, and dockyard hands were not in possession of this equipment.

Plans for Retaliation

If the Germans initiated gas warfare our own plans for retaliation were based on two contingencies :—

(*a*) The initiation of gas warfare in conjunction with invasion, which would probably involve attacks on the civil population.

(*b*) The use of gas by the Germans without invasion simply as part of a plan to demoralize our civil population.

Contingency (a)—Invasion

If our own civil population had been attacked, retaliation in kind on the German population would have been undertaken on the largest possible scale at the first favourable opportunity. We should never have begun, but we should certainly have replied in full measure. Had the first use of gas against the civil population coincided with invasion and had this occurred in the autumn of 1940, immediate retaliation on the German population would have been unjustified owing to our limited resources then available. In these circumstances gas would have been used primarily in regard to the military situation, attacks being directed on disembarking troops and those on landing beaches, on aircraft de-planing points, enemy ports and docks supporting the expedition, and on enemy aerodromes for emplaning airborne troops. Detailed plans were drawn up for this eventuality, and mustard gas air weapons (bombs and spray containers) were deployed to specified R.A.F. stations and maintenance units to implement them. No gas weapons were sent to Northern Ireland and, had it been necessary to employ gas against an enemy force invading the Irish Republic, the work would have been undertaken by bomber squadrons based on England. By October, 1940, a small number of phosgene bombs were becoming available, but owing to danger and difficulties in storing them they were not deployed but held centrally until likely to be required.

In using gas in defence of the United Kingdom high altitude spray would have involved dangers to our own civil population and therefore spray attacks would have been mainly low spray against specific targets, except possibly at night or where danger to our own population was unlikely to be involved. The use of phosgene against targets in this country, whether from air bombs or Livens projectors, would also have been restricted for similar reasons. In any case it was recognized that the Livens Projector was of limited value, owing to lack of mobility, and the 5-inch U.P. rocket was still in the development stage. In the defence, widespread contamination would have been undesirable in view of the fluid nature of the probable course of operations, and mustard gas shell of the base ejection type, which were then becoming available, would have been the main Army weapon for ground contamination purposes. A careful record of any contamination would have had to be kept.

A requirement was stated at the time for nose gas artillery shell, but although production of nose gas was eventually established no shells were ultimately so charged.

Contingency (b)—The Gassing of our Civil Population without Invasion

In these circumstances retaliation would have been immediate. The views of the Air Staff on the methods of using gas against a civil population remained the same as in 1938, *viz.*, that gas, H.E. and incendiary were complementary weapons. The incendiary bomb extended the destructive effect of H.E. while gas, particularly mustard gas, in conjunction with these added seriously to the strain on civilian morale, further harassed the A.R.P. workers, and caused a diversion of effort and resources to purely defensive activities. The primary gas weapon against civilians was the air bomb. Spray might be used in certain circumstances but would be a secondary weapon applied to suitable areas with a dense population for its harassing and demoralizing effect.

In the light of these views no detailed plans for using gas against the civil population of Germany were prepared in advance. When the decision had been taken for such retaliation the R.A.F. would have selected those objectives in their current plans where an attack would have reacted most immediately on the civil population. The psychological effect of the inclusion of gas in attacks on centres such as Berlin would have had to be considered in the light of the strategical situation existing at the time. Production of gas bombs was planned to form 25 per cent. of the average bomb lift of the bomber force employed on night operations. Spray operations would have been on a limited scale until the early part of 1941 owing to scarcity of the right types of spray container and aircraft. The shortage of respirators for the civil population in Germany and the reliance placed on gas-proof shelters as a first line of defence indicated that a combination of phosgene and H.E. bombs might prove very effective.

Measures to Prevent the Outbreak of Gas Warfare

During the winter of 1940–41, although our retaliatory capacity was rapidly increasing and the most active steps were taken under the personal instructions of the Prime Minister to rectify the position, the immediate future was a matter of anxiety to the War Cabinet. It would have been gravely to our disadvantage had gas warfare been initiated at this stage. Consideration was given to the possible deterrent effect of the Prime Minister making a public announcement that, although we should never use gas ourselves unless it had first been used against us, we actually had in store very large quantities of deadly gas with which we could immediately retaliate. On the whole it was thought better to say nothing until we had evidence that attack was imminent, since, apart from the fact that there would have been too much bluff in any such statement, the Germans might have said that we had threatened them with gas warfare ; this might have encouraged them to invent a pretext.

The Position in the Middle East

The Army in the Middle East was entirely devoid of any offensive means of waging chemical warfare. Small quantities of R.A.F. gas weapons existed as shown in Appendix II. Action was taken to strengthen the offensive potential of our Middle East Forces—both Army and R.A.F. ; R.A.F. weapons were to be completed to estimated requirements, from new production, by the end of the year, and 50 per cent. of production of all Army chemical weapons was to be sent to the Middle East until five months supply had been built up. The normal standard of stocking with three months supply was considered inadequate owing to the length of the voyage round the Cape.

The effect on gas of Middle East conditions needed study and this was to be undertaken with the assistance of members of the anti-gas laboratory as part of their duties on arrival there.

Chemical warfare staffs were incomplete but were brought up to establishment by the end of the year. On the defensive side there were equipment deficiencies in units and no reserves, and the organization for decontamination was unsatisfactory.

Measures in relation to Possible Use of Gas by the Italians

In November, 1940, the Commander-in-Chief, Middle East, stated that there were reports to the effect that the Italians were contemplating the use of gas against the Ethiopian patriots and that such action might well squash the rebellion unless we retaliated. It would have had a reassuring effect on the Abyssinians if we could have informed them that we would retaliate.

It was decided, however, that we should neither make nor threaten retaliation in kind against the Italians if they made use of gas in Ethiopia and, also, that we should make no public mention of such use if it occurred. Retaliation might have led to widespread gas warfare for which our offensive preparations were inferior to those made by the Germans. As already stated above, the Government's policy was to make no threats of retaliation unless we had evidence that an attack was imminent, because they might have been seized on by the Germans as an excuse to initiate gas warfare. The results might have been similar had we given publicity to its use by the Italians while not retaliating ourselves, since the Germans might have thought that we were actuated by fear of consequences.

Measures were taken, however, to warn the Italians by underground means that not only would we retaliate but that we possessed new and potent forms of gas which would be used against them.

At a later date (March, 1941), when the Greek Minister in London stated that he had reliable information that the Italians had landed in Albania two battalions equipped for chemical warfare and asked the British Government to make a statement regarding retaliation, similar views prevailed and the request was refused. The previous underground warning to Italy was considered adequate.

SECTION 4.—1941 AND THE ENTRY OF JAPAN INTO THE WAR

1941 was a period of extensive preparations on all sides in the United Kingdom to meet the threat of gas attack, either alone or combined with invasion, and of the building-up of our retaliatory capacity, both at home and overseas. The effect of the German onslaught on Russia was to introduce new factors in relation to the probability of their initiating the use of gas, and the entry of the Japanese into the war at the end of the year extended the area in which we had to make preparations to meet the threat of gas warfare.

In the United Kingdom, training of the Home Forces during the autumn of 1940 and winter of 1940/41 was progressive and was made more realistic by the use of toxic smoke generators and of artificial air spray under field conditions. On the whole, the Services were " gas conscious ", but the problem remained of keeping them on their toes as time passed and both gas attack and invasion hung fire. Similarly, with the civil population, there was a risk that if the public were prematurely roused they would relapse into a state of indifference before the danger materialized. This factor had to be borne in mind in relation to the timing and intensiveness of public propaganda. Initially, the spring was the danger period for which preparations had to be keyed up, but later, as neither gas warfare nor invasion materialized, the autumn became the peak period for preparedness.

Our Own Capacity to Wage Gas Warfare

As far as the Army was concerned, the position remained unsatisfactory throughout the year.

By the spring of 1941 the supply of minor defensive weapons (bombs, ground 6-lb., chemical mines and bulk contamination vehicles) was adequate, the total reserve requirements having been very nearly completed.

Supply of other weapons was, however, far from satisfactory. The building-up of stocks of gas shell for the 25-pounder was slow, owing to problems of design and technical difficulties ; although empties for charging began to come forward in increasing quantities during the year, the production of fuzes lagged. Further, the mark of 25-pounder base ejection shells manufactured during this year were on the border-line for stability and were subject to limitations in use, particularly in hot climates. 6-in. Howitzer gas shell began to come forward in June, 1941, but no 5·5-in. became available during the year.

By the spring, reasonable reserves of drums for the Livens projector, charged with both mustard gas and phosgene, were available, but there was a deficiency in bursters and propellant charges, and, although this had been overcome by the autumn and reserves were then adequate, this equipment was obsolescent and the chemical warfare companies armed with it would have been of little use in modern conditions of mobile warfare. The intention was to withdraw this equipment as soon as the 5-in. U.P. rocket was available, and to hold it as a trench store for possible use under static warfare conditions. This decision had been taken late in 1940, but owing to technical difficulties in design, no rockets fit for operational use became available during 1941. Towards the end of the year a requirement was stated for the 4·2-in. chemical mortar to supplement the armament of the chemical warfare troops, but the mortar did not become available until the following year.

The position at the end of 1941, therefore, was that we had 34 chemical warfare companies but no adequate offensive weapons for their equipment. (*See* Chapter II, Section 3, and Chapter IV for the history of chemical warfare troops and of their armament.)

Had gas warfare eventuated in 1941 in any theatre where we were in land contact with the enemy, our gas offensive would have had to be left largely to the R.A.F.

Supplies for the R.A.F. were more satisfactory, although, to begin with, production of phosgene bombs lagged. All phosgene surplus to other essential war requirements had been charged into 250-lb. bombs during the winter of 1940/41, but the numbers available were small. It was not until February, 1941, that firm demands for this type of bomb were formulated. Expansion of phosgene production had been based only on Army requirements which had been stated at the beginning of the war, and there was no surplus available for the R.A.F. New production capacity, therefore, had to be established. In any case, " teething " troubles and technical difficulties were encountered with the plants already under construction. The fact that no 5-in. U.P. rockets for the Army became available for charging during the year enabled the majority of available phosgene to be allocated to the R.A.F., but this was fortuitous.

Stocks of mustard gas were still being built up and there was sufficient to charge empty weapons (both Air and Army) as they became available. Early in the year, at the instigation of the Air Ministry, negotiations were put in train with the South African Government for the production of mustard gas there, as well as for the manufacture of as many air gas weapons as possible, in order to provide an alternative source of supply for the Middle East and Far East theatres.

By the spring there had been a small increase in stocks of both mustard gas and phosgene air weapons, and production of empties was improving. In the summer of 1941 the 65-lb. L.C. bomb was introduced as the standard R.A.F. mustard gas bomb in preference to earlier types owing to its greater tactical flexibility and simplicity of manufacture. By the end of the year it was coming forward in good quantity.

The potential mustard gas expenditure of the R.A.F. increased considerably, partly owing to the planned expansion of the bomber forces, and partly through the introduction of the 65-lb. bomb and the acceptance of a new spray weapon (I.S.C. 400 lb./S.L.3), known as the " Flying Cow ", since these weapons carried a larger quantity of gas for the same aircraft effort. (For details of air gas weapons *see* Chapter IV, Section 2.) The position was therefore reviewed in the autumn. To begin with, the Air Staff based their expansion requirements on an arbitrary assessment that up to 50 per cent. of our total bomber effort might be exerted in the form of gas. Expansion of gas and gas weapon production to maintain this effort was, however, of such magnitude that it was likely not only to impair our general war effort, but also to restrict the development of aircraft output. On further consideration, this planning was thought to be excessive, and the policy applied for relating our gas requirements to the expanding bomber forces was :—

Gas Bombs :		At sustained rates of expenditure.
Mustard ..	15 per cent. of total effort of all bombers	
Phosgene ..	6 per cent. of total effort of all bombers	
Spray :		
400-lb. S.L.3 ..	5 per cent. of total effort of heavy and medium bombers.	
S.C.I. 500 lb...	10 per cent. of total effort of light bombers	

Reserve stocks were to be based on three months' expenditure at sustained rates, or to be sufficient to enable production to be worked up to requirements before reserve stocks were exhausted, whichever was the longer period.

Stocks of gas weapons and existing planned production in the United Kingdom in December, 1941, would have enabled the assessed average effort to have been maintained for at least a year.

Appreciation of Enemy Intentions

The general factors, as previously outlined, governing the possible use of gas against the United Kingdom by the Germans continued to hold good.

It was thought that the use of gas before invasion was unlikely, even during preparatory bombing, since, if invasion were launched, considerable local tactical advantage would have accrued to us had we been free to use gas to meet the initial attack on the beaches, and the Germans were unlikely to hand us the opportunity for this.

Further, after the entry of Russia into the war, the Germans were likely to use gas on all fronts at once, otherwise the advantage of surprise would have been lost. This lowered the probability of the launching of air gas warfare against the United Kingdom during the winter months when weather conditions were unfavourable in Russia. It was also assessed that the German Air Force, with their Russian commitments, could not produce a sufficient scale of attack on this country to make the introduction of gas warfare by aircraft worth while, and they were therefore likely only to use it against us in support of a major operation such as invasion. A scare raised by the United States representative in Berne in December, 1941, that gas attack was imminent was discounted for these reasons.

Among probable invasion targets for gas would have been those in connection with isolating the Kent area from the rest of the country, as it was felt that the Dover Straits area was the only one where there was any real chance of the Germans maintaining a secure line of communications, and that they would make a determined effort to seize and isolate the vital Kent area by using airborne troops and possibly gas.

As time passed, however, it became open to doubt whether militarily the use of gas in invasion would be to the German advantage. In the battle subsequent to the first landing, the use of gas by both sides was not likely to be materially advantageous to either side and its employment by us against German embarkation and disembarkation operations would have had considerable nuisance value. Further, it was thought that the Germans might have appreciated that, in the event of failure or partial failure of invasion, our subsequent retaliation on their civil population would have had serious results for them, as they were reported to be not well protected against gas. On the other hand, the urge to add every possible factor favourable to their initial attacks on the beaches might have outweighed all other considerations, particularly had they assessed a high value to the surprise factor in relation to any assumed lack of preparedness on our part.

However, we could not afford to take any risks and our preparations, defensive and offensive, proceeded apace.

In relation to the use of gas by Germany in an invasion of the United Kingdom it is interesting to note that all their leading personalities both military and scientific when interrogated after the war stated that the Germans never planned to use gas, except in retaliation. As evidence of this they drew attention to the fact (duly proved to be correct) that no stocks of gas were ever moved outside the borders of the Reich. Their statements are considered to be probably true but it must be remembered that at the time of their interrogation the chance of being sent to Nuremburg was ever present in their minds.

Defensive Preparations

On the whole the state of equipment and general preparedness of the Fighting Services was satisfactory, although there were some respirator deficiences, particularly in the Home Guard where only 20 per cent. had received Service respirators. Further, the link up between the Services and the civil organizations needed to be perfected.

Anti-gas precautions in the United Kingdom were the subject of a review, with particular reference to the civil side, by the Lord Privy Seal in February, 1941, and of a further report to the War Cabinet at the end of April, 1941. Generally speaking defensive preparations were satisfactory, but further action was necessary to meet concentrated attack on nodal points. Railways, factories and utility services were warned to bring their precautions up to date, cleansing and first aid arrangements for the public were overhauled, and measures for the protection of foodstuffs were made more stringent. Steps were taken to ensure co-ordination and mutual support, both between the Services and the civil organizations within a particular area, and also between areas. Arrangements for close liaison between the Fighting Services, Regional Commissioners and local authorities were made and included the exchange of information on chemical and medical gas intelligence at various levels, the pooling of military and civil cleansing and decontamination facilities where this was expedient, and co-operation for the recognition and disposal of unexploded gas bombs and for the speedy circulation of information about the structure and appearance of enemy gas bombs as soon as this was established.

Although cloud gas attacks projected from ships at sea were not very probable they were a possibility and a special alarm system was set up in areas likely to be affected (*see* Chapter VI, Section 2).

An assessment, covering both phosgene and mustard gas, was made of the likely effects of a gas raid on the maximum probable scale, assumptions on bomb distribution being based on the bomb census of raids on Birmingham in November, 1940. The assessment was worked out for a large city, a dock area and a nodal point in communications. A review of the scale of our defensive preparations in the light of this assessment showed that, if mustard gas alone was used, the task of decontamination after a prolonged night raid would be well within the power of the appropriate services. The most serious problem would be the obstruction of key points and vital centres of communication by contaminated debris caused by H.E. bombs followed by mustard gas. Bulldozers would have been useful for clearing this contaminated debris away, although not many of them were then available. As far as attacks with phosgene were concerned the assessment showed that, with one proviso, the probable effects should not be serious. The proviso was that the public were properly trained in putting on their respirators quickly.

It was the public apathy in relation to possible gas attack which was a major concern of the Government. Long immunity from gas attack had led to carelessness on the part of the public in relation to precautionary measures such as carrying the respirator, and yet the Government's desire not to raise unnecessary scares still pointed to a policy of avoiding publicity.

In February, 1941, unostentatious propaganda was carried out as part of the campaign to make the public invasion-conscious, in order to familiarize them with anti-gas precautions without alarming them, and in such a manner as to avoid the appearance of any marked development in Government policy which might give the Germans a pretext for starting gas warfare. By March, 1941, it was evident that this restricted propaganda policy was not having the desired effect on the public, and in any case it did not appear to be serving its purpose abroad. German spokesmen and broadcasters were reported as having commented on our pre-occupation with the gas question and on the tendency of our press to ascribe to Germany the intention to attack, while remarking that all the activity seemed to be on the British side. In view of the danger of the heavy casualties which might have resulted from initial large scale attack on an unready civil population, considerable risks were attendant on the continued restriction on public instructional measures and publicity. A statement of Government policy was, therefore, deemed necessary. This was broadcast by the Home Secretary on 28th March, 1941, to the effect that, although we regarded the gas weapon with abhorrence and would under no circumstances be the first to use it, we could not rely on the enemy's undertakings not to use it as proof that he would in fact not do so, and that it was therefore necessary to take full defensive measures for the safety of the civil population which must co-operate in obeying instructions given to it. This announcement was followed by a press campaign and by the issue of a leaflet of instructions to householders as well as by an increase in *tempo* and scale of civilian gas exercises. Further publicity was kept going at a steady level. All possible measures—short of making failure to do so a legal offence— were taken to induce the public to carry their respirators and to practise using them.

The Procedure to be adopted on the First Use of Gas by the Enemy

Among the measures adopted in relation to the general public was the preparation in advance of a series of broadcasts to be given in the event of the enemy using gas, and also of leaflets, containing information and instructions to householders, which were held ready for issue by the G.P.O. at District Offices.

By May, 1941, a procedure had been evolved for confirming the first use of gas by the enemy, for authorizing the release of broadcasts and leaflets, and for obtaining War Cabinet sanction to adopt retaliatory measures ; the responsibilities of various authorities in this connection were specified.

A proposal that, should gas be used in connection with an invasion, the C.-in-C., Home Forces should be given authority to retaliate immediately without further reference to the War Cabinet was refused. After the War Cabinet had sanctioned the employment of gas the C.-in-C., Home Forces would have ordered and controlled its use in Great Britain (and in Ireland the G.O.C.-in-C. British Troops in Northern Ireland), but, in the case of air gas attack, requirements would have been passed to Bomber Command, in the same way as for bombing attacks, who would have decided the method of meeting the requirement. The Chiefs of Staffs would have ordered and controlled the use of gas against targets overseas such as Berlin.

Government Policy Regarding Declarations involving Threats of Retaliation

The Government attitude towards making any declaration involving a threat of retaliation remained the same as in 1940. No threat of retaliation was included in the Home Secretary's statement on defensive policy on 28th March, 1941.

Although in some quarters it had been felt that inclusion of such a statement would have acted as a deterrent, the balance of opinion was against it on the grounds that the Germans must be well aware that we should retaliate and that any statement to this effect would be judged by them as an indication that we intended to take the initiative and might, therefore, provoke them to adopt this course.

In July, 1941, a telegram was received from Reuter's correspondent in Moscow (which might have been inspired by the Russian Government) to the effect that the Germans intended using poison gas against the Russians, and which suggested that it would prove an effective deterrent if we announced that we would retaliate on the Germans if they did so use gas. The War Cabinet, however, considered it premature to give such an undertaking.

Plans for the Use of Gas in Retaliation

In general, plans for retaliation remained the same as in 1940. In anti-invasion plans for action if gas were used against us, more emphasis came to be placed on the use of mustard gas bombs which were now regarded as more effective than high spray, although this latter was considered a useful adjunct for attacks on enemy embarkation ports to interfere with the movements of reinforcements and to complicate supply movements. Against efficient A.A., it was felt that low spray was likely to cause us aircraft casualties out of all proportion to the effect produced. When using S.C.Is. the pilot was unable to take full evasive action when actually on the spray run and it was this factor of aircraft vulnerability that in part led to the development of the type of spraying bomb known as the "Flying Cow." Against targets in enemy territory gas bombs were considered to be the more suitable weapon.

In February, 1941, the decision not to send gas weapons to Northern Ireland was reversed. The original arrangement under which gas tasks would have been carried out by Bomber Squadrons based on the United Kingdom had pre-supposed that an invasion of Ireland and an invasion of the United Kingdom would not be simultaneous. The G.O.C.-in-C., British Troops in Northern Ireland had pointed out that in any case such arrangements made the close tactical co-operation necessary between the Army and the R.A.F. difficult, and requested that stocks be provided for his air component. It was, therefore, agreed that in order to ensure that gas could be used quickly in Ireland under conditions of general invasion an outfit of air gas weapons for the air component should be sent to Northern Ireland as soon as they could be spared. Gas shell were also to be sent for Army use.

The Position in the Middle East

Army capacity for retaliation remained small. No chemical warfare units could be sent to the Middle East as their obsolescent Livens equipment was totally unsuitable to that theatre. By the end of the year stocks of 25-pr. base ejection gas shell had been built up to about 25 per cent. of the target and the target for 6-inch howitzer gas shell had been completed. Had gas warfare eventuated in this theatre retaliation would have had to be left mainly to the R.A.F.

By December, 1941, stocks of mustard gas air weapons in the Command or ready for immediate shipment from the United Kingdom were sufficient to maintain the assessed average effort for six months. The weapons available (30-lb. and 250-lb. bombs) were less efficient than the new 65-lb. bomb. Stocks of this new bomb could not be sent overseas ready charged owing to short tropical storage life ; empty bombs and charging drums had to be sent separately. An initial shipment of 5,000 of these bombs was made in December, 1941. No bombs charged phosgene were available during the year for shipment overseas. Stocks of gas weapons in the Middle East in December, 1941, are given in Appendix II. The defensive position was satisfactory.

World-wide Position on the Outbreak of War with Japan

With the entry of Japan into the War, we had to be prepared for world-wide gas warfare. Japan was not a signatory to the Geneva gas protocol of 1925. The Chinese had claimed on a number of occasions that the Japanese had used gas against them and, although it was not known whether the Japanese had in fact done so, it was felt that they would use it if they wished.

Generally speaking, except in India itself, defensive equipment in overseas theatres was adequate. All troops were equipped with respirators, eyeshields, ointment and capes, and reasonable reserves of defensive clothing and equipment had been built up during the first half of 1941. In India the equipment position could hardly have been worse. Resources had been barely sufficient to equip formations, before they proceeded overseas from India, with the basic essentials of personal equipment ; deficiencies in unit scales and reserves were completed under War Office arrangements on arrival in the theatre of war. In India itself the cupboard was bare ; production lagged behind expansion of the forces. War and P.A.D. scales of equipment were negligible and even training scales of respirators were deficient.

All units in all theatres had unit anti-gas instructors trained in Army schools in the United Kingdom, Middle East or India. Little stress had been laid on the offensive aspects at the Middle East school, and the school in India taught defensive chemical warfare only.

The chemical warfare staff organization in the Middle East—both at G.H.Q. and in formations—was fully established (*see* Chapter II, Section 2), and there were technical officers at Malta and Gibraltar. Further East the position was less satisfactory ; in Iraq, each division had a G.S.O. III trained in defensive chemical warfare and camouflage, and there was a G.S.O. II at Force Headquarters, but there were no staff or technical officers at other formations ; in India there were no staff or technical officers at any formation except for a G.S.O. II and a technical officer at G.H.Q., India. In the Far East there was a scientific officer on the Air Staff, and some divisions had a G.S.O. III (C.W.).

The position in the R.A.F. was generally similar to that in the Army, except that in India there were no qualified staff or scientific officers to handle C.W. at Air Headquarters, nor were there any Command gas schools. A very limited number of R.A.F. officers had attended the Army Gas School there.

In Australia and New Zealand all troops were in possession of respirators but full scales of personal and unit anti-gas clothing were incomplete.

Our retaliatory capacity in the United Kingdom, Middle East and Far East can be seen from Appendix II. In India, Australia and New Zealand there was no offensive chemical warfare equipment of any kind—either Air or Army.

The extension of our liabilities to the Far East further accentuated the unsuitability of the chemical warfare troops armed with Livens projector and no C.W. groups could be sent overseas until rearmed with the rocket and the mortar.

Measures were taken forthwith to reinforce the Far East. An Anti-gas Laboratory from the United Kingdom and staff and technical officers were *en route* to Malaya when Singapore fell and were diverted to India. Similarly, by the time that increment staffs for Burma had reached India, the evacuation had reached the stage when their posting was inexpedient. The despatch of an initial consignment of 25-pr. chemical shell for India was put in hand at once, and her requirements of chemical warfare troops and air gas weapons were examined in conjunction with G.H.Q. and Air Headquarters, India.

In view of the weakness of the position in overseas theatres, the despatch of a mission consisting of a General Staff officer, a technical and research expert and a production expert, was arranged by the War Office in consultation with the Ministry of Supply. This mission left the United Kingdom in January, 1942, and toured West Africa, the Middle East, India, South Africa, Australia and New Zealand to examine the position at first hand and to advise on offensive and defensive measures for chemical warfare. They returned *via* Canada and United States. This mission also acted for the Admiralty and the Air Ministry.

On the civil side abroad provision had been made for respirators for the entire civilian population of Malta, Gibraltar, and such civilians as would remain in Aden under attack, and also for the civil populations in the principal towns of Palestine and Cyprus. In Egypt they were provided for British civilians, and the Egyptian Government had over a million respirators and some protective clothing for Egyptian key personnel and essential services. In Trans-Jordan and in most of the African Colonies respirators were provided for key personnel. In almost all cases respirators were held in bulk for issue if chemical warfare broke out.

Half a million civilian respirators for issue to the civil population were shipped to Singapore but these did not arrive before its fall and were diverted to India where they were held as an Imperial reserve. South Africa, New Zealand and Australia were virtually unprepared for chemical warfare.

The provision of respirators for the civil population in Colonies exposed only to the risk of attacks by warships or seaborne aircraft was considered unnecessary. The risk of gas being included in such attacks was remote, owing to the space required to store and transport gas. Respirators for key personnel of essential services, however, were thought desirable. Provision was ultimately made for key personnel in Ceylon, Mauritius, the Seychelles and the Indian Ocean bases.

In India the civilian population was entirely unprotected against gas. Even had it been possible to provide, within a reasonable time, the vast number of respirators involved, the problems connected with their care and with training in their use would have been insurmountable in the case of the uneducated majority of the native population. An anti-gas organization on the United Kingdom model was set up within the Civil Defence Services and stores were provided. This organization was implemented during 1942/43 to varying degrees in major towns according to the scale and type of attack anticipated. No wide scale anti-gas precautions were ever undertaken in India owing to the impracticability of protecting the civil population as a whole. As little publicity as possible was given to gas and to anti-gas measures. The protection of native populations in other parts of the Empire presented similar problems. In the Middle East the main concern was with the native dock population at ports ; respirators or no respirators, it was felt that this population would remove itself on the first use of gas.

SECTION 5.—1942 TO "D" DAY, 1944

Offensive and Defensive Capacity to wage Gas Warfare in the United Kingdom and the European Theatre

OFFENSIVE CAPACITY

The supply of Army gas weapons remained unsatisfactory until the end of 1942. Although charged chemical shell started to come forward in good quantity, the lag in supply of fuzes was not overcome until the autumn. Production was also chasing increased requirements as a result of revised demands arising out of our world-wide commitments.

In January, 1943, it was decided to abolish the chemical warfare troops (*see* Chapter II, Section 3). During 1942 the rearmament of these troops with the 5-in. rocket and 4·2-in. mortar had been slowly but painfully achieved. Owing to the many " teething " troubles encountered in the development of both these weapons, the lack of spare parts and tools and, in the case of the rocket, deficiencies in components of the firing equipment, these troops were never fully operationally fit.

The decision to abolish the chemical warfare troops was conformed to throughout the Empire, and from 1943 onwards it was accepted that the role of Army weapons (artillery and infantry support weapons) was harassing, and that reliance would be placed in the R.A.F. to put down heavy neutralizing concentrations of gas in forward areas in support of Army operations as decided on between the Air and Army commanders in the theatre of war (*see* Chapter V).

By the spring of 1943 Army capacity to retaliate in this harassing role was satisfactory. The six months reserve of artillery ammunition was almost complete and stocks of mustard gas and phosgene bombs for the 4·2-in. mortar were coming forward well. Charging of gas bombs for this weapon ceased in September, 1943, owing to high priority requirements for H.E. and smoke bombs ; stocks were adequate to tide over the first few months of chemical warfare until experience should determine the actual expenditure of gas *versus* H.E. and smoke. The manpower situation did not admit of the retention of a large capacity for charging 4·2-in. chemical bombs at the expense of other vital and urgent stores.

At the end of 1942 production of all Air and Army C.W. weapons was reviewed in the light of the manpower stringency. The leeway in the production of gas and gas munitions of the early years of the war had been caught up ; production was therefore reduced to a level which allowed for expansion to full wastage rates within six months of gas warfare breaking out, and at the same time admitted of the provision of improved types of munitions and the completion of reserves. This resulted in a considerable reduction during the first quarter of 1943 in the labour force employed. In April, 1944, a further reduction in the labour force was considered, but in view of the fact that production capacity was closely linked to the planned scale of retaliation and any reduction would therefore have reduced our state of readiness, action was postponed until after " D " day. We could not afford to relax until it was clear that the Germans would not use gas in an attempt to repel our invasion of the Continent.

The R.A.F., from early 1942 onwards, were in a sound position to wage gas warfare in the European theatre. Reserve stocks of air gas weapons were sufficient to carry over until production could be expanded to maintain a sustained effort based on the policy formulated at the end of 1941. By the spring of 1943, although production capacity remained unchanged, our ability to carry a heavy scale of attack into the heart of Germany was increased by reason of expansion of the bomber force, whose total lift was 70 per cent. greater than in the spring of 1942. Further, the Pathfinder Force and new navigational aids had increased the efficiency of attack.

In implementing retaliation some four days' notice would have been required to bring forward sufficient stocks of mustard gas bombs to maintain attacks once they had been initiated, owing to the difficulty of transportation of these weapons in large numbers. Initial attacks could have been made within 48 hours from dispersed stocks held at operational bomber stations or maintenance units in the United Kingdom. All phosgene bombs continued to be held in central storage seven days behind operational units ; their use in initial attacks would have been dependent on sufficient warning.

In December, 1943, in view of the expansion of the bomber force and the increasing ability of the bombers to carry a greater number of gas weapons, with the consequent and rapid increase in air gas weapon requirements if they continued to be based on the 1941 policy, *i.e.*, approximately 25 per cent. total gas lift, a more realistic method of computing gas lift was adopted and applied in particular to strategic gas bombing. A very detailed appreciation was made of the effective concentrations of gas it would be necessary to put down on a major European city in an " all out " gas raid. A new basis of computing requirements was reached which took into consideration the relative requirements of H.E. and incendiary as opposed to gas attack as well as the characteristics of both phosgene and mustard gas, *viz.*, the exploitation of the surprise factor for maximum effect with phosgene, and the duration of the immobilizing effect of a very heavy scale attack with mustard gas.

A calculation was made of the number of gas bombs required for the total number of sorties necessary to achieve the requisite concentrations in a single raid on an average size city, *e.g.*, Frankfurt, allowance being made for a proportion of bombs not reaching the target area. The minimum number of bombs necessary for an attack of this nature was estimated at 3,600 × 500-lb. bombs in the case of phosgene and 12,800 × 65-lb. bombs in the case of mustard gas. The new policy for strategic gas bombing was based on making two such attacks with phosgene and six with mustard gas every month, supplemented by continuous harassing with gas during normal H.E. and incendiary attacks in order to enforce anti-gas measures. For this purpose provisioning allowed for two-thirds of all strategic bombing sorties flown (other than those carrying full gas loads) each to carry a 500-lb. container with three 65-lb. mustard gas bombs and for the remaining one-third to carry a 500-lb. phosgene bomb. Total weapon requirements for strategic bombing were calculated on the above basis.

The revised calculations showed that as at June, 1944, there would be little difference between the new basis and the old basis of provisioning and that there was, therefore, no need to devote greater resources to the manufacture of gas weapons. In certain cases, for a few months until production expanded, it would have been necessary to use a proportion of obsolescent weapons in lieu of the latest types, *e.g.*, 250-lb. and 30-lb. mustard gas bombs in lieu of the 65-lb. bomb. The operational aspects of the new policy are discussed later in this section under " Plans for Retaliation ".

Provision for 2nd Tactical Air Force for " Overlord " and subsequent operations was made on an appropriate basis. 500-lb. phosgene bombs and mustard gas spray equipment were available. Although a 500-lb. mustard gas bomb, which was intended as an airburst bomb with a dual role of ground contamination and direct anti-personnel attack, became available from June, 1944, onwards, it suffered from certain deficiencies in functioning, details of which will be found in Chapter IV, Section 2.

Stocks and production capacity within the Empire on " D " day are shown at Appendix III.

Naval Preparedness

There was no change in Naval policy. The Navy continued to regard gas as an ineffective weapon against ships as compared to H.E. The Navy had no gas bombs of their own but Naval aircraft were cleared to carry R.A.F. bombs should necessity have arisen. Naval aircraft would not have been used for spraying gas. A small quantity of phosgene shell of 8-in. calibre and below had been available since 1941 in case H.M. ships should have been required to bombard shore objectives with gas in support of a landing, with the object of enforcing on the enemy the wearing of the respirator.

Allies

The U.S.A. Air Forces in the European theatre were not in a position to take offensive action with their own gas weapons until early in 1944, by which time adequate stocks had been built up ; before that date certain of their aircraft could have carried certain types of British gas weapons had it been necessary. By " D " day, however, they were adequately equipped. Co-ordination between the British and the Americans in relation to provisioning for gas warfare is discussed in detail in Chapter II, Section 9.

Little was known of *Russian capacity*. There was some evidence that they had increased their preparations in the spring of 1943 but details were lacking. Although probably not so well prepared offensively as defensively there was no reason to believe that their capacity would have been dangerously insufficient.

Defensive Preparedness

Although interest was stimulated by the increased use of real war gases (*see* Chapter VII), the state of training in the Services, both in the United Kingdom and elsewhere, continued to be variable. This was largely due to the difficulty of creating any enthusiasm in training against a weapon which might never be used. In the active theatres of war, where we were in land contact with the enemy, the difficulty of maintaining a generally high standard of training was accentuated by the small amount of time which could be allocated to such training when troops were out of the line : they naturally had to concentrate first on learning to use new weapons and equipment which they were shortly to use in action. Nevertheless, the standard of training in anti-gas measures was such that it was thought that once the formidable effect of the first impact was over undue casualties through inefficient training should not occur.

Provision of defensive equipment in the United Kingdom was satisfactory. There was some concern in 1942 owing to the lack of a detector for the odourless blister gas which it was thought the Germans might have (*see* Chapter III, Section 3.—Nitrogen mustard gas), and immediate development of a detector (the pocket vapour detector) was put in hand, but issues were not completed until early 1944. Provision of light type respirators lagged in 1943 owing to a hold up in haversack production, but 21 Army Group anti-gas equipment was complete by January, 1944, and reserves had been built up simultaneously with issues. The issue of anti-verm clothing to troops continued until September, 1942, when the policy to make a general issue of this clothing was suspended owing to administrative difficulties. In 1944, for the " D " day operations, all troops were provided with one suit of this clothing. (*See* Chapter VI—Defensive Equipment.)

The loss of our rubber resources in the Far East made it essential that respirator wastage should be reduced to the minimum. In April, 1942, it was proposed that the order to the public in the United Kingdom invariably to carry their respirators should be relaxed, owing to the high wastage occurring through constant carriage. It was felt that, provided adequate measures were taken for maintaining anti-gas efficiency, the knowledge that the public were not always carrying their respirators would be outweighed by factors of greater strategic importance in any German decision to initiate the use of gas. Should the risk have increased at any time, immediate warning could have been issued through the B.B.C. and the Press. The War Cabinet, however, decided that the moment was inappropriate for a change in policy because the Government were about to issue a declaration that we should use gas against the Germans if they used it against the Russians.

By July, 1942, this objection no longer applied and the injunction to civilians always to carry their respirators was relaxed, with certain provisos for the Civil Defence Services. Special measures were taken to ensure that the public could have their respirators tested periodically if they took advantage of the facilities provided. This relaxation of the order to carry respirators at all times was paralleled in the Fighting Services where respirators were only to be carried in field service order, on guards and pickets, when ordered by commanding officers, or on reinstatement of the previous instructions should the risk of gas attack become imminent. The small risks involved in this change of policy under the prevailing conditions were accepted in the interests of the war effort as a whole.

The acceptance, at the end of the year, of the light type respirator in gradual replacement of the General Service type throughout the Army also contributed to a saving in rubber owing to the elimination of the connecting tube.

Towards the end of 1942, the shortage of manpower in the United Kingdom was so acute that the production of all unessential items of equipment had to be eliminated. The stock position of anti-gas equipment was good ; the whole position in respect of anti-gas equipment was therefore reviewed, particularly in relation to the scales of reserves which were thought to be excessive. The basis for provision of reserves in all theatres was consequently reduced by the elimination of certain reserves.

Stocks of anti-gas clothing were considered to be sufficient and it was decided that production in the United Kingdom could be discontinued, subject to arrangements being made whereby it could be restarted quickly in emergency, and to periodical review in relation to the probable date of exhaustion of stocks and the time required to get production under way again. Respirator production was reduced by 10 per cent. in each of the first three months of 1943 with the proviso that production of General Service and light type respirators should not fall below a certain figure. All capacity for the G.S. type respirator was gradually being turned over to the light type. Curtailment of production of other items of anti-gas equipment was effected by the elimination of the production lag reserve.

These measures led to a considerable saving in manpower and materials. When the position was reviewed in March, 1943, it was found that further cuts could be made in the main items, *viz.* respirators, eyeshields, anti-gas ointment and anti-dim outfits, as a result of a further revision of Service requirements. A saving in wastage was effected in April, 1943, by the withdrawal of second and third line reserves of all items of anti-gas clothing and equipment to ordnance charge and by the imposition of strict economy measures on equipment in use. On calculated wastage rates a " short-fall " of several million anti-gas capes was to be expected by the end of 1944, but it was hoped that actual wastage would be considerably less and would go some way towards closing the gap. No steps were taken at this stage to resume production.

The above measures were not based on any diminution of the probability of the outbreak of gas warfare. They arose from manpower considerations. For security reasons information regarding this policy was restricted to those immediately concerned.

On the civilian side anti-gas equipment in the United Kingdom was complete including reserves. As a result of the drive to tighten up precautions in 1941 there was an all-round improvement in general preparedness and this was maintained, although difficulties were encountered, particularly in connection with personnel for the decontamination services, owing to manpower stringency and the calls of industry and the Fighting Services as opposed to Civil Defence. Owing to the lull in raiding, there was some slackening of interest in gas questions on the part of the public, but the civilian population was as well trained as could be expected. They were reminded from time to time that the possibility of gas still existed. Special attention was given in 1944 to all aspects of anti-gas preparedness in the " D " day ports and the civilian areas surrounding them.

Allies

The U.S. Forces in the European theatre were fully equipped personally from their own resources but were supplied by us with certain items of decontamination and other equipment under the arrangements for co-ordination of supply.

Russian defensive equipment had been deficient in 1942 but it was believed that her anti-gas measures for both Army and civilians were by now of a high order and that any gaps in her 1942 equipment had been made up by 1944.

Offensive and Defensive Capacity in Overseas Theatres and the Dominions

General

The building up of stocks of Air and Army gas munitions overseas proceeded within the limits of shipping availabilities and priorities as agreed by theatres.

By May, 1943, holdings of chemical shell and stocks in transit were surplus to reserve targets in all theatres except North Africa where reserves were not complete until the end of the year. This resulted from a recalculation of reserve targets based on new activity factors and reserve periods, and revised orders of battle.

There was doubt about the storage properties of some shell chargings, and the stability of certain marks of 25-pdr. shell in high temperatures was known to be poor. The position, however, was not bad because the mark of shell in current production in the United Kingdom by 1943 was ballistically stable and charged with the type of gas suitable to the tropics (*see* Chapter IV). Much of the overseas stocks could be rectified locally and further shipments of shell made as required.

For air weapons the R.A.F. 1941 Provisioning Policy was applied to overseas theatres.

On the defensive side there was a progressive improvement during 1942–43 in those theatres where equipment deficiencies existed. The War Office had been unaware of the seriousness of shortages—more especially in India—until the visit of the liaison mission referred to in Section 4. Local production of respirators and of a number of items of anti-gas clothing and equipment was initiated or expanded in the Dominions and India and this was supplemented by shipments from the United Kingdom although these were limited by shipping priorities. Staff deficiencies where they existed were also made good (*see* Chapter II, Section 2).

On the civil side our inability to provide protection for the non-European populations in overseas theatres in some degree offset our growing capacity to conduct air gas operations on the Continent of Europe. Even if it had been practicable to make more extensive issues to civil populations abroad, it would not have been possible in view of the labour shortage and the extra shipping space involved.

In August, 1942, a chemical warfare group of two rocket companies and one mortar company arrived in the Middle East. The 5-in. rocket was considered an unsuitable weapon for the Middle East theatre : the 4·2-in. mortar was preferable in every way. At the end of the year the two rocket companies were to have been converted to mortar companies for employment in an infantry close support role, using H.E. and smoke, but brigade support units came on the ground before conversion could be implemented.

The air gas weapon position remained satisfactory. 250-lb. phosgene bombs became available by the spring of 1942 and stocks of 65-lb. mustard gas bombs (empty) with charging drums continued to be built up during the year.

It had been intended to stock up North Africa on the same basis as the Middle East as soon as conditions were sufficiently stabilized(1) but, in the event, shipping priorities did not admit of this until towards the end of 1943.

By this time we were established in Italy and the strategic situation had changed. We were not planning any large-scale gas offensive from the Mediterranean or Southern European bases. It was not considered desirable, for geographical reasons and others referred to later in this section, to divide our air gas effort against the Continent between the United Kingdom and the Mediterranean. Any small advantages which might have been gained by conducting air gas warfare from bases in Italy would have been outweighed by difficulties of transporting air gas weapons there. Stocks in the Mediterranean theatre were enough to sustain for four months a harassing scale of attack, and sufficient to impose anti-gas measures, as a supplement to the major gas attacks conducted from strategic bomber bases in the United Kingdom. During this period stocks could have been replenished from the United Kingdom or South Africa to continue the same scale of attack.

Defensively, as far as the Services were concerned, the position was good, except that reserves of anti-gas equipment in North Africa were not completed until August, 1943. On the civil side the situation in the principal towns of Cyprus and Palestine, and also in Malta, was unchanged. In Egypt it was the the opinion of the British Technical Supervisor to the Ministry of Defence that the measures taken by the Egyptian Government would prove futile and that gas attack would result in wild panic. In Syria and the Lebanon anti-gas measures were completely inadequate. In Iraq and Persia similar conditions prevailed although an effort was made to provide respirators for essential services. The Anglo-Iranian Oil Company were provided with some respirators by the Tenth Army.

PERSIA AND IRAQ COMMAND

No chemical warfare troops were sent to this theatre in 1942, but the British Chemical Warfare Group sent to India was, until its abolition, regarded as an Imperial reserve which covered this theatre. Adequate stocks of chemical shell were built up. Owing to difficulties in storage conditions the majority of air gas weapons for this command were held in the Middle East.

Note (1).—No British gas munitions were taken with operation " Torch " (Anglo-American invasion of North Africa). Some gas munitions for the U.S. Forces were phased in early, direct from the United States, and sooner than was intended.

India

Stocks of chemical shell started to arrive in April, 1942, and continued to be built up until by mid-1943 holdings were in excess of revised establishments. In June, 1942, a training cadre, together with training ammunition and equipment (rocket and mortar), arrived from the United Kingdom and the conversion of field companies, Indian Engineers, to chemical warfare companies began. Initially, it was intended to provide three chemical warfare groups, Indian Engineers (nine companies), but this was later reduced to two groups (six companies) owing to stringency of Engineer manpower. All companies of Indian chemical warfare groups were to be armed with the 4·2-in. mortar, since the 5-in. rocket was considered generally unsuitable in this theatre where supply problems on extended L.s of C. of limited capacity necessitated restricting types of arms and ammunition to essentials. A British chemical warfare group, which included two rocket companies, had arrived in India and provided sufficient reserve of the rocket weapon for use in specific areas which lent themselves to its employment. The Indian chemical warfare groups reverted to engineer duties early in 1943, in conformity with the home policy to abolish such troops, and this provided a welcome relief for the heavy manpower commitments of the Indian Engineers.

No production of war gases or gas weapons was established in India owing to limitations imposed by lack of raw materials, tools and production capacity. The manufacture of empty 65-lb. air bombs was projected but ultimately dropped. Production of mustard gas and phosgene was considered as a long-term project but never implemented ; technical difficulties and lack of certain raw materials made the project not worth while since foreseen requirements could be met from United Kingdom and South African production.

Shipment of stocks of air gas weapons from the United Kingdom began in July, 1942, but it was well into 1943 before adequate stocks were built up, including stocks of the 65-lb. bomb and drums of mustard gas for charging them.

At no time, however, before the end of war against Germany could more than a harassing scale of attack have been maintained against Japan from South East Asia, simultaneously with gas operations in Europe.

On the defensive side deficiencies in anti-gas clothing were gradually made good from Indian manufacture, and respirator production was stepped up. A large number of respirators was sent out from the United Kingdom to India to bridge the gap between her own production and her serious deficiencies. Although shipping priorities had delayed the completion of reserves of items of equipment which could not be manufactured locally the position was generally satisfactory by 1944. Steps were taken to turn respirator production over to the light type but this did not begin until 1944. The main handicap was to training, owing to the non-arrival, through losses at sea, of adequate quantities of training stores of the real war gas type. The weaknesses in our defensive equipment under tropical conditions, notably in respect of impregnated clothing and ointment, are discussed in Chapter VI.

In the winter of 1943–44 all anti-gas equipment of forward troops of 11 Army Group (South East Asia Command) was withdrawn to ordnance charge under special arrangements for rapid re-issue in emergency (including air dropping) and for its use during periods of training. The risk involved, which also inevitably entailed a falling off in the standard of training, was acceptable in the interests of mobility under the conditions of climate and terrain on the Burma Front, and was not felt to be great since the chances of the initiation of gas warfare by the Japanese on this front was even more remote than elsewhere.

AUSTRALIA

Australia was gradually stocked up with air gas weapons and chemical shell from the United Kingdom. A beginning was made in 1942, with the assistance of trained instructors from the United Kingdom, on training one C.W. company on a special organization incorporating both the rocket and the mortar, but this was abandoned in 1943 in conformity with United Kingdom policy.

By 1944 the Americans were in a strong position to wage gas warfare in the South West Pacific and Australia could have added her quota. Had gas warfare eventuated in both the Japanese and the German theatres at the same time, Australian air forces would have been provided with United States gas weapons, but barring this eventuality she would have continued to be provisioned with British weapons from the United Kingdom and South Africa.

A chemical warfare experimental station was established by Australia in 1942. This was subsequently augmented by a team of British and American workers in 1943 and made a considerable contribution to our knowledge of tropical gas warfare and brought to light certain defects in our defensive equipment. (Details will be found in Chapter II, Section 9 ; Chapter III ; Chapter IV and Chapter VI.)

Manufacture of G.S. respirators was established and turned over to the light type in 1943. This was discontinued in 1944 but could have been re-started and expanded had it been necessary. Civilian respirators were provided by 1943, partly by local production of face-pieces which were matched by containers sent from the United Kingdom, and partly by the placing of large demands on the U.S.A.

NEW ZEALAND

New Zealand Forces were mainly equipped with anti-gas equipment from the United Kingdom. There was no intention of providing respirators for the civil population. Some civilian respirators for key personnel were produced locally. No stocks of air gas weapons were sent to New Zealand. Some 25-pdr. chemical shell were sent there and, in the unlikely event of gas being used in connection with a landing there by the Japanese, the policy was to retaliate by using these for harassing.

SOUTH AFRICA

Two mustard gas plants were established in South Africa—one of which was sent out from the United Kingdom. These came into operation during 1943. Phosgene capacity also existed. With some assistance from the United Kingdom there was a fairly good potential capacity for gas weapon production. Although some 65-lb. bombs and charging drums and S.L.3. spraying bombs were made there, no large scale production of empties was embarked upon since combined United Kingdom and U.S.A. capacity was considered adequate. Had gas warfare started, and, particularly, had we been involved in a gas war simultaneously with both Germany and Japan, South Africa's potential would have been a valuable reserve for supply to the Mediterranean or the Eastern and Australian theatres.

Canada

Canada's contribution to chemical warfare was out of all proportion to her size. Apart from being generally self-supporting, Canada put a considerable effort into research work. The organization was centred on Ottawa whence control was exercised on the extra mural work which was carried out at Montreal. Canada, in conjunction with the United Kingdom, established proving grounds at Suffield, Alberta, which in effect became an imperial centre, and which had perhaps the best combination of laboratories, experienced staff, and practically unlimited range facilities of any C.W. establishment in the world. The closest liaison was maintained with the U.S.A. The mustard gas capacity which was created was for the American type (*see* Chapter III, Section 3).

Liaison with Allies

Russia

In July, 1941, Lt.-Col. H. Paget, R.E., who was an expert in chemical warfare, was sent to join our Military Mission in Moscow, to establish liaison with the Russians on chemical warfare matters. This officer came up against a blank wall and returned after six weeks as the Russians would not participate in any discussions. He had been provided with an open letter of introduction from the Royal Society to the Russian Academy of Science. It was thought afterwards that this line of approach had not been the best owing to the suspicions existing between the Russian civil and military authorities.

In March, 1942, when replying to the Prime Minister's offer (page 39) to make a declaration to the effect that we would retaliate should the Germans use gas against the Russians, M. Stalin asked for certain material assistance and offered to send an expert to London. This was accepted and on 29th April, 1942, a civilian delegation arrived, consisting of M. Kasatkin, Deputy People's Commissar of Chemical Industry, and M. Sevastianov, a chemical engineer and interpreter. There was no military member of the delegation and it is doubtful whether the Soviet General Staff had been consulted regarding, or were even aware of, the visit. The War Cabinet directed that all technical information be given to the delegation regarding designs and types of offensive and defensive gas weapons and equipment, but that details about stocks and production capacity were not to be disclosed. Visits were arranged to research and development establishments and to factories producing C.W. equipment, and discussions were held at which our views on technical and research aspects were frankly explained. The Russians put forward no views on possible technical developments of gas warfare which were not already known to us, and either would not or could not provide any details of Russian military policy or preparedness for gas warfare. The impression gained of M. Kasatkin was that he was a good chemist but not a chemical warfare expert, particularly in relation to its military application. Our Military Mission in Moscow were likewise unable to obtain any details of Russian C.W. preparedness.

The Russian delegation stressed the fact that to be effective any assistance must be immediate. They thought that the months June to September, 1942, were critical for the success or failure of the German Eastern campaign and stressed the necessity for immediate C.W. counter-measures to be complete. We were unable to meet their demands for the types of war gases they requested, since we only manufactured them on a small scale for experimental purposes, but some shipments of certain raw materials were arranged. Their major demand was for three million respirators. This was difficult to meet as production was only sufficient for our own very large requirements, including large demands from India, and the production position was aggravated by the rubber shortage. This very large number of respirators had not been included in M. Stalin's original request to the Prime Minister ; it only materialized on arrival of the delegation. There was some suspicion that our favourable reactions to their requests for support had encouraged the Russians to take all they could get on seeing the chance of getting something for nothing. The request was not accompanied by any details of the number of respirators already held in the Russian Army, nor could this information be obtained from our Military Mission in Moscow. The Russians maintained that the respirators were urgently required to complete the equipment of front line troops and were not for replacement of out-of-date or inferior respirators of their own. In the end the request was accepted as genuine and, in view of the strategical importance of the Russian front, the War Cabinet agreed to allocate more rubber to respirator production and to ship one million respirators (complete with haversack and anti-dim outfit) immediately from our reserves, to be followed by a further half a million during the summer. Demands for bleaching powder and equipment for testing for war gases were met in full. We also let them have a million sets of anti-gas Ointment No. 2, and two million eyeshields.

Before leaving, M. Kasatkin said he intended on his return to raise the question of establishing close relations between the British and Russian Armies on chemical warfare matters. We particularly welcomed this as we were concerned at our complete inability either to confirm any Russian allegation of the German use of gas, or to ascertain whether any such use was on a large scale or was only an isolated incident. We would have liked to have had our own military experts with our Mission in Moscow who could have been given facilities for independent confirmation. Although the matter was pursued, after M. Kasatkin's return to Russia, first through military and later through Foreign Office channels, nothing came of it.

Subsequent to this the only liaison with the Russians was through the Soviet Military Mission in London which had a chemical warfare and flame expert—Lt.-Colonel Sidorsky—on its staff. Exchange of information was fairly free in respect of German intelligence, but when confronted with other questions, particularly regarding Russian offensive equipment, Sidorsky could " not remember ". On one or two occasions he personally showed some evidence of desiring to talk more freely but was carefully watched by his associates. He and his associates visited the Army Gas School and the Experimental Station at Porton and attended a number of demonstrations.

We must thank the Russians, however, for two very valuable pieces of intelligence, *viz.* the first specimens of the new German respirator container—the F.E.42, and a copy of the German Offensive Manual " Gasscheisen aller Waffen ", both of which were captured on the Russian front and passed to us through our Military Mission in Moscow.

Liaison between the Russians and Americans developed similarly. A Russian C. W. Mission to America was hospitably shown round and given much information, but the Russians did not reciprocate.

The United States

Liaison between the British and Americans was detailed and extensive and is dealt with separately in Chapter II, Section 9.

Appreciation of Enemy Intentions

Germany

The use of gas against the United Kingdom unaccompanied by invasion and before the outbreak of gas warfare elsewhere seemed unlikely, since this would have meant the sacrifice of the element of surprise in its initiation on some occasion strategically more profitable to the Germans.

Early in 1942, there had been indications that they might initiate its use in their spring offensive against Russia. As this did not materialize, it was thought, in the autumn, that, in view of the course the war was taking, the likelihood of the Germans starting gas warfare had lessened. They would certainly have used it in retaliation had gas been used against them, but Allied air superiority was such that they were unlikely to initiate the use of gas themselves except in circumstances where they felt that it would be the final stroke which would bring the war to a quick and victorious end. Invasion of the United Kingdom would have been one such circumstance but, at this stage of the war, it could only have followed the elimination of Russia. There was also a third condition under which the Germans might have used gas, namely, a " mad dog " act by Hitler when he saw himself faced with imminent military disaster.

On post-war information this last contingency seems to have been unlikely since Hitler appears to have been consistently averse to gas. At the time, however, the incalculable factor of Hitler's character could not be excluded.

During 1943 there was little evidence to prove any German intention to initiate gas warfare. Soviet staff fears in the spring that the German might use gas against them combined with reports of the tightening up of German defensive preparedness by the movement of gas stocks to their Eastern front, and with rumours that Hitler would only have accepted defeat after all means of warfare, including gas and bacteriological warfare, had been tried, led to the conclusion that the chances of Germany initiating gas warfare, although still small, were somewhat less remote. In the summer their anti-gas preparations were further intensified, partly owing to the fear that the Russians might start gas warfare against them. When retreating in Italy they destroyed or removed all Italian chemical warfare stocks, and information was also received that the Germans had removed French C.W. material from French dumps. From these facts and from the knowledge that the Germans were nervous about the possible use of gas by the Allies in invasion it was inferred that the Germans' main motive was not to provoke the Allies to initiate gas warfare. In the light of after knowledge on the conclusion of the war this inference was proved correct. During the latter part of the war the Germans had greatly feared the large Anglo-American gas potential combined with our air superiority. At the time, however, we had to recognize the possibility that the Germans' intensive defensive preparations might mean that they were themselves preparing to take the initiative when it suited them.

Other factors which were then thought to militate against the Germans initiating gas warfare were that they would appreciate that such an act was likely to prejudice any negotiations from which they might secure some political advantage, and that in the event of their defeat, a possibility which was by then widely appreciated by them, their use of gas would increase the Allied retribution which they had come to fear.

Overall, it continued to be felt that Germany was unlikely to use gas unless she was convinced that it would give her a decisive advantage which might result in an early and more materially favourable outcome of the war.

"D" day was obviously an occasion to which this might have applied. The outcome of the war depended on the success or failure of our invasion of the Continent. It had always been appreciated that the surprise use of gas against amphibious operations would afford considerable tactical advantage to the defender ; it might have delayed the Allied build-up and penetration, and thus given the German counter-attack better prospects of success. On the other hand the Germans would have had difficulties in bringing gas into action at the time and the place of invasion. They knew we were ready to use gas if necessary, and it was doubted whether the Germans would ever feel confident that Allied retaliation would not rapidly redress any initial advantage in the use of gas against our bridgeheads. They also knew that sooner or later we should have retaliated against their towns and cities, and they were bound to fear our great air superiority in this connection. Finally it was doubtful whether the Germans were in a position to stand up to the additional heavy administrative and material burdens of a gas war against three major powers. It was therefore felt that although the Germans might consider the use of gas more seriously than ever before in an attempt to repel invasion, it was improbable that they would in fact initiate it.

Italy

During 1942 there were no indications that Italy had any immediate intention of using chemical warfare unless forced by the Germans. Fear of retaliation was thought to be a more weighty factor with the Italians than with either the Germans or the British, as their home defence organization against gas was not of the same high standard. Geographically, before we were established in Tunisia in 1943, many parts of Italy were less liable to air attack from our existing aerodromes.

Japan

Japanese intentions were difficult to assess in 1942 as there was little concrete information regarding their state of equipment, particularly of their air arm. They were known to have used tear and nose gases against the Chinese on a small scale on a number of occasions, and prussic acid bombs had been carried by forward troops in Malaya and Burma and had been used against us on one or two isolated occasions during the retreat. The temptation existed for Japan to use gas against the unprepared native populations of the countries she was then threatening. It was probable that she would have feared retaliation on her home islands since her civil population were not equipped for gas defence on the same scale as ourselves or other European nations, but, for the time being, neither we nor the Americans could effectively attack her there. It was felt that she would use gas if she thought it to be to her advantage, irrespective of whether gas warfare had started elsewhere or not, but the favourable effect on Indian public opinion of refraining from its use might stay her hand.

During 1943/44 more information regarding Japanese capacity for gas warfare became available largely through documents and equipment captured by the Americans in the South West Pacific and to a smaller extent from those captured by ourselves in Burma. This information indicated that Japanese gas warfare development was on stereotyped lines and that it was unlikely that she had anything novel up her sleeve. Post-war information confirmed this (*see* Chapter VIII). As the Japanese came to realize the possibility of the bombing of certain parts of Japan, there was evidence of improvement in anti-gas measures there. Although she still could have gained immediate advantage by attacking the Indian population with gas, this would have been counter to her policy of appealing to the Indians against us. It was felt also that she was even less likely to initiate gas warfare in Burma owing to the difficulties of maintaining a sustained rate of attack at such a great distance from her centres of production. Overall, in view of the rapidly increasing Allied air superiority, it seemed unlikely that she would initiate a type of warfare in which, in the long run, she would be unable to compete on equal terms. There was no reason to suppose that Japan would have concerted her efforts with Germany if Germany had initiated gas warfare in Europe. The considerations influencing a decision by Germany or Japan to initiate gas warfare had so little in common that it was unlikely they would decide to do so either jointly or simultaneously.

The Strategic Position

It was appreciated that, in countries where anti-gas preparations were good and morale was high, the effect of heavy and sustained gas attack would be only slightly greater than a similar scale of normal bombing, although the psychological effects might be considerable, particularly at the outset. In these circumstances, the real importance of gas warfare lay in its surprise value. In unprepared countries and where the standards of intelligence and morale were low, the initial effect of gas attack was likely to be widespread panic, leading to a state of complete disorganization as long as attacks were maintained. The use of gas under such conditions had been of great strategic value to the Italians in Abyssinia in 1935.

Europe

In Europe, therefore, the use of gas against Germany in 1942 would not have been likely to produce any new or startling turn of events, because there were no targets where gas attacks were likely to produce decisive results. As the Germans had liabilities on the Russian front, we should have been able to put in a heavier scale of air gas attack than they could have brought to bear on us, and it was felt that German attack, on the scale probable, would not seriously impair our civilian morale and the effect on industry, transportation, etc. would be little greater than from a similar scale of normal bombing. Even if by freeing air forces from the Russian front Germany used gas against us on a larger scale, it was doubtful whether she would have gained any great material results. Bombing with H.E. and incendiaries could have upset the life of the country just as much, if not more.

Had gas warfare started in 1942 we should have gained little advantage from it in the European theatre. With the passing of the initiative to the Allies in 1943 our views about the advantages and disadvantages of the outbreak of gas warfare were dominated by one factor. This was that in amphibious operations the use of gas against the bottlenecks of beaches and newly captured ports would have enabled the enemy to impose very considerable delays, which might have spelt the difference between success and failure. The outbreak of gas warfare prior to or on "D" day would have been gravely to our disadvantage. When, early in 1944, the American Chemical Warfare Service were seriously considering whether it would not be to their advantage to initiate gas warfare in the South-West Pacific, the possible repercussion on Allied strategic plans in Europe was the most powerful argument against such a course of action.

The effect of gas on the German civil population was likely to have been greater towards the end of 1943 than in 1942 since morale was lower, particularly in urban areas. The use of gas in addition to H.E./Incendiary attack would have further stretched their already strained civil defence services. Initial attacks would therefore probably have caused considerable panic, but the overall effect, once the initial surprise had worn off, was impossible to forecast. It was unlikely that the resultant demoralization would have been sufficient to bring about unconditional surrender. Germany's satellites were ill-prepared for gas warfare and would have feared it ; it would have increased their desire to get out of the war but probably not to the point of capitulation.

The Russian Front

It was felt that the outbreak of gas warfare would bring few advantages to the Russians. As the Russians advanced, the use of gas against their L. of C. might have assisted the Germans in temporarily stabilizing their eastern front.

Middle East

In the Middle East during 1942, although we had good capacity for retaliation from the air and the supply of Army gas weapons gradually improved, it was considered that the outbreak of gas warfare in this theatre would be greatly to our disadvantage. Generally speaking, the desert was not a good area for the effective use of gas, except against defended localities such as Tobruk. The scale of air attack which we could have delivered against Italy from existing bases was negligible. On the other hand, the Axis could have attacked Malta, Cairo and Alexandria. Civilian A.R.P. arrangements in Egypt were poor ; and gas attack would have produced chaos among the native population and would probably have raised acute internal security problems.

The Mediterranean

In the Mediterranean in 1943, once we were established in Italy, our air superiority would have enabled us to exploit the use of gas, and by using it we would have been able to impose a greater delaying effect on the enemy than he could have imposed on us. The use of gas would, however, have meant reduction in our total H.E. bomb load which was likely to offset any advantages otherwise gained. Further, the effect on morale of the Italian civil population and on the unprotected native population of the Middle East and North Africa would have been adverse, and it was likely that the morale of guerrillas and resistance groups in the Balkans and elsewhere would have been similarly affected owing to our inability to supply them with the anti-gas equipment they would certainly have demanded.

Japan

In the war against Japan even a light scale of gas attack against India would have had the most deplorable effects on civilian morale in that country which would have greatly impeded the war effort there with possibly far-reaching results. Initially we were totally unequipped for offensive gas warfare, and although the position improved both offensively and defensively during 1942–43, we were still not in a favourable position to wage gas warfare in the South-East Asia Theatre. In any case, there were no vulnerable targets against which we could have retaliated.

In the South-West Pacific both Australia and New Zealand were virtually unprepared for gas warfare to begin with, but by end of 1943 this had been rectified. The scale of possible gas attack on Australia would have been unlikely to be of the same intensity as was to be expected in Europe, owing to the extended nature of the Japanese L. of C., but, against this, the useful targets, *e.g.*, docks, aerodromes, factories, were fewer and the weight of attack could have been concentrated. In the case of New Zealand the intensity of any gas attack was thought to be unlikely to be more than harassing.

Although the United States had built up a fair capacity (mustard gas only) to wage gas warfare in the South-West Pacific by the end of 1943, the Allies were still unable to deliver effective air gas attacks against the Japanese homelands.

Policy on Declarations involving a Threat of Retaliation against Germany

Up to 1941 it had been thought that any such declaration involved too large an element of bluff, but, with our own increasing capacity and the United States potential behind us, this phase was passing. In March, 1942, there were strong rumours that the Germans might use gas against the Russians in their spring offensive. In view of the importance of the Russian front, the Prime Minister sent a message to M. Stalin that if the Germans used gas against the Russians we would immediately retaliate by using gas against Germany, and offered to announce our intentions publicly if Stalin thought this advisable. M. Stalin agreed and suggested that the declaration be made not later than the end of April or beginning of May.

By then, however, the strategic position had changed. The rapid advance of Japan westwards laid India open to gas attack, and we were not ready to wage gas warfare in that theatre. It was open to doubt whether the promised declaration should not be postponed. Further, if we made a declaration and gas warfare ensued, we might have been considered partly responsible for the start of gas warfare. Another cause of concern was that the Russians might be tempted to accuse the Germans of having used gas against them in order to bring our counter-measures into play, thereby causing the Germans to withdraw some air effort from the Russian front to deal with the situation and thus relieving pressure on the Russians. We should have had no means of independent check on Russian statements.

On balance, however, it was deemed expedient to make the declaration at the time requested by the Russians, for the following reasons :—

(*a*) It seemed unlikely that the Japanese would be influenced by events in Europe in initiating gas warfare against India.

(*b*) Politically it was desirable to accede to the Russian request for an immediate declaration.

(*c*) Whether or not we made a previous declaration, we had already promised to retaliate if Germany used gas on Russia.

(*d*) There was some chance that a declaration on our part might deter the Germans, since, if they were engaged in a heavy scale of gas attack on Russia, we would have been able to direct a heavier weight of attack on Germany than she could have directed on us.

The text of the declaration is given in Appendix I. It was so worded that we were safeguarded from initiating gas warfare against Germany. It also provided for our being satisfied that gas warfare had actually been started by the Germans before beginning retaliation.

In the spring of 1943, a recurrence of Russian nervousness that the Germans might initiate the use of gas against them led to an offer by the Prime Minister to renew our declaration of the previous year ; this was accepted by M. Stalin, and a statement re-affirming our intention to retaliate, should the Germans use gas against the Russians, was issued on 22nd April, 1943. This warning was strengthened in June, 1943, when President Roosevelt announced that the use of gas by any Axis power against any of the United Nations would be regarded as having been committed against the United States itself and would be followed by the fullest possible retaliation.

Shortly before " D " day, consideration was given to the issue of another statement by the Prime Minister and the President of the United States to the effect that we had no intention of using gas, but would retaliate on German towns and cities if any toxic substance or form of gas was used against us. This was decided against. The Germans knew we would retaliate and were well aware of the threat from our great air superiority. Apart from the fact that the issue of any statement might have suggested to the enemy that an attack was imminent, a further announcement would have tended to give the impression that we were frightened of gas and might thus have influenced the Germans into using it rather than deterring them from doing so. Had we had firm intelligence that the enemy intended to use it against invasion it might have been worth while making a fresh announcement, but not otherwise.

Plans for Retaliation

The decision to retaliate in any theatre was reserved to the War Cabinet in all circumstances irrespective of whether chemical warfare had broken out in other theatres or not. Co-ordination of policy with the United States is dealt with in Chapter II, Section 9. The procedure for confirming the first use of gas by the enemy, which had been developed in the United Kingdom in 1941, was co-ordinated with the U.S. Forces there and also with all overseas theatres where systems were set up on the United Kingdom model but adapted to local conditions (Chapter II, Section 7).

As the danger of invasion waned and the strategic initiative passed to Allied hands, following our victories in the North of Africa, the advance into Italy and the turn of the tide in the war in Russia, more specific plans were made for exploiting gas in relation to our bombing policy against German industrial and strategic objectives.

It was intended that strategic gas bombing should be directed only against enemy territory and not against enemy-occupied territory. As the main offensive would have been against Germany, gas attacks against Hungary, Bulgaria and Roumania would, although these countries were open to attack, have been diversionary and at the expense of the main effort. The majority of suitable targets in Germany were within range of strategic bomber bases in the United Kingdom and the main attack would have been launched from there and supplemented by the use of gas with normal H.E./incendiary attacks from Mediterranean bases as a harassing measure.

Our views on the strategic employment of gas had not changed fundamentally since before the war. Gas was still visualized as supplementary to H.E. and incendiary attack and not a decisive weapon in itself unless enemy morale was at such a low level that " all out " gas attack was likely to tip the scales. Although heavy attacks with chemical weapons would have caused considerable casualties and disorganization and would have lowered morale, they could have inflicted little permanent damage on property and could not, of themselves, have critically reduced the enemy's industrial potential, except by reducing manpower by infliction of casualties. Heavy and concentrated attack with mustard gas might have compelled the enemy to evacuate the target area, and in the case of a large city thereby make it particularly vulnerable to subsequent incendiary attack. Unless a successful incendiary attack were delivered within the vulnerable period, the mustard gas attack would only have affected war industries during the period of evacuation and decontamination, although this might well have extended to a good many days according to weather conditions and other factors. With phosgene, the maximum effect would have been obtained by first using incendiaries in sufficient quantity to bring out fire fighting personnel and then some H.E. to ventilate the buildings. Phosgene was most valuable as a surprise weapon against the unprepared. Heavy attacks would have exploited this factor initially and, having established in the minds of the enemy the risks of encountering this gas, would subsequently have been made spasmodically as and when it appeared that an element of surprise might be achieved. Phosgene could not have caused any serious or lasting damage to industrial plant and, therefore, would not have seriously impeded the enemy's war industry.

In any case, the effectiveness of an " all out " scale of gas attack was unproven, whereas the effectiveness of our incendiary and H.E. bombs was established. We continued to regard these weapons as the principle means of achieving the aim of strategic bombing forces, *i.e.*, the long term destruction of the enemy's war economy and the under-mining of morale. It would have been unwise to curtail their use had chemical warfare started. Moreover, although our gas warfare production capacity was linked to our retaliatory plans, the supply of gas weapons was limited. Owing to the manpower shortage the balance of our production as a whole would have been upset if gas warfare production had been increased beyond its existing planned limits.

The basis of our plans for heavy scale gas attacks was, therefore, that, unless enemy morale were to reach such a critically low level that heavy scale gas attack alone promised to be decisive, the gas effort was best concentrated on a limited number of such attacks a month (which were within our capacity) supported by attacks in which normal weapons constituted the major part of the total bomb lift. In addition, gas (principally mustard gas) would have been included in all incendiary/H.E. attacks, to harass and to enforce anti-gas measures, thereby increasing their efficacy.

As the effect of gas attack on cities is in direct proportion to the density of the population in the area attacked, the most densely populated areas would have been chosen as targets. For phosgene, selection of targets would have been governed by the conditions prevailing at the time, *e.g.*, meteorological conditions, hours of darkness, enemy defences. In the opening stages, to exploit surprise, as many targets as practicable would have been attacked with phosgene in the shortest possible time. Mustard gas attacks would have been co-ordinated so that two or more would have followed in quick succession to create conditions which would have enforced evacuation of the target area, and would subsequently have been followed by incendiary attack. Arrangements were made to co-ordinate with the U.S. Air Forces in the European theatre the selection of targets in the light of the progressive destruction of German cities.

The Americans tended to consider gas a decisive weapon in itself, and this difference in viewpoint is referred to in more detail in Chapter II, Section 9.

In January, 1944, the use of gas against German towns was considered as retaliation for attacks with flying bombs, when these eventuated. Apart from the moral and political considerations involved in view of our being a signatory to the Geneva Protocol of 1925, it was not considered that, in the existing strategic position, it would have been to our advantage to initiate gas warfare, particularly because we knew there would be repercussions on our amphibious operations against the continent. Further, gas would not have been more effective than H.E. in the bombing operations against the flying bomb sites which were then under construction.

After the flying bomb attacks had been started, when we were considering any possible means of stopping them or mitigating them before the capture of the ramps, the question of gas attack was reviewed, but was still thought to be strategically unprofitable (*see* Section 6) and unlikely to deter the Germans from using the bombs. We could not, of course, have initiated gas warfare without the agreement of our Allies, including the Dominions and Russia.

Plans in the Event of the Germans initiating Gas Warfare against "Overlord"

Retaliation in the first instance would have been limited to air gas attack from bases in the United Kingdom, since the policy was that no gas munitions would be taken overseas. The Army gas effort was, in comparison, insignificant and the taking of ground weapons would not have been worth-while. Political and security considerations were against it, and there would not have been "room" for them in the bridgehead.

Spray was not immediately available because the use of S.C.Is. would have involved the " locking up " of aircraft that could have been used otherwise in the opening phases ; air gas action in the initial stages therefore would have been limited to gas bombing. To meet the high initial effort sufficient gas bombs (mustard and phosgene) were brought forward to airfields by " D "—8, and stocks in depots were held ready for immediate issue.

It was appreciated that if the Germans had decided to use gas against " Overlord ", there were three occasions on which he might choose to initiate it—either before the assault, or from the beginning of the assault in order to increase his chances of preventing landings taking place, or thirdly, at a later stage when he considered that the concentration of troops in the small bridgehead would present a very advantageous target. A very likely time was thought to be between " H "+8 and " H "+24, when the beaches would be covered with stores and personnel and offer an ideal target for gas.

Detailed plans were drawn up to cater for the first two contingencies. It was impracticable to appreciate the tactical considerations of the third contingency in advance.

It was intended that, in general, gas would be employed by the tactical air forces as an adjunct of normal H.E. bombardment, rather than as a weapon in itself, and that gas bombs should form a part of the normal bomb load, in proportion of not less than 20 per cent., against all targets in the tactical area except those considered unsuitable for gas, *e.g.* dispersed airfields, beach defences, batteries in the assault area. Lists of suitable targets for such attack were drawn up from lists of targets scheduled for normal bombing. Owing to the grave danger to French civilians the use of phosgene would have been limited, and each target for attack with this gas would have been chosen with that risk in mind. Lists of targets suitable for gas bombing showed separately those targets where this risk was involved. The risk in the case of mustard gas was not so great and mustard gas could therefore have been used where, on this count, phosgene was contra-indicated. In addition, targets particularly suitable for intensive gas attack were selected and listed according to whether gas was first used before or on " D " day. The bomb load for engaging such targets would have been 75 per cent. mustard gas and 25 per cent. H.E.

The types of target considered most suitable for attack are given at Appendix IV. Gas would have been included in pre-arranged attacks on request from 21 Army Group, and would have been at call for opportunity attacks on enemy troops and forming-up areas.

Apart from the infliction of casualties and from retaliation as a matter of policy,(1) the object in using gas would have been to delay the movement of enemy reserves. Had the Germans initiated gas warfare shortly before " D " day, the targets selected for gas attack were areas where Panzer formations and H.Q. had been located. If the Germans had first used gas against us on " D " day, intensive gas attack would have been directed against movements of German reserves in areas nearer the bridge-head, and against enemy columns on roads, and would have been extended on " D "+1 against road and rail movements of the more distant reserves.

Note (1).—As well as fulfilling the threats made publicly by the Prime Minister and the President, there was an important psychological factor in retaliation—namely, the effect on our own troops of being able to see for themselves on their own front that the enemy was being given a full dose of his own medicine.

The above refers to the policy for 21 Army Group and the Allied Expeditionary Air Forces. The use of gas against more distant communications and against industrial areas would have been the responsibility of the Strategic Bomber Forces, whose general policy for the strategic use of gas has already been discussed. Plans to implement this policy had been drawn up.

Under S.H.A.E.F. there was uniformity of policy for all British and U.S. forces. This included the co-ordination of procedure for reporting the first use of gas by the Germans, which is referred to in detail in Appendix XVI.

Comparison of " D " day plans for retaliation with those made in 1940 before the German invasion of the Low Countries (*see* Section 2), shows that in 1940 our main air gas effort would have been devoted to gas spray, whereas for " D " day, although spray equipment had been provisioned for the Tactical Air Forces, spray was not immediately available in the early phases of the assault. Because of the inherent difficulties in the ground handling of gas-filled S.C.I. or spray tanks, the vulnerability of the aircraft so employed, and the necessity for immobilizing aircraft which otherwise might have been used for other purposes, it was considered that gas spray should be used only in exceptional circumstances. Low spray was still recognized as a potentially useful weapon against troop movements and concentrations by day, but high spray was no longer considered effective as it did not produce a sufficiently high number of casualties for the effort expended. One opinion was that a very small high spray effort by night against, for example, troops bivouaced in the open or moving up under cover of darkness, would produce a harassing effect out of all proportion to the bomb load or the number of aircraft employed. On the other hand, formation flying at night was impracticable, and it was difficult to establish a line along which the plane could fly when discharging spray. Since gas was never used it is not possible to determine the extent to which the diversion of even a small air effort to such a task would have produced worth-while results.

Sixty days expenditure of ground gas weapons was ear-marked and held ready in the United Kingdom. Expenditure was calculated at the following rates :—

Artillery.—10 per cent. of F.F.C. intense rates in proportion three-quarters mustard gas and one-quarter tear gas.

4·2-in. Mortar.—40 per cent. of F.F.C. intense rates in proportion one H.E., three smoke, four phosgene and eight mustard gas.

For defensive equipment the basic policy laid down by Supreme Headquarters Allied Expeditionary Force was that the amount of anti-gas equipment carried was to be kept to a minimum, subject to the guiding principle that troops were not to be separated from their personal anti-gas equipment except for a definite period of operations for which the risk could be accepted. It was generally felt that the scales were weighted against the Germans initiating gas during operation " Overlord ", since it would have been to their ultimate disadvantage owing to our great air superiority.

In order to reduce the weight on the man, it was at one time decided that the leading waves of the assaulting brigades would not carry respirators. Finally, however, the risk was considered too great, since the temptation to the Germans to use gas against the beaches might conceivably have been strong enough to override prudence.

It was therefore decided that all troops landing should take personal equipment (cape and light respirator with eye-shields, ointment and cotton waste), plus unit establishment of local alarms (the rattle) carried in unit vehicles. Once ashore, respirators were to be collected at the discretion of Army commanders if it was quite clear that the enemy did not intend to use gas. All ranks wore anti-verm clothing. Vehicles carried vehicle anti-gas equipment and, where this did not include bleach, a 2-lb. tin was taken.

With the exception of beach groups, no units took unit anti-gas equipment (except the rattles and bleach referred to above), first line reserves, or decontamination reserves. (For details of scales and reserves *see* Chapter VI, Section 4.) A general transport company from the later build-up was held in a concentration area in the South of England pre-loaded with unit equipment and the reserves in question for all troops ashore by " D "+5. Had gas been used unexpectedly in the early stages, this G.T. company would have been given complete priority over to France. In these circumstances, other units would have had to take complete anti-gas equipment and it was recognized that this would have entailed displacement of other equipment. As this did not occur, unit equipment and the reserves for the whole force were phased in as part of the stores programme between " D "+4 and " D "+30 and held in bulk by Ordnance. Beach groups landed from the outset with full unit equipment and first line and decontamination reserves. All decontamination reserves included anti-verm clothing.

Second and third line reserves (for details of scales and reserves *see* Chapter VI, Section 4) were held in specially selected depots in the United Kingdom, at full scales for all troops engaged in operations, ready for issue at short notice either in the United Kingdom or overseas. Vehicles thereby rendered surplus supplemented second and third line transport in formations, but continued to be earmarked for these reserves if re-issued.

Special arrangements for base, port and beach areas were laid down ; the responsibilities of Naval, Military and Air commanders were clearly defined in this respect. Anti-gas organization in such areas overseas was the special responsibility of the military commander. For base sub-areas in the United Kingdom, including ports and hards, detailed instructions agreed to by the War Office, the Home Office, and Ministry of Home Security, were issued separately.

R.A.F. defensive preparations were equally thorough and orders for personnel entering France during the assault phase were similar to those for Army personnel. As with the plans for the offensive, there was uniformity of policy between British and U.S. Forces.

SECTION 6.—THE WAR IN EUROPE—" D" DAY TO " VE" DAY

Offensive and Defensive Capacity to Wage Gas Warfare

Shortly before " D " day, in the light of the generally satisfactory stock position of gas and gas munitions, the labour force employed on gas warfare preparations had been reviewed, but measures for a deliberate run-down had been postponed until after " D " day. Actually, a reduction was by then automatically occurring, owing to the diversion during the early part of 1944 of the activities of the Ministry of Supply production and research staffs from gas warfare to the provision of high priority demands for smoke materials, insecticides (D.D.T.) and other items.

By October, 1944, the industrial and research staffs employed on chemical warfare were some 40 per cent. less than in April, 1944, and production generally was at a " trickle " level to cater for expansion should gas warfare start. Stocks of Army weapons were in excess of calculated reserve requirements and shell charging ceased in November, 1944.

By the summer of 1944, stocks of air gas weapons (about four months' supply) were less than had been anticipated in December, 1943, when the provisioning programme had been reviewed. On the policy of increasing production to meet fully planned expenditure within three months of the outbreak of gas warfare, there would have been a gap between output and expenditure after about four to five months. By early 1945, there would have been a gap after about six months, since although the target of air reserves was nearing completion, the Ministry of Supply estimated that, with the labour conditions then prevailing, full expansion equivalent to planned expenditure was unlikely to be attained in less than eight months. This caused little concern, since the planned rate of expenditure was based on a fairly heavy and continuous scale of attack, which was thought unlikely to have to be sustained at the stage then reached in the war in Europe. As far as the Japanese war was concerned, the use of gas in the South East Asia Command would have been limited to Army and Air Forces in a tactical role, since Japanese cities were out of range of bomber aircraft in that command. This limitation would not have applied after the re-deployment of the strategic bomber forces on the conclusion of the German war. The disposition of our reserves of gas weapons would then have had to be re-considered. Allied capacity to wage gas warfare in the Asiatic and Pacific theatres is dealt with in Section 7.

In March, 1945, it was decided that production of gas and the charging of all gas weapons should be discontinued, on the understanding that, as was done when previous reductions had been made, the skilled operatives released should be earmarked for immediate return if production had to be resumed. Although " trickle " production would have enabled more rapid initial expansion in the event of gas warfare, its retention would not greatly have affected the time taken to attain peak production, owing to the time factor involved in recruiting and training large numbers of operatives without previous experience of the work.

By the end of the war the total of the Ministry of Supply labour force employed on the production of gas and the charging of gas weapons was approximately 10 per cent. of that employed at the peak period at the end of 1942. Research staffs had also been considerably reduced.

The question of closing the mustard gas plants in South Africa had been considered in the summer of 1944, but it had been decided that "trickle" production should be maintained until the close of the war in Europe as an alternative source of supply for the Far East and Australia in the event of the initiation of gas warfare in North West Europe. By the end of January, 1945, however, all available empty weapons and storage facilities in South Africa had been filled, and it was agreed that production should cease and the plant be put on to care and maintenance. After the conclusion of hostilities in Europe, British factories could have fully matched their output to the potential British expenditure of mustard gas against Japan, and in July, 1945, the South African plants were closed down.

Stocks of gas weapons at the end of the war were materially unchanged from those shown at Appendix III, although there were some increases, notably in 65-lb. and 500-lb. mustard gas bombs and in chemical shell. The total gas and gas weapon production effort for the whole period of the war and its cost is shown at Appendix V.

Production of essential items of anti-gas equipment continued to the extent necessary to meet the scheduled requirements of the Services until the end of the Japanese war. The production of all anti-gas equipment finally ceased in September, 1945. The total production of anti-gas equipment from 1939 to 1945 and its cost is given in Appendix XXI.

In all, the cost to the country of gas preparedness—defensive and offensive—amounted to some £110 million for equipment alone. This does not include the maintenance of research establishments at home or abroad (abroad these maintenance costs were a charge or part-charge on His Majesty's Government), nor does it include the cost of units, staffs, and Service establishments, including training establishments. At the peak period in 1942 the manpower bill on British resources for the Services, including the Chemical Warfare troops when they existed, and for the Ministry of Supply and Admiralty (1) labour force on production of offensive and defensive gas stores was upwards of 40,000. Ministry of Home Security commitments are not included in this figure.

Appreciation of Enemy Intentions

When the Germans did not use gas against us in the early stages of "Overlord" it was felt that the chances of their initiating gas warfare had decreased. Evidence continued to be received until the very end of the war that the Germans were maintaining their preparedness for gas warfare. The possibility that this might indicate offensive intentions could not be ruled out, but it was generally felt that these measures were of a precautionary nature. This was proved true after the war when it was revealed that the Germans had greatly feared that the Allies would take the initiative and had taken considerable pains to keep stocks of gas munitions, and even training stores, out of our hands both in Italy and North West Europe in order not to provide us with any excuse to start gas warfare, and to obviate the risk of its outbreak through some chance incident.

Note (1).—The Admiralty were responsible for production of anti-gas oilskin clothing—*see* Chapter II, Section 1.

From the military standpoint, had they used it before the breakout leading to the closing of the Falaise Gap in August, 1944, it was felt that they could not have expected to achieve more than a temporary and tactical advantage from its use, although the tightly packed concentrations in the bridgehead area were an ideal gas target, and its use at this juncture might have involved us in serious difficulties. Had we then known of their development of Tabun (nerve gas, *see* Chapter III, Section 4) our anxiety would have been intense. During the retreat after Falaise, when their supply lines were disorganized and increasingly exposed to Allied air attack and sabotage, the Germans were in no position to initiate gas warfare effectively and stood to lose more than they would have gained.

As time passed the German Air Force became less and less capable of waging gas warfare on any appreciable scale, and as the German Army neared its own soil, although from the point of view of supply of materials it was in a better position, many of its formations were composed of ill-trained and ill-equipped troops, and any temporary tactical advantage occurring from the initial use of gas was likely to be rapidly off-set.

By February, 1945, the most profitable moment for initiation of gas warfare on their Eastern Front had passed with the establishment of Russian bridgeheads across the Oder. It was known that German gas warfare stocks and personnel were being dispersed from areas threatened by the Russian advance, and that they were making extensive preparations for evacuation at a number of war gas factories. From this and other evidence it was deduced that the Germans had no intention of initiating gas warfare on that Front. Also, Berlin, shaken by Allied bombing and full of ill-protected evacuees, presented a highly vulnerable target for gas.

With the crossing of the Lower Rhine the last opportunity had passed in which they could have hoped for any substantial advantage from the initiation of gas warfare in the defence of Western Germany. The fact that they had not started gas warfare on that occasion, nor previously in defence of the Oder, was confirmation that the use of gas was not part of their strategy to defend the Reich. After the Rhine crossing, had Hitler ordered the use of gas, its application would only have been haphazard and sporadic, and on such a limited scale that it could not have seriously affected Allied operations.

At all stages in the war from " Overlord " onwards, overshadowing all other considerations, there lay the deterrent afforded by our great air superiority. Although for this reason and from the military standpoint the initiation of gas warfare by the Germans was felt to be most improbable, one possibility was present in our minds right up to the final disintegration of the Reich, namely, that either Hitler personally, or the " Party," as a last desperate resort, might order the use of gas without regard to military or humanitarian considerations. As the factors prompting Hitler to such a decision were unlikely to be reported by normal Intelligence sources, we had to maintain a state of constant readiness for this eventuality. It seemed probable that the German General Staff might have obstructed any such order on military grounds, but this factor was given less weight after the failure of the plot to kill Hitler in July, 1944, owing to the increasing control being exerted by Himmler and the S.S. (Special Guards). In the final stages it was probable that any such order would only have been carried out by a minority of Army commanders in the field.

Another contingency we had to be on guard against towards the end was that, in the general confusion, enemy units might have used gas shell in error and in ignorance of the nature of the charging, or that some units might deliberately use them on their own initiative as a final act of desperation. On our own side orders were issued for the greatest care to be taken in using enemy ammunition from captured dumps in order to avoid an accidental and unwitting initiation of gas warfare.

It was not until April, 1945, when stocks of German chemical munitions were unearthed for the first time(1) that we discovered that they possessed the new and potent war gas, Tabun, which, although the respirator afforded protection against it, presented formidable difficulties in recognition. Had the Germans initiated gas warfare at any stage there can be little question but that the results of the first impact would have been far more serious for us than was appreciated at the time.

It is interesting to note that the reason why the possession of this new gas did not influence the Germans to initiate gas warfare was that they thought the Allies also possessed it. They formed this view owing to the cessation early in 1943 of all reference to the substance D.D.T. or to organic phosphorus insecticides in American scientific journals. As they had themselves discovered Tabun during research work on phosphorus insecticides, they ascribed this silence to a similar discovery on the part of the Americans, whereas it was in fact the military potentialities of D.D.T., which were then coming to be realized, that caused the complete ban on publication of material regarding all insecticides.

The Germans themselves were ill-protected against Tabun in that they had not solved the problems of recognition nor found an antidote for the treatment of casualties. This did not worry them until they thought we also had discovered the gas. In the end their discovery of the new gas acted as a boomerang, particularly in the later stages of the war when our assumed possession of it was allied to our air superiority.

The Strategic Position

Had the Germans initiated gas warfare on any occasion subsequent to "D" day we could, with our great air superiority, have drenched German cities with a far greater weight of gas than they could have brought to bear on us, and, other things being equal, this should have reacted to our advantage.

This does not mean, however, that we should have welcomed the outbreak of gas warfare in as much as it was likely to have shortened the war or eased our task in bringing it to a final conclusion.

Operationally, had surprise been achieved, the advantage in the first use of gas would have lain with the initiator. Thereafter it would have lain with the defence rather than the attack.

Note (1).—Stocks were uncovered in a number of widely separated places within a few days of each other at Hundstadt in the U.S. Zone 20 miles N.W. of Frankfurt (Blue Ring shell only), at Espelkamp in the British Zone 25 miles N.E. of Osnabrück (gas shell of sundry chargings including Tabun and chemical mines charged mustard gas), at Orrel (Munsterlager) in the British Zone (aircraft bombs of sundry chargings including Tabun), and at Wildflecken in the U.S. Zone (sundry shell, mines, rockets, and generators, but no Tabun). Ultimately 34 German depots containing chemical ammunition or bulk stocks of gas were located of which nine were in the British Zone.

In North-West Europe, before the breakout from the bridge-head area in Normandy, gas attack on our restricted supply lines would at least have caused serious administrative repercussions. Once fluidity had been achieved movement would have been hampered through the contamination of communications and demolitions.

Later at Arnhem and Nijmegen, in the waterborne operations in Holland, and at the crossing of the Lower Rhine gas attack would have had a cramping effect.

Had gas been used before the landings in the South of France, it would have favoured the enemy there rather than ourselves. It would have increased the difficulties of the assault, and the build up would have been hindered if the ports of Toulon and Marseilles had been extensively contaminated. Subsequently it would have been likely to have delayed our advance up the Rhone Valley in view of the narrow front to which we were confined. Also it would have tended to reduce the amount of assistance the French Resistance Groups could give us.

On the Eastern Front, although the Russians had an overwhelming superiority in the air, and in artillery and mortars, and were believed to be well equipped for the offensive use of gas, it was felt that they would have stood to lose more than they would have gained had gas warfare started. It would have tended to slow down the *tempo* of the Russian advance owing to the contamination of important objectives, road and rail junctions, and possibly water supplies, and would have imposed on their lengthening L. of C. the increased strain of moving large C.W. stocks.

Had the Allies initiated gas warfare, and thereby gained the advantage of the initial surprise, it seems improbable that it would have benefited us greatly in the long run. Had a stalemate occurred in Normandy in the bridge-head area, it might have assisted us to break out ; in Italy it might have assisted an initial break through of the German defences, but in both cases when fluidity had been achieved, gas warfare would, in general, have been to our ultimate disadvantage. In addition, had we taken the initiative, we should have received a rude awakening when the Germans retaliated with Tabun.

In its effects on civil populations there were elements of doubt on both sides. In our own case, after five years of war, public morale was less resilient, particularly in those areas affected by flying bombs ; considerable potential interference with the war effort was already involved by these attacks, and gas attacks would have further accentuated this. The effect of gas on the largely unprotected civil populations in liberated areas and battle areas on the continent would have created additional problems.

In the German case, although the effects on morale were likely to have been considerable, even amounting to panic in the areas immediately affected, there was no reason to believe that the German authorities would have had any greater difficulty in holding down the cowed German population than they had already had during the past months of intensive high explosive and incendiary bombing.

As regards the Far East, it was felt improbable that the Japanese would be affected by events in Europe in any decision to wage gas warfare. The position in the Far East is discussed in more detail in Section 7.

Before "D" day we were convinced that the outbreak of gas warfare would be to our disadvantage. Whether gas warfare after "D" day would have led to an earlier conclusion of the war in Europe is a matter for speculation. On the one hand lies our overwhelming air superiority and our capacity to have swamped the German cities with gas. The German fear of our taking the initiative can be used to support the contention that the use of gas might have been decisive in shortening the war. On the other hand the operational disadvantages to us referred to above, and the incalculable factor of the effect on civilian morale both in Germany and the United Kingdom, in the latter case accentuated by the use of a gas of greater lethal potency than any hitherto known, support the contrary view.

Policy of the Allied Expeditionary Forces in France and Germany

Plans for retaliatory action were kept under constant review at all stages of operations.

The policy not to ship gas munitions overseas before the outbreak of chemical warfare held good until 10th January, 1945, when the ban on shipment to the Continent was lifted by Supreme Headquarters, Allied Expeditionary Force. At that time, although there was no change in the general appreciation about the German intentions to use gas, odd items of information were received which could have been interpreted to indicate that the Germans might be considering its use. Also, the stage of the war was being reached when, if ever they were going to do so, they might use it.

As a precautionary measure, therefore, the gas munitions shown in the footnote (1) were shipped to the Continent at the end of January under the most stringent security conditions, since, had the Germans got wind of this, the repercussions were likely to have been serious. The munitions were sent through a quiet route, as far West as possible, both for security reasons and to obviate the danger of casualties through a chance hit by a bomb. Ostend was the port selected and temporarily cleared for the purpose.(2) On arrival they were stored in an A.O.D. in a sparsely populated area between Ostend and Bruges.

On the defensive side the fundamental principle held good that a man should not be separated from his personal equipment except for a definite period, operation or locality. In general, personal equipment was carried in unit transport, but on occasions it varied from on the man at rest to separation (*i.e.*, in B Echelon) by 12 miles when in action. Overall, every man should have received his respirator within from six hours to twenty-four hours of the order to issue, except for troops in M.T. whose respirators were always with them.

Note (1).—25-pdr. and 5·5-in. gas shell—14 days at 10 per cent. contact rates in proportion 3 mustard to 1 tear gas (total : 50,000 and 12,000 of each calibre respectively). 4·2-in. mortar—14 days at 30 per cent. contact rates, in proportion 2 mustard gas to 1 phosgene (total : 28,000).

500 lb. L.C. Aircraft Bomb—4 days, in proportion 2 mustard gas to 1 phosgene (total : 1,500).

Note (2).—The necessity for stringent precautions when routing gas munitions through ports in active theatres of war where bombing is likely was evidenced by the disaster at Bari in Italy in December, 1943, of which details are given at Appendix VI.

In due course, anti-verm clothing came to be treated as ordinary clothing and was " wasted out ". Unit equipment, first line reserves and decontamination reserves, phased over during " Overlord " continued to be held in Ordnance charge and about 25 per cent. were phased up to Advanced Base early in 1945. Shipment from the United Kingdom continued to the extent necessary to maintain personal equipment and the reserves referred to above. Second and third line reserves continued to be held in the United Kingdom and were never sent to the Continent.

At the crossing of the Lower Rhine, where conditions fundamentally paralleled those of the " D " day operations of " Overlord," indications of German intentions to use gas were not strong enough to warrant the issue of a general order that respirators would be carried, and the decision was left to formation commanders. In some cases they were carried on the man, and in others they remained in transport.

SECTION 7.—THE WAR WITH JAPAN—1944 TO "VJ" DAY

Appreciation of Enemy Intentions

It continued to be thought that the chances of the Japanese initiating gas warfare were remote, although she was in a position to retaliate. With the long range bombing of the Japanese mainland by the Americans we could have carried gas warfare to the ill-equipped civil population of Japan. We possessed tactical air superiority in the battle areas, and our gas warfare equipment, organization, and supplies were superior to the Japanese.

At one time it was felt that a contingency which had to be guarded against was the indiscriminate use of gas in an emergency by a Japanese commander on his own initiative, owing to the relative autonomy of their task force commanders and the dispersed character of Japanese operations.

As with Germany, the major deterrent was the increasing Allied air superiority and our large gas potential. Reliable information during the latter part of 1944 and in 1945 consistently endorsed that the Japanese High Command were apprehensive of the Allies initiating gas warfare—particularly the United States, who were not bound by the Geneva Protocol, and amongst whom, the Japanese knew, was a school of thought, supported by the Press, in favour of waging gas warfare against them. They took active measures not to give the Allies a pretext for starting it. These included orders to commanders not to initiate the use of gas at their own discretion, the withdrawal of gas shells and bombs from forward areas, the destruction of stocks threatened with capture, and orders precluding the use of certain smoke bombs without permission from higher authority.

The possibility, however, existed that in the final defence of their own country the Japanese might resort to the use of gas as a last desperate effort and at whatever cost to the civil population, but there was no evidence that they ever, in fact, contemplated so doing.

Post war information regarding the Japanese C.W. effort and intentions will be found in Chapter VIII.

Allied Capacity to Wage Gas Warfare in the Asiatic and Pacific Theatres

Although conditions in South-East Asia were in some respects more favourable for the use of ground weapons, the bulk of retaliatory gas attack would still have come from the air. Stocks of air gas weapons in this theatre remained until the end of the war substantially the same as those shown in Appendix III (July, 1944), largely owing to shipping difficulties, and our capacity for immediate retaliation was small. General policy gave priority to the European front and the bulk of our gas munitions were located in the United Kingdom in case they should be required for use against Germany. In any case there seemed little point in diverting large stocks of gas bombs to South-East Asia so long as suitable targets for British bomber forces were lacking (*see* page 59).

Had the Japanese initiated gas warfare in this theatre before the conclusion of the German war (a remote contingency), the strength and rate of build-up of gas munitions would have been dependent on developments in Europe. Our resources were insufficient for global gas warfare simultaneously on both the European and Eastern fronts, and expansion to meet such a development would have required an increase in both manpower allocation and production plant. There was a potential reserve capacity in South Africa, but this would have required a little time to get into its stride. Even with support from the large U.S.A. capacity a co-ordinated full scale effort would have taken time to develop. Stocks of gas munitions for the United States strategical air forces in the China–Burma–India theatre were larger, but below that required to sustain the high percentage of air effort which they visualized devoting to gas in their general policy for air gas bombing. (*See* Chapter II, Section 9.)

As regards ground weapons, the stock position of artillery shell in the South-East Asia Command was good, but the shell themselves were not universally well suited for use in all conditions of terrain in that theatre. Some 4·2-in. mortar ammunition was available, but this weapon was not part of the normal armament in the theatre. Developmental work on a gas bomb for the 3-in. mortar was put in hand, and details of this and certain other development items, which did not reach finality before the end of the Japanese war, are referred to in Chapter IV.

On the defensive side, quantitively speaking, equipment was adequate. Qualitively, the lack of impregnated clothing and the inferiority, under tropical conditions, of immediately available ointments constituted a weakness rather than a danger, but as long as these deficiencies persisted we could not be considered to be adequately equipped for gas warfare, or in a position fully to exploit the offensive superiority we could have developed on conclusion of the German war. The measures taken to rectify the position are referred to in Chapter VI. For impregnated clothing we became dependent on American resources to implement the limited provision decided on, but impregnite and impregnating equipment were not forthcoming by the time the Japanese war closed down.

In the South-West Pacific the Americans had large stocks of mustard gas munitions and in the initial phase could have delivered a heavy scale of attack with mustard gas, but a decisive scale of attack would have been difficult to maintain because the majority of the chemical bombs prepared for and allotted to this theatre were, owing to shipping difficulties, still in the continent of America even when the war ended.

The Increased Effectiveness of War Gases under Tropical Conditions in Relation to the War with Japan

The increased offensive potential of mustard gas and certain other war gases under tropical conditions is discussed in detail in Chapter III.

This increased potential led to the view being held, particularly in certain quarters in America, that gas, exploited ruthlessly and on a vast scale against the Japanese, could achieve decisive strategic and tactical results, and was likely to bring the war to an earlier conclusion with a saving in loss of Allied lives. This view was also subscribed to by the Lethbridge Mission—a War Office mission which visited the Asiatic and Pacific theatres towards the end of 1943 to study on the spot the special requirements of the war against Japan both in equipment and organization.

During 1944–45 the pros and cons of the use of gas against Japan were exhaustively examined on both sides of the Atlantic. The views expressed evidenced the fundamental differences in the viewpoints of the United States Chemical Warfare Service and of the British on the role of gas warfare. The U.S. Chemical Warfare Service regarded gas as a conclusive weapon of warfare itself, whereas the British regarded it as a supplementary weapon only to be used if it seemed certain to achieve the end more quickly and more cheaply than other weapons.

We were never approached officially by the United States about any military factors which might have influenced their decision to use gas other than in retaliation. The fact that they had not signed the Geneva protocol gave them greater freedom of action in this respect than we ourselves possessed. There was considerable support in the American press for the use of gas against the Japanese, and this became intensified at times when American forces were suffering casualties in the Pacific, *e.g.* Tarawa, Okinawa and Iwojima.(1)

The strategical and political factors governing the use of gas both in the Pacific and Asiatic theatres are discussed below ; apart from these factors, and before the end of the German war, we could not be certain that if the Americans had initiated gas warfare in the Pacific, it would not have spread to Europe and Asia, and the Allies were not in a position to sustain global gas warfare on a scale sufficiently large to have had decisive effects.

On conclusion of the war in Europe we could have developed the capacity, but in the considered British view the initiation of gas warfare in the Pacific would have been strategically unprofitable. This, however, did not prevent us from contributing our full quota towards obtaining the precise data required for planning gas warfare in tropical theatres which is referred to in Chapter II, Section 9. We had to be prepared for all eventualities.

Administratively, problems connected with the storage and maintenance in a tropical climate of the vast quantities of gas munitions involved and with their provision to forward aerodromes under conditions in the Pacific would have been considerable, although not insurmountable.

Arguments for and against the Use of Gas against the Japanese in the Pacific

The arguments in favour of initiating the use of gas against the Japanese relied in the main on the contention that the enhanced power of mustard gas in the tropics, particularly when considered in relation to the inferiority of the Japanese protection against it, would be such that, if used on a vast scale with overwhelming air superiority, complete centres of resistance could be overcome by the action of air gas bombardment alone, thereby either enabling such centres to be by-passed or rendering the defenders so debilitated as to ensure a " walk-over ", and thus obviate a costly amphibious operation.

Note (1).—It is interesting to note, however, that in November, 1944, the answer to a Gallop Poll on the question " If it means an early end to the war in the Pacific would you approve the use of poison gas by the Allies ? ", was 71 per cent. against. In January, 1945, to the suggestion to use gas in retaliation for the execution of American bomber pilots by the Japanese, the answer was 52 per cent. opposed.

It was recognized that any question of the use of gas would involve some reduction in the efficiency of assault troops owing to the use of protective equipment and its maintenance under difficult conditions of supply, but it was felt that the Japanese retaliatory effort would be restricted and that any loss of efficiency involved would be outweighed by the far greater loss imposed on the enemy in the vast scale(1) of attack visualized.

It was also maintained that where gas was used as a decisive weapon both strategy and tactics would be built around it and that in these circumstances the relatively light additional logistical burden on ground troops could be accepted.

The opposing view, while acknowledging the potentialities of mustard gas and admitting that, weight for weight, it should be more effective than H.E. against such intermediate objectives as the small islands and atolls in the Pacific where no escape from severe gas bombardment was possible, nevertheless thought that any advantage thereby gained would be more than outweighed by the subsequent disadvantages of being faced with defensive gas measures in major amphibious operations of critical importance in bringing the Japanese war to a rapid conclusion, such as the capture of the Philippines or the ultimate invasion of Formosa and Japan itself. Under conditions of gas warfare the mobility of the assaulting forces would be materially reduced and a substantial loss of overall fighting effectiveness would be incurred.

If the enemy maintained only a harassing scale of gas attack an extra daily maintenance load would be imposed on the L. of C. of the assaulting force. If a heavy scale were maintained the enemy could effectively delay the rate of build-up of our assault, and also the construction of airfields and advance landing grounds on which the maintenance of air superiority might depend.(2) The carrying of large quantities of anti-gas equipment could only be achieved at the expense of the maximum rate of build-up, on which success or failure of the operation depended, and this impediment would have to be accepted irrespective of the weight of enemy reaction. Sustained and heavy scale of attack would further impede the build-up owing to the constant wearing of anti-gas equipment and the necessity for frequent decontamination. The landing beaches would present an ideal gas target.

An amphibious assault must be planned to the minute and, owing to the difficulties in timing referred to below, it could never be assessed with certainty that our own troops would not meet the after effects of our own gas. Therefore, even assuming that no enemy gas retaliation was forthcoming, our own use of gas would necessitate the carrying or even wearing of anti-gas equipment, and compel the allotment of space in assault craft for such equipment and stores.

Note (1).—In interpreting the term "vast scale" in relation to the scale of attack considered appropriate as preparation for an assault the United States Chemical Warfare Service visualized an initial attack with non-persistent gas on a scale of six hundred 1,000-lb. bombs a square mile, followed by attack with mustard gas on a scale of 50 to 112 short tons a square mile (or 1,786 to 4,000 100-lb. mustard gas bombs) during the four days immediately prior to the assault. At this time the potentialities of certain non-persistent war gases (AC and CK) were somewhat over appreciated, particularly in the United States (*see* Chapter III). In any case, in the British view, although the possibility was recognized that a sudden heavy non-persistent attack, especially at night, might have a profound effect on Japanese morale, the effects of such attacks, after the first shock, were thought to be of a decreasing order of value as gas discipline improved.

Note (2).—Initial air superiority could be achieved by carrier borne aircraft, but such air support has a limited "life."

Although the drenching of the beach defences with mustard gas could be expected to render them temporarily untenable, the silencing of coast defence batteries could not be guaranteed and the question of timing in relation to the assault presented a difficult problem, particularly where the operation was mounted from a distant base. Gas as a weapon is dependent on weather conditions, and it was impracticable to calculate precisely how long the effect of mustard gas would last on the target area. If the assault came in too soon the gas would cause casualties to our own troops, if too late, the enemy would have time to re-occupy the area. In any case gas would not destroy weapons, nor prevent determined men from accepting the risks and continuing to operate their weapons to the limit of their physical capabilities.

The means by which the gas bombardment could be put down before the assault also presented a special problem. In the Pacific much reliance for air support had to be placed in carrier based aircraft. To have used such aircraft for gas bombing would not only have exposed the carrier force to grave dangers, but was likely to affect adversely the initial establishment of air superiority over the assault area.(1) Further, bomb stowage in aircraft carriers is limited and the employment of carrier-based aircraft in gas attack would have been expensive in carriers. For the preparatory gas bombardment, therefore, it was essential that sufficient land-based bomber aircraft should be able to operate within range of the assault area.

In assessing the possible scale of retaliation, although it was to be assumed that local air superiority over the beach head would be provided initially by carrier based aircraft, it was to be expected that the enemy would be able to carry out some air attacks against the assaulting forces and craft and that these would include gas.

To sum up, although the use of gas in such an operation offered possibilities of achieving rapid and decisive results, the handicaps imposed on the assaulting force, unless full success was achieved, were such that the consequences might be serious or even disastrous.

The above considerations applied in general whether the use of gas had been initiated either prior to or on the occasion of a critically important amphibious operation. If gas had been used before such an operation, then the advantage of surprise would have been lost and we should have had to face the use of gas against us at an earlier stage.

As a result of the detailed study of the conditions peculiar to the use of gas in the Pacific, certain general conclusions were drawn which were applicable to amphibious operations as a whole.

Note (1).—In a clash between carrier task forces, each force commander must take as his primary objective (which is both an offensive and defensive one) the sinking of the enemy's carriers as early in the action as possible. Chemical bombs are ineffective against enemy vessels as compared to H.E. bombs and therefore only useful against shore targets. The bomb space in a carrier is limited and gas bombs could only be carried at the expense of H.E., which would reduce the ability of the carrier for self-protection ; when aircraft are bombed up with gas the carrier force would be at a disadvantage if located by an enemy carrier force. In amphibious operations the protection of the task force as a whole and the initial establishment of air superiority would be first priority tasks for aircraft carriers.

These were that :—

(i) The feasibility of decisively employing gas in support of an amphibious operation depends upon the availability, within range, of land bases for a sufficient number of bomber aircraft.

(ii) Gas is not a suitable weapon for employment in support of an amphibious assault as its characteristics are in conflict with two essentials for success :

(*a*) Its effects are uncertain in duration, depending on the vagaries of climate, and thus conflict with the necessity for accurate timing.

(*b*) It is essentially delaying, whereas amphibious operations depend for their success on the maximum mobility and rate of build-up.

(iii) The landing beach is an ideal target for enemy gas attack, being confined in area, open in character and a vital bottleneck on the attacker's line of communication.

In regard to the tactical aspects of the use of gas in land fighting in the Pacific area, gas offered a means of causing eviction from stubbornly-held fortified positions which were tedious and costly to reduce by normal weapons. Against this was the greater harassment caused by the wearing of anti-gas equipment in the tropics, as compared to temperate climates, and the additional commitments arising from its maintenance over difficult supply lines, in particular where human porterage was involved which was entirely unprotected against gas. Further, the general nuisance value to the individual soldier would have been great when superimposed on some of the worst living and fighting conditions met with during the war, *e.g.* New Guinea. Admittedly these factors would have applied to the enemy as well as to ourselves, and would have been of less significance in our own case where we possessed air superiority. All the same they were additional burdens we should have been loathe to shoulder.

Politically there was little doubt that public feeling against the use of gas was widespread. In their public utterances the Prime Minister had referred to the unprovoked use of gas as an outrage and the President had classified it as "this inhuman form of weapon." To have gone back on this would have prejudiced our relations with the peoples of the world, particularly those in the Far East who might have been exposed to gas attack. As we became responsible for the protection of native and friendly populations of cities from which the Japanese were driven, the political and military repercussions of enemy gas attack upon them might have been considerable, particularly had the responsibility for its first use rested with the Allies.

To sum up, apart from the political disadvantages of initiating gas warfare, it could not be guaranteed that, ultimately, the conditions of gas warfare might not be to our disadvantage, despite the profits likely to accrue in its use against the stepping stone targets of the small islands and atolls.

Military and Political Factors governing the Use of Gas in S.E. Asia

Throughout the campaigns which led to the recapture of Burma, notably in Arakan and at Kohima, many costly and stubborn actions were fought to evict the Japanese from their strongly fortified bunkers and fox-holes. These positions were often small in area and difficult to pick out from the air ; 25-pdr. H.E. shell made little or no impression on them except at point blank range, and, although effective use came to be made of the 5·5-in. gun howitzer, where it could be brought up, the potentialities of gas as a quicker, easier and less costly means of eviction were considerable. Formation commanders were much concerned by the cost in men when using existing weapons and, at one time, some desire grew up to use gas. This, however, was almost invariably counterbalanced by a reluctance to accept the overall loss of mobility involved in wearing anti-gas equipment during the attack, since the duration of effectiveness of our own gas could not be calculated to a nicety of timing sufficient to preclude wearing the equipment.

Tactically, for the smaller day-to-day individual operations, there would have been considerable scope for the use of ground gas weapons, particularly where the target was difficult to locate or to engage from the air. The developmental work carried out in the United Kingdom in this connection is referred to in Chapter IV. Even so, the main weight of attack would have come from the air, particularly where complete centres of resistance could be isolated and wholesale eviction eventually caused by the extensive use of mustard gas. In jungle areas the major air gas task would have been the contamination of tracks and paths to complicate supply and harass reinforcements.

Broadly, however, although there were many occasions on which gas might have been of great value, military exigencies were not such that they would have warranted the initiation of gas warfare.

As to supply problems, the maintenance of anti-gas equipment in Burma would have been an additional load on the tenuous L. of C., but with the increasing use of air supply it is likely that this problem would have been overcome as were the other supply problems of this theatre.

On the long term view the conditions of gas warfare would have been disadvantageous to us in the early stages of any amphibious operations, and, even assuming our complete air supremacy, considerable harm could have been inflicted by the use of ground gas weapons against us in the first few days of a landing by upsetting our time-table.

In one other case the use of gas would have been seriously to our disadvantage : against the "Chindits" (the Long Range Penetration Groups) during their operations in 1943 and 1944 behind the Japanese lines.

Political considerations were also against the use of gas in this theatre. Irrespective of which side had initiated its use, our own use of it against targets in countries whose government we intended to reassume would have been likely to have had an adverse effect politically, which would have been all the greater had we been the initiators. Possible harm to the civil population in Burma and Malaya—and also in Siam—would have placed limitations on tactical and strategical use of gas from the air. Targets other than in Japanese occupied territory were not within range of R.A.F. bombers in this theatre. Only the United States Strategical Bomber Force in China could have reached suitable targets.

Furthermore large numbers of prisoners of war—many of them Indian Army—were held in Malaya and Siam. Had we initiated gas warfare they would have been open to reprisals by the Japanese. The reaction on public opinion would have been unfavourable and, in India, likely to have had political repercussions.

The effects of even an occasional and very light scale gas attack on the industrialized areas of Calcutta and in Bengal were likely to have involved grave interference with the war effort owing to the wholesale desertion of labour which it was likely to have caused. Towards the end of the war the probability of such an attack waned rapidly as we advanced into Burma, but it nevertheless remained as long as Japanese aircraft were in range, and was another factor to our disadvantage when considering the use of gas in this theatre.

CHAPTER II.—ORGANIZATION

NOTE : *The term " Chemical Warfare " initially denoted " Gas Warfare " only. From* 1942 *onwards it came to be used loosely to cover Gas, Smoke, and Flame Warfare, in which sense it was always used by the United States Forces. In this volume, where gas warfare only is referred to, the term " Chemical Warfare" is, as far as possible, avoided, but, of necessity, has to be used where it forms part of the title of a Committee, unit, or appointment, etc. The sense in which the term is used should be clear from the context in the light of this note.*

SECTION 1.—ORGANIZATION TO CONTROL POLICY, RESEARCH, DEVELOPMENT, SUPPLY AND PRODUCTION

Introductory

During the 1914–18 war, although gas policy matters were handled by the relevant branch of the Staff as required, the War Office role had been largely that of an intermediary between the Gas Directorate of G.H.Q., The B.E.F. in France and the Ministry of Munitions which, from 1917 onwards(1), was responsible for all offensive and defensive research and for the production of all gas stores.

After the Armistice gas warfare research and experiment was in abeyance. It was restarted towards the end of 1920, by which time the Ministry of Munitions had been abolished, and the War Office, in addition to the policy functions of the General Staff, assumed those previously carried out by that Ministry. These latter functions became the responsibility of the Master General of Ordnance, which he discharged through the Director of Artillery. This organization continued until 1938 when the Department of the M.G.O. was absorbed into the Department of the Director General of Munitions Production, which had been formed in 1936 for administration of Ordnance factories and business affecting the production of munitions. In the summer of 1939 the D.G.M.P's organization was merged into the Ministry of Supply, and henceforward the War Office was no longer responsible for the administration of gas warfare research and experiment or for the production of such stores. The D.G.M.P's responsibilities for the Ordnance Services and consequently for provision, storage, issue and maintenance of gas stores passed to the Quarter-Master-General.

The above shows broadly the major distribution of responsibilities as between the War Office and the Ministry of Supply as they stood at the beginning of the war. This remained fundamentally unchanged throughout the war.

It should be noted that the source of technical advice moved outside the War Office, and for a time the balance of the " machine " was disturbed ; this was a factor underlying the development of organization at the War Office which is discussed later in this chapter.

Note (1).—Until 1917, the Army Medical Department of the War Office had been responsible for defensive research and manufacture of respirators and other defensive equipment. From 1917 onwards there was unified control of all research and production. (" Official History of the War—Medical Services—Diseases of the War—Vol. II," pages 324 *et seq.*)

Organization for the formulation of Major Policy and for Inter-Service Co-ordination

Before the war, major policy in connection with preparations for gas warfare was formulated by the Committee of Imperial Defence. There was, however, no sub-committee whose special responsibility it was to co-ordinate plans and policy between the Services and to keep the position under constant review.

In the summer of 1939, a special sub-committee was set up to progress outstanding matters resulting from the Anglo-French conversations on gas warfare. It was hoped that this sub-committee would develop into a permanent body to advise on gas warfare policy, but it was dissolved with the majority of the sub-committees of the Committee of Imperial Defence on the outbreak of war.

From then on, until September, 1940, policy was formulated by the submission of papers by Service Chiefs, or of joint papers, to the Chiefs of Staff Committee which was a War Cabinet Committee. There was no specific machinery for co-ordination of gas warfare matters between the Services until the formation in September, 1940, of the Inter-Service Committee on Chemical Warfare (I.S.C.C.W.) as a sub-committee of the Chiefs of Staff Committee. The circumstances leading to its formation are referred to in Chapter I, Section 3, and its terms of reference are given in Appendix VII.

The chairmanship of the committee was held by an R.A.F. Officer(1), the R.A.F. being the major gas arm, with representatives at Director's level from the War Office, Air Ministry and Admiralty, and the Controller of Chemical Defence Development, Ministry of Supply, other officers attending as required by the subject matter under discussion. The secretaries were provided from the Chiefs of Staff Secretariat which was a sub-division of the War Cabinet Secretariat.

It is not possible to give a description here of the working of the War Cabinet and Chiefs of Staff organization to show in detail how a decision was reached. Broadly, the I.S.C.C.W. was the " working party " who thrashed out the many problems affecting the joint gas effort of all Services and submitted their recommendations to the Chiefs of Staff for a policy decision, which was subject to the overriding authority of the Prime Minister and the War Cabinet. The I.S.C.C.W. was not, however, responsible for strategic planning or for the preparation of Intelligence appreciations affecting the probability of the outbreak of gas warfare. These appreciations were carried out, as necessary, by the applicable sections of the Joint Staff Secretariat (under the Chiefs of Staff) *viz.* the Joint Planning Staff and the Joint Intelligence Committee. In fact, the appreciations of these two committees normally formed the background for policy decisions and for the work of the I.S.C.C.W.

The War Office—General Staff

The development of the organization to handle gas warfare policy is linked with that which led up to the eventual formation of the Arms Directorates of the General Staff which largely had their roots in certain portions of the Staff Duties Directorate. This development is shown diagrammatically at Appendix VIII, where the development in the organization above Director level is also shown.

Note (1).—Air Vice Marshal (later Air Marshal) Sir Norman Bottomley was Chairman of the Committee continuously from January, 1942, until the end of the war.

It is both undesirable and impracticable to divorce the General Staff organization handling gas warfare from that handling " special weapons " ; details affecting individual weapons are given in the relevant parts of the three volumes of this history.

Pre-War

Before the war Gas Warfare was the responsibility of the Director of Staff Duties. Gas questions were handled by S.D.2 together with other matters in the normal course of staff business affecting fighting efficiency (other than training) regarding personnel, arms and equipment. Major policy was the responsibility of the D.M.O., and the D.M.T. was responsible for training and tactical doctrine.

S.D.2, however, did not have a qualified officer to handle gas matters—in fact until the end of 1937, when a G.S.O.II was appointed in M.T. Directorate, there was no appointment specifically for gas warfare in any branch of the General Staff.

As a result, policy frequently emanated from sources outside the branch of the Staff officially responsible for its formulation. These sources included the Director of Artillery (the technical and research authority in the M.G.O's Department) ; the Inspector of Royal Engineers, during 1936/37 (owing to the R.E. being the parent Arm for gas measures (*see* Section 3 and Section 4)) ; the Commandant of the Anti-Gas Wing of the Small Arms School, or, from the end of 1937 onwards, the M.T. Directorate, whose influence was further strengthened in March, 1939, by the formation of the Army Gas School and the appointment of its Commandant as *ex officio* Inspector of Anti-gas Training (Section 5).

From the Outbreak of the War to June, 1942

It was not until November, 1939, that a gas warfare section—S.D.9(*c*)—was set up under the D.D.S.D. (R.E. and C.W.) (Brigadier C. R. Ling, R.E.) in the Staff Duties (Weapons) Directorate which had broken off from the original Staff Duties Directorate. In addition to gas warfare, the D.D.S.D. (R.E. and C.W.) had certain R.E. responsibilities and also dealt with policy and scales for infantry and horsed cavalry equipment and " B " and R.A.S.C. vehicles. There was a direct link between R.E. and gas warfare since the C.W. troops were raised on this Corps (*see* Section 3).

In October, 1941, an Engineer-in-Chief was established, under the Q.M.G., who also acted as adviser to the C.I.G.S. on the General Staff aspects of engineer matters and in this connection assumed certain functions previously carried out by the D.D.S.D. (R.E. and C.W.). In so far as gas warfare was concerned he discharged his responsibilities through the Deputy E.-in-C.(B) whose executive section—E.4.—became responsible for the technical administration of chemical warfare units and was also the source of technical advice on gas matters within the War Office, and for technical liaison between the War Office and the Chemical Board of the Ministry of Supply.

This S.D.9(*c*)—E.4. organization continued until June, 1942. During this period certain other appointments were intimately connected with the formulation of gas warfare policy. On the outbreak of war the Inspector of Anti-Gas Training (Brigadier I. S. O. Playfair) became Inspector of Chemical Warfare (while still combining this appointment with that of Commandant, Army Gas School) and became adviser to all General Staff Directors over the whole range of gas warfare policy, doctrine and training. In July, 1940, the duties of this appointment were further extended to include those of Director of Chemical Warfare at G.H.Q. Home Forces, (this joint appointment, I.C.W.–D.C.W. was taken over by Brigadier (later Major-General) Crawford). From November, 1940, the appointment of I.C.W.–D.C.W. was divorced from that of Commandant, Army Gas School. In October, 1941, the appointment of Inspector was abolished on the creation of the appointment of D.E.-in-C.(B), this latter combining his duties with those of D.C.W., Home Forces (Brigadier Crawford, initially. After a few months he handed over the joint appointment (D.E.-in-C.(B)–D.C.W.) to Brigadier R. M. H. Lewis, R.E.)

1942

The gradual trend towards the formation of the Arms Directorates from 1941 onwards can be seen from the charts at Appendix VIII.

In June, 1942, the S.D. (Weapons) Directorate was abolished(1) and a new second class Directorate, the Directorate of Weapons and Vehicles,(2) was created. This Directorate was responsible for chemical warfare (which included smoke equipment for use in forward areas, and flame warfare less static flame installations), field camouflage, infra-red equipment,(3) and infantry equipment, small arms, " B " and R.A.S.C. vehicles. Smoke weapons had previously been the responsibility of Directorates controlling the user arms, and development of individual user requirements had proceeded independently. Flame had been ill co-ordinated between the War Office and the Petroleum Warfare Department of the Ministry of Fuel, under whose aegis development had been taking place largely in direct liaison with G.H.Q., Home Forces in regard to static flame installations, and with the Home Guard in relation to flame appliances for that force(4).

The D.E.-in-C.(B) and E.4 continued to act as the technical advisers on gas warfare and smoke weapon matters.

The detailed lay-out and responsibilities of the W.V. Directorate, when formed in June, 1942, is shown in Appendix IX. It was still largely a weapon, vehicle and equipment directorate and its function as the sponsor branch for the subjects it dealt with was not fully recognized.

1943

In the latter part of 1942 and early in 1943, developments, which are discussed below, led to a reorganization of the W.V. Directorate on the assumption by the D.W.V. of certain new responsibilities and the shedding of some old ones.

Note (1).—The reorganization within the General Staff as a whole at this period should be noted and can be seen from Appendix VIII. The responsibilities for Artillery (other than A.A.) of the D.S.D. (Weapons) passed to the D.R.A.

Note (2).—Brigadier (later Major-General) G. Brunskill was appointed D.W.V. and subsequently D.S.W.V. which appointment he held until after the end of the war.

Note (3).—For details *see* Volume II, Part V.

Note (4).—For detail of organization for flame warfare *see* Volume II, Part III.

Liaison between the Army (Home Forces), the War Office, and the Chemical Defence Research Department (Ministry of Supply), over smoke stores was still unsatisfactory. User reports on both experimental and approved gas and smoke stores were not receiving proper attention. There was a lack of co-ordinated action to obtain improved stores as a result of user experience. In addition, individual arms were still stating their individual development requirements of smoke weapons unilaterally to the Ministry of Supply. This not only led to considerable confusion, but also to difficulty in the allocation to the various weapons of those smoke producing materials which were scarce. Furthermore, the C.W. technical officers in the General Staff of lower formations were beginning to play a part in the smoke screening of ports abroad, and there was need for the examination and collation of their reports and experiences if the best use was to be made of the information for future development

The D.E.-in-C.(B), as the technical adviser in the War Office for gas warfare, was becoming involved in preparations for smoke warfare, and although officially a technical adviser, was also in practice in some cases stating requirements.

This situation led to the suggestion that the General Staff and E.-in-C. branches, W.V.1(*c*) and E.4, should be combined under one Director in the General Staff. From every point of view this proposal seemed practicable.

At the end of 1942 the Chiefs of Staff decided that the manning of, and administration of A.A. smoke screens at home and abroad should be the responsibility of the Army (with certain provisos about employment of local labour in certain specified overseas areas), and that the responsibility for endorsing material requirements for smoke defence plans at home and abroad should rest with the War Office. This decision led to the transfer of the administration of all A.A. screens in the United Kingdom from the Ministry of Home Security to the War Office.(1) This involved increased responsibilities for the D.W.V. and added force to the proposal to combine W.V.1(*c*) and E.4.

Another development was the decision to abolish the C.W. Groups R.E. (*see* Section 2). This decision finally severed the link that had hitherto existed between the R.E. and gas warfare, and there was little justification for the E. in C. continuing to be charged with a special gas role.

Other items where more vigorous G.S. action was required were in connection with the rapidly increasing and urgent demands from all theatres of war for coloured smoke weapons and for flares,(2) and with the necessity for the guiding hand of the General Staff in relation to flame warfare weapons and their development by the Petroleum Warfare Department(3).

Finally, there was no officially accepted " sponsor " directorate for each of gas warfare, flame, smoke, camouflage and deception. The lack of co-ordination on smoke matters has already been referred to above. The

Note (1).—*See* Volume II, Part IIA for detail of history of organization for smoke. From this date on the War Office became responsible for guiding design and development and arranging provision of every type of smoke producing weapon other than those required exclusively by the Admiralty (smoke floats) or Air Ministry (smoke screens laid from the air).

Note (2).—*See* Volume II, Part II (B).

Note (3).—*See* Volume II, Part III.

position in respect of gas warfare was better, but the division of responsibility between the D.M.O., D.M.T., D.W.V. and E.-in-C. was ill defined. Co-ordination on camouflage and deception(1) was imperfect because the D.M.T., E.-in-C., D.W.V. and D.R.A. were each handling various aspects individually.

The above factors led to a reorganization of W.V. Directorate in March, 1943. The appointment of D. E.-in-C.(B) lapsed—the incumbent remaining Director of Chemical Warfare, G.H.Q., Home Forces only. E.4 were absorbed into W.V. Directorate, the S.O.R.E.II of E.4 becoming the G.S.O. I.(2) of the new chemical warfare section W.V.1.

The lay-out and responsibilities of the reorganized Directorate are shown in Appendix IX. In so far as administration at the War Office of gas, smoke, and flame warfare were concerned, all three had become merged into " chemical warfare."

It will be noted that the D.W.V. was no longer charged with responsibilities in connection with infantry weapons and equipment, and assault weapons and anti-tank mines. The former passed to the Director of Infantry, who was established in the General Staff concurrently with the reorganization of W.V. Directorate, and the latter to the Director of Royal Engineers.

This reorganization was in effect only an interim step to the establishment of a new and enlarged Directorate in July, 1943. The Weapons and Vehicle Directorate had become over-loaded, and expansion to meet the increased volume of work as well as the assumption of further responsibilities led to the establishment of a first class Directorate—The Directorate of Special Weapons and Vehicles. The new responsibilities included jungle warfare equipment,(3) tropicalization of all equipment(3) and, in the case of camouflage, the co-ordination of policy and production questions for the carrying out of cover plans. Henceforward until the end of the war this Directorate was a " sponsor " branch for all " arms " not within the special sphere of other Arms Directorates. The fact that S.W.V. Directorate was the sponsor branch, did not mean that it usurped the special responsibilities of the D.M.O., D.M.I. and D.M.T., but that it was the nodal point round which the organization to deal with the subjects concerned revolved.

1944–1945

During 1944/45, in view of the manpower position, no new appointments were allowed to be created in the War Office without a corresponding decrease elsewhere and the pressure of work on all staffs was very considerable. Early in 1944 an arbitrary cut in all War Office staffs was made and there was some small reorganization of the S.W.V. Directorate to meet this. The organization of S.W.V. Directorate from February, 1944, onwards is shown in Appendix IX.

Note (1).—*See* Volume III, Part VI for detail of organization for camouflage, and deception or, as it is now known, " visual warfare ".

Note (2).—Lt.-Colonel (later Brigadier) O. H. Wansbrough Jones.

Note (3).—*See* Volume II, Part V.

The Directorate acted as rear link to the Lethbridge Mission, which toured India, Burma and the South West Pacific areas in the winter of 1943/44 to obtain first hand knowledge of the development requirements for the war against Japan, in particular of the requirements for jungle warfare. This necessarily gave rise to a considerable volume of work for all S.W.V. sections. On its return, a G.S.O. I of the Mission was appointed G.S.O. I of S.W.V.6 (Jungle Equipment) to progress the Missions recommendations, the appointment being compensated for by the reduction of a G.S.O. III in S.W.V.8 (Camouflage) where there was some falling off of work due to operational reasons. In April, 1945, S.W.V.6 was merged with S.W.V.1 to become S.W.V.1 (*a*) as all main problems of items of jungle clothing and equipment had, for practical purposes, been solved.

In July, 1944, the Chemical Warfare Section, S.W.V.2, assumed two new responsibilities—namely for the screening and co-ordination of requirements for incendiary projectiles([1]), and for anti-insect measures([2]) (in particular anti-malarial stores), which in the latter case involved giving assistance in the production of user instructions.

As the history of vehicles forms the subject of another book in this series this aspect of the activities of S.W.V. Directorate is not dealt with here.

General Staff Committee

Before the war a number of chemical warfare committees were set up by the General Staff from time to time to assist in the co-ordination of gas warfare matters and in the formulation of policy. They were generally low level committees which dealt with such matters as were put to them by branches from time to time rather than committees which actively and continuously progressed policy.

On the outbreak of war the " General Staff Chemical Warfare Committee " was established as a standing War Office committee on the initiative of the D.S.D. The Committee was under the Chairmanship of D.D.S.D. (R.E. and C.W.) and its terms of reference were " to review gas warfare policy, organization and equipment, both offensive and defensive, in the light of changing conditions." Branch representatives were frequently at a low level (G.S.O. II).

This Committee met periodically (on an average about five times a year) until January, 1942. It cannot be said to have actively progressed gas warfare policy. Rather it was called together to discuss piecemeal particular problems as they arose and it became a convenient medium for consultation between the controlling section—S.D.9 (*c*)—and other branches of the Staff. It also afforded a meeting ground with representatives from the Ministry of Supply. Although it performed a useful function in this latter connection it by no means met the full requirements of liaison between the policy and the research and production sides.

Note (1).—*See* Volume II, Part III.
Note (2).—*See* Volume II, Part IV.

It became redundant in July, 1942, on formation of the "General Staff Policy Committee on Weapons and Equipment." This Committee, which was on Directors' level under the Chairmanship of the D.S.D., was set up with the object of formulating G.S. policy on research, design and production of all weapons and equipment which the Army was likely to require. In the past the procedure had been rather that a weapon had been invented or produced haphazardly and the Army had then utilized it in the best way it could. Gas warfare weapon and equipment policy was henceforward dealt with by this Committee.

This Committee was the precursor of the "Organization and Weapons Policy Committee (O.W.P.C.)" which was formed as an Army Council committee in July, 1943 under the Chairmanship of the D.C.I.G.S. Membership consisted of the A.C.I.G.S., D.M.O., D.M.T., D.S.D., the Arms Directors and the Scientific Adviser to the Army Council. Membership was personal to the Directors and in their absence their Deputies could attend but could not vote. Henceforward all major General Staff decisions on organization and equipment policy were taken by this committee, subject to approval by the C.I.G.S. or the Secretary of State. Equipment and weapon development requirements were formulated, and circulated to the Ministry of Supply with the full weight of the Army Council behind them. It is to be noted that the formation of the O.W.P.C. coincided with the final evolution of the Arms Directorates.

At the same time as the "General Staff Policy Committee on Weapons and Equipment" was established in July, 1942, an Army Council Committee, "The Weapons Development Committee," was set up as a liaison committee between the War Office and the Ministry of Supply for the general co-ordination of munition development in accordance with G.S. policy, and to bring to light any new lines of development the Ministry might have in mind. Being a high level inter-ministerial committee it could deal with questions of priorities with the necessary authority.

Factors Underlying the Development of General Staff Organization

Before the formation of the G.S. Policy Committee on Weapons and Equipment, the general procedure by which policy was formulated was that the branch primarily concerned with the particular aspect of the subject on which a decision was required either obtained the opinion of other branches by direct minuting, or thrashed out the problem at committees, such as the General Staff Chemical Warfare Committee, or called an *ad hoc* committee of those concerned. The case was then put to the Director primarily responsible, *e.g.*, D.M.O. for major policy, D.S.D. for organization, D.M.T. for training and tactical doctrine, who either gave a decision or put the matter to higher authority according to its importance.

With this organization there was a tendency for decisions to be taken piecemeal, and, as expansion occurred and the volume and complexity of subject matter increased, the problems of co-ordination became more and more intricate. Directors found themselves handling matters which were not strictly within the briefs of their Directorates, and proper co-ordination became increasingly difficult.

Furthermore, the rapid advances in scientific development made it vital that, while leaving the scientists a free hand in fundamental research, they should be given sound direction on the broad requirements with which weapons and equipment should comply, if their activities were to be directed to the most useful ends, and also that the General Staff should put its house in order, to control the many developments as they occurred. It was not only necessary that the soldier should be interpreted to the scientist, but *vice versa*, and in this connection the appointment of a Scientific Adviser to the Army Council in 1942 is to be noted.

It was natural that, as long as co-ordination in the War Office was imperfect, liaison with the Ministry of Supply, who were responsible for research and development, presented difficulties. The authorities in the Ministry of Supply had a number of " masters " to refer to, and they did not always speak with one voice.

The above factors, as well as those connected with the administration of a vastly expanded Army operating in many theatres of war and closely integrated with Allied Forces, had a major bearing on development of General Staff organization. They led to the establishment of the Arms Directorates as " sponsors " for particular subjects, with clearly defined co-ordinating responsibilities. These Directorates were linked to a policy forming body at the highest level—the O.W.P.C.—which not only promulgated policy but also formulated development requirements for the guidance of the scientific and technical authorities.

The organization to handle gas warfare matters is only one case in point in connection with the above general development. Before the war, in 1936, the research side had complained that they received insufficient direction from the General Staff. They frequently felt that they were groping in the dark and that their developments might prove useless on account of being out of line with General Staff views. The General Staff reply to this was to increase its representation on the Chemical Defence Committee (*see* pp. 70–71, " Organization to Control Research ") and raise the level of its representatives. Early in 1940 there was a similar complaint. Again the General Staff considered that representation from the Ministry of Supply on the newly formed General Staff Chemical Warfare Committee met the case. At this stage the General Staff were somewhat reluctant to be at all dogmatic about probable developments in warfare once the stalemate on the Maginot line had been broken.

In some respects in the early days of the war co-ordination and G.S. direction were better effected in regard to gas warfare than was the case with some other subjects ultimately sponsored by D.S.W.V. The earlier organization had the advantage of the closely knit C.W. Staff organization with its tentacles down to divisions (*see* Section 2). S.D.9 (*c*) was closely linked with this and with the offensive gas organization referred to in Section 3 of this Chapter. The I.S.C.C.W. came on the ground early in the war for inter-service co-ordination, and the eventual establishment of E.4 oiled the machinery for interpretation between the scientists and the soldiers. On the other hand the fact that S.D. 9 (*c*) was in a large degree the co-ordinating point within the War Office led to its performing functions which were not legitimate to a Staff Duties (Weapons) branch. Further, the set-up in which the Inspector of Chemical Warfare and the D.E.-in-C. (B) combined these functions with that of

Director of Chemical Warfare, at G.H.Q. Home Forces, frequently led in practice to policy being formulated outside the General Staff. Not infrequently the Gilbertian situation arose of the same officer having failed to put through a policy in one capacity, changing his office and putting it through *via* other channels. Also there was more than one " voice " to speak to the development side. As long as such oscillatory machinery existed for policy making, co-ordinated and authoritative guidance was not in all cases forthcoming. It was not until the re-organization of the W.V. Directorate in March, 1943 that fully satisfactory liaison with the Ministry of Supply was established in detail, and supported by authoritative policy announcements formulated by the O.W.P.C.

The closest personal contact was maintained between S.W.V.2 and the Chemical Defence Department and the Chemical Board of the Ministry of Supply, the G.S.O.1 of S.W.V.2, for example, becoming Chairman of a Chemical Board Sub-Committee.

Development in other Branches of the War Office

The developments and re-organization which occurred in other branches of the War Office to meet the pressure of war and changing circumstances do not come within the scope of this book. No special organization was set up in other branches to deal with chemical warfare matters, such matters being handled *inter alia* by the relevant Directorates and branches of the Military Secretary, the Adjutant General (including Medical(1),) and of the Quarter-Master-General.(2)

The War Office " screened " the provisioning of all anti-gas stores except those peculiar to a particular Ministry, *e.g.*, civilian respirators which were solely the responsibility of the Ministry of Home Security. Allocation between the Services and Ministries was effected by the Inter-Departmental Committee on Anti-Gas Stores (I.D.A.G.).

Organization to Control Research

The organization for research, which grew up under War Office auspices during the inter-war period, and which was transferred to the Ministry of Supply in 1939, had been under the administration of the Director of Artillery in the Master General of Ordnance's branch. It consisted of an advisory committee—the Chemical Defence Committee—and the Chemical Defence Research Department(3) under a Chief Superintendent.

Note (1).—The Director of Pathology was responsible for the medical aspects of Chemical Warfare.

Note (2).—The Director-General of Army Requirements was responsible for the co-ordination of all War Office requirements for equipment, stores, and material obtained from the Ministry of Supply.

The Director-General Army Equipment (later Deputy Quartermaster-General Army Equipment) was responsible for provision, receipt, storage and distribution.

Q.M.G. Directors, *e.g.* Director of Supplies and Transport, Director of Transportation, were responsible for gas matters which affected their respective services.

Note (3).—Up to 1931 the Committee and Department were known as the Chemical Warfare Committee and the Chemical Warfare Research Department respectively. In that year, they were re-christened for reasons of policy since we did not wish to be accused of developing offensive gas warfare during the disarmament era. Henceforward the term " chemical defence " covered all defensive and offensive gas matters.

The Chemical Defence Committee was composed of eminent scientists (chemists, physiologists, physicists, etc.), representatives of chemical industry, and the Chief Superintendent of the Research Department. The various interests in the Admiralty, War Office, Air Ministry, Home Office and India Office were represented as required. Sub-committees, *e.g.*, respirator, physical, physiological, medical, chemical, meteorological, each under a chairman appointed by the main Committee, were set up for detailed examination of particular aspects, matters of general research policy being referred to the main Committee. The Secretariat for the Committee and Sub-Committees was provided from personnel in the headquarters of the Research Department.

The function of the Chemical Defence Committee was to act as an advisory body on all matters pertaining to gas warfare which might be referred to it by the Admiralty, War Office, Air Ministry and any other Ministry concerned, and also to raise questions of special importance without waiting for direct reference. It was charged with advising broadly the lines on which chemical defence could be developed, and with advising and assisting in the maintenance of an efficient organization for the conduct of research, in which connection it received reports on experiments and investigations and made recommendations for further action, bearing in mind the special requirements of the Services. It was also charged with giving advice on the best means of utilizing the country's scientific resources for the development of chemical defence, and on the best method of mobilizing both these and industrial resources in the event of a national emergency.

The Research Department had its headquarters in London. With the general advice of the Committee, it controlled the work of research centres at universities and similar institutions, where investigations of a fundamental nature were carried out, and also the work of the Chemical Defence Experimental Station at Porton, which was the main centre for applied research and for field trials as well as for fundamental work that could not, for various reasons, be carried out extra-murally. It also controlled the research establishment at Sutton Oak whose particular function it was to translate laboratory methods of making war chemicals into full-scale production processes.

The organization and functions of the Committee and the Department remained fundamentally unchanged, after transfer to the Ministry of Supply, until the end of the war, except for expansion and changes in designation. The Chemical Defence Committee was renamed the Chemical Board. The organization of the Chemical Board and of the Department at the end of the war is shown at Appendix X.

Organization for Development and Production

Up to the summer of 1939, the D.G.M.P. at the War Office had been responsible for research, development and design, approval and specification, provisioning, production, inspection, and supply of all gas stores and munitions. All these became functions of the Ministry of Supply, with the exception of the responsibility for provisioning and for the co-ordination and presentation of demands, which passed to the Q.M.G.

It is impracticable in this book to show the development of the complex organization to handle these matters in the Ministry of Supply *pari passu* with development at the War Office. Rather is it more profitable to examine the general procedure by which a gas or smoke store gets into production.

A demand may arise for a particular piece of equipment either in the field or as a result of staff representation. It is first considered by the relevant Director at the War Office, *i.e.*, for gas and smoke stores, the Director of Special Weapons and Vehicles, and passed through the General Staff machinery for stating development requirements. The G.S. requirement would then be passed to the Ministry of Supply with some indication of priority and of the scale of provision which would be necessary if ultimately accepted.

In the Ministry of Supply it would pass to the Director General of Artillery. In so far as gas and smoke stores and munitions are concerned, the Director General of Artillery holds a similar position in the Ministry of Supply to that of the D.S.W.V. at the War Office, *i.e.*, he is a nodal co-ordinating point. D.G. of A. after examining the requirement would pass it to the Chemical Defence Experimental Station at Porton through the Controller, Chemical Defence Development, and to the Chief Engineer Armament Design. These two authorities have respectively the assistance of the Chemical Board and the Ordnance Board, the latter being concerned where questions of ballistics are involved. Development proceeds, primarily under the aegis of the Chemical Defence Department, early contacts being made with the production side to ensure that long-term factors, such as availability of materials, receive due consideration during design. Pilot designs are then produced leading to the production of a prototype, or, where a prototype is not necessary, a small-scale production order may be placed to clear questions on the suitability of design for large-scale manufacture. This is inclusive in the production of a prototype. Trials—including Service trials where necessary—take place at appropriate stages in this process of development.

It is then passed back to the War Office for acceptance and, if accepted, for detailed provisioning and the placing of demands. After acceptance by the War Office, designs are approved and specified by D.G. of A. It then goes on to the production chain and thence to the inspection chain, to ensure that production complies with specification, and is finally delivered to Army ordnance depots. In some cases the War Office places a trial production order before final acceptance for provision on a wide scale.

When a new item arises in the first place from fundamental research or scientific development, information about it is passed from the Ministry of Supply to the War Office Director concerned, and subsequently if a requirement is found for it the matter is dealt with through the same channels.

The above is a brief description of the somewhat complex process of bringing a piece of equipment into Service use. The many changes in the organization of the Ministry of Supply during the war were largely directed to speeding up the procedure and perfecting it so that each vital interest was consulted at the appropriate stage. Unless this occurs it is probable that at some advanced stage it would be necessary to re-trace steps, with consequent delay, owing to an unforeseen inability to comply with some detail of specification, or the specification might even be such that it would be impracticable for the production authorities to give effect to it.

Production of oilskin anti-gas clothing was carried out under Admiralty auspices since they were the major users of oilskin for general purposes. In this connection, they acted as the agents of the Ministry of Supply and development procedure was similar to that described above, through the relevant Director General at the Ministry of Supply.

In the development of air gas weapons, the Ministry of Aircraft Production stood in the same relation to the Air Ministry as did the Ministry of Supply to the War Office. In this case, however, while the Ministry of Aircraft Production was responsible for all " ironmongery " aspects and for design and final approval, the Chemical Defence Department of the Ministry of Supply acted on an agency basis for the Ministry of Aircraft Production in so far as research and development were concerned.

The channels for development of other equipment for which the D.S.W.V. at the War Office was responsible were similar, but the authorities concerned varied ; these are referred to in the parts of this book dealing with each particular type of equipment.

SECTION 2.—CHEMICAL WARFARE STAFFS

The Inter-War Period

In 1936, the Corps of Royal Engineers was made the parent arm for all gas measures. At the time, one of the objects of this was to provide a source of qualified officers in war, but, in the event, staff officers were drawn from Arms other than R.E., the primary concern of this Corps being the Chemical Warfare Troops (*see* Section 3 of this Chapter).

A logical outcome of this policy was that chief engineers and commanders, R.E., were charged with the responsibility of acting as advisers in gas warfare to their commanders. Consideration was given to the provision of staff officers to assist them but was not implemented until later. In 1938 the following appointments were included in War Establishments :—

G.H.Q. 1st Echelon ..	One G.S.O. I and one G.S.O. III for gas warfare on the staff of the Chief Engineer.
H.Q. Corps ..	One G.S.O. II for anti-gas and air raid precautions duties under the Chief Engineer.
H.Q. Division ..	Captain, R.E., for anti-gas and air raid precautions in H.Q. Divisional Engineers.

At the end of 1938 anti-gas staff officers were appointed to the Headquarters Commands in the United Kingdom. These were the only gas warfare staffs that existed during the inter-war period.

September, 1939, and onwards—Organization of C.W. Staffs

At the beginning of the war the appointments in War Establishments shown above were transferred to the General Staff side. Concurrently, the gas warfare advisory functions of C.Es. and Cs. R.E. lapsed.

In November, 1939, a basic organization for gas warfare for the Army in the field was evolved. This included, in addition to General Staff officers (C.W.), the appointment of chemists as advisers. Such chemists had been included in formation staffs in the 1914–18 war and had proved their value, but there had been no provision in peace for such appointments in a future war and after the war broke out arrangements had to be made to find and train suitable chemists for these appointments. They became available for G.H.Q., B.E.F. appointments in February, 1940. Three officers for corps appointments arrived in France in May, 1940, but were unable to join their headquarters owing to the confused situation.

The basic organization is shown below. Only those appointments which are starred were authorized in the case of the B.E.F. ; no Army H.Q. was formed.

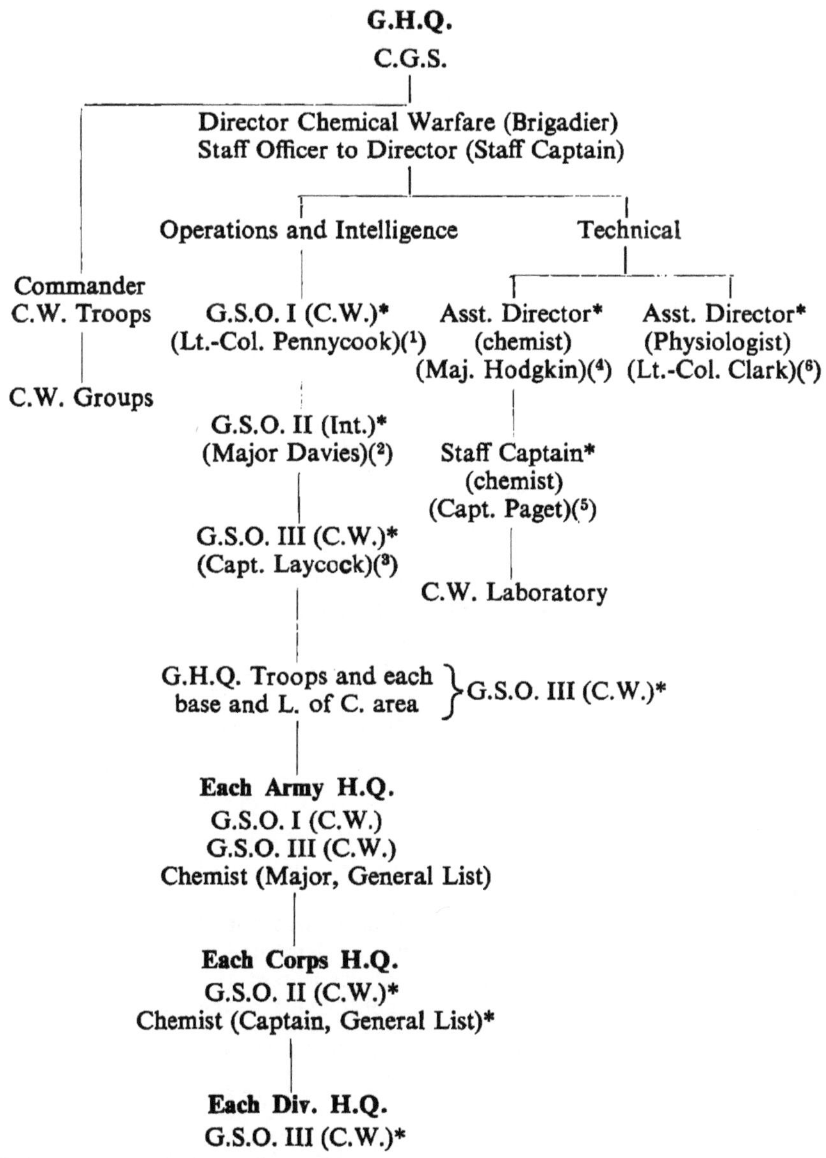

Note (1).—Subsequently Brigadier G. H. C. Pennycock, The Cheshire Regiment, D.C.W. 21 Army Group.

Note (2).—Subsequently Brigadier E. F. Davies, Royal Ulster Rifles.

Note (3).—Subsequently Major-General R. E. Laycock, Royal Horse Guards, Chief of Combined Operations.

Note (4).—Subsequently Brigadier A. E. Hodgkin, R.E., Deputy Chief of Civil Affairs, Second Army.

Note (5).—Subsequently Colonel H. Paget, R.E., W.O. representative on British Army Staff and agent of the I.S.C.C.W. in Washington.

Note (6).—Lt.-Col. A. J. C. Clark was Professor of Medicine at Edinburgh University. He was nominated as Physiological Adviser but was not called up until April, 1940.

Passive Air Defence continued to be a responsibility of C.W. staffs in the B.E.F. It was not until after Dunkirk that "C.W." and "P.A.D." became divorced both in staff administration and in training (*see* Section 5 of this Chapter).

Broadly, with certain modifications in the title, grading and duties of appointments, the organization outlined above was in due course applied in all theatres of war and throughout Empire forces.

After Dunkirk this organization, adapted to suit immediate requirements, was implemented in Home Forces in the United Kingdom. At G.H.Q. Home Forces there was a Director of Chemical Warfare who combined this appointment, until January, 1943, with one at War Office (Section 1 of this Chapter). In his capacity as D.C.W. at G.H.Q. he had a G.S.O. II (C.W.) to assist him and an Assistant Director (Technical), Major. An Assistant Director (Physiological), Major, was earmarked but remained at the experimental station at Porton, visiting G.H.Q. when required. At each Command Headquarters and Corps Headquarters there was a G.S.O. II (C.W.) and a Technical Officer, Captain—the Technical Officer appointment at Commands not being implemented until June, 1941. At each Division and District H.Q. there was a G.S.O. III (C.W.).

In October, 1943, when H.Q. 21 Army Group was being formed out of G.H.Q. Home Forces, a Director of Chemical Warfare (Brigadier E. H. C. Pennycook) was appointed, and by January, 1944, a staff of a G.S.O.I. (Lt.-Col. des C. Chamier), a Technical Officer, Major (Major R. C. Pink, R.E.) and a Technical Officer, Captain, were on the ground. A physiologist (Professor J. H. Gaddam, Professor in Pharmacology at Edinburgh University) was on call if required but was never called up. The analogous appointments in G.H.Q. Home Forces wasted out. The D.C.W. 21 Army Group was responsible for the co-ordination of operational policy with G.H.Q. Home Forces, where he had a G.S.O. III. For "Overlord" and up to the time that Twelfth (U.S.) Army Group became operational two U.S. Chemical Warfare Service officers were attached to the staff of D.C.W. 21 Army Group.

At Army and Corps there were G.S.Os.II trained in C.W. and Technical Officers (C.W.), and G.S.Os.III trained in C.W. continued to be included in the establishment of divisions.

In the Middle East(1) the staff organization was not fully established until the end of 1940, and in India(2) it was not implemented until 1942. In Allied Force Headquarters in North Africa there was a joint U.S.–British C.W. staff section,(3) the staffs in lower formations conforming to the basic organization.

Note (1).—Colonel G. A. C. Peter, R.A. was initially G.S.O. I and later D.C.W., G.H.Q. Middle East. Major T. W. J. Taylor, R.E. (later Scientific Adviser to the Supreme Allied Commander South East Asia) was Technical Officer (Major); he was relieved by Major (later Lt.-Col.) J. P. Hall, R.E. Lt.-Col. W. A. Robinson R.A.M.C. was Physiological Adviser.

Note (2).—There was no appointment of Director of Chemical Warfare in India. Lt.-Col. D. J. C. Wiseman, 15th Punjab Regt., was G.S.O. I, G.H.Q. India and Major (later Lt.-Col.) N. Hartshorne, R.E. was Technical Officer. Lt.-Col. P. C. Scudamore, The Rajput Regt. was Inspector of C. W. (Training). Lt.-Col. (later Colonel) R. S. Dickie, R.A.M.C. and subsequently Lt.-Col. T. Twigg, R.A.M.C. held the appointment of Physiological Adviser. Towards the end of 1944 a G.S.O. I, Lt.-Col. F. J. Hodson, Oxfordshire and Buckinghamshire Light Infantry, was appointed in South East Asia Command. At Allied Land Forces, S.E.A. there was a G.S.O.II and Technical Officer (Major), the staff of lower formations conforming to the home model, with the exceptions that corps technical officers were ultimately pooled under H.Q. Allied Land Forces, and G.S.O.s III of divisions were trained in camouflage as well as C.W.

Note (3).—The Chief Chemical Officer was Colonel (later Brigadier-General) C. S. Shadle, U.S. Chemical Warfare Service, Lt.-Col. G. des C. Chamier, King's Own Yorkshire Light Infantry was G.S.O.I. and Major K. J. B. Earle, R.A. was Technical Officer. There were also four American officers and a British G.S.O. III.

Provision of C.W. Staff

Except for the initial delay in filling the technical appointments in B.E.F. sufficient technical officers (C.W.) were found and trained in time to be available to fill appointments as they occurred.

The title of " Technical Officer (C.W.) " replaced that of " Chemist, Major or Captain," in May, 1941, as it more aptly described their functions. The title of " Assistant Director (Technical) " similarly lapsed and became that of " Technical Officer (C.W.), Major."

Technical officers (C.W.), had originally been commissioned into the General List. They were carefully selected from university and industrial chemists with good honours degrees, and usually research experience, and, initially, given direct commissions, without O.C.T.U. training, after a period of technical C.W. training at Porton. The fact that they had no parent arm led to a number of administrative difficulties. In particular they had to work in close contact with, and, where under the command of their H.Q., control the activities of, anti-gas laboratories. These were R.E. units whose officers were similarly qualified scientifically to the technical officers (C.W.) and held commissions in the Royal Engineers. Also the technical officers (C.W.) formed a general part of the organization that included the chemical warfare companies which were also R.E. units. With effect from January, 1941, existing technical officers were transferred to the R.E. and after that date all entrants were commissioned into this Corps.(1) Latterly the giving of direct commissions was stopped and potential technical officers had to do the normal O.C.T.U. training.

General Staff officers (C.W.) were provided from the output of C.W. staff courses held at the Army Gas School, Winterbourne Gunner. In 1943 the appointments of G.S.O. II (C.W.) at corps and G.S.O. III (C.W.) at division, and districts in the United Kingdom, were changed to G.S.O. II (Ops.) and G.S.O. III (Ops), trained in C.W. The reason for this was that chemical warfare was not a whole-time job and this change gave these officers a better standing and the ability to achieve more. The continuance of gas peace had led to chemical warfare staff officers being used more and more on miscellaneous duties not connected with any particular branch. It was still necessary, however, to have qualified officers to keep divisions prepared, and to give advice on the use of smoke. Their new status kept these officers more in touch with operations, which would have become essential had gas warfare started. A similar change was made in due course throughout Empire forces.

Duties of G.S.Os. trained in Chemical Warfare

The duties of G.S.Os. trained in C.W., as they stood at the end of the war, are given in Appendix XII. Initially these officers had been staff officers for gas only, but finally their duties embraced gas, smoke, and flame.

The circumstances under which smoke and flame came to be included in the duties of General Staff officers trained in C.W. were the same as in the case of technical officers (C.W.) which is discussed below.

Note (1).—There were some exceptions to this general rule ; certain technical officers were seconded from units or the general list, and not transferred to the R.E.

Duties of Technical Officers C.W.

Technical officers (C.W.) were the technical advisers to the G.S.Os. trained in C.W. Although they received staff pay they were not given the status of General Staff officers and were not supposed to perform executive staff duties. They were advisers only on technical aspects. In effect, however, in the latter part of the war both in Central Mediterranean Forces and, in particular, in 21 Army Group, these officers did, in fact, function as staff officers and technical advisers to their commanders. They were still not graded as such.

As with G.S.Os., their duties were initially concerned solely with gas. With the advent of A.A. smoke screening of ports in the Middle East and North Africa in 1942, these officers became increasingly involved (as did all chemical warfare staffs) in smoke. They had the technical knowledge and qualifications to apply to the subject and became the advisers to the staff in this connection. They were, in fact, the only officers available who had the necessary qualifications and knowledge to take on the job. Not only did they act as advisers, but they carried out much executive work in connection with the lay-out of smoke screens, the development of smoke materials from indigenous sources and the improvisation of smoke producing equipment. In the Middle East, North Africa and Malta, in particular, the successful development of smoke screens with local and improvised equipment was in many cases directly due to the drive and initiative of these officers.

In the United Kingdom, up to 1942, A.A. smoke screening had been the responsibility of the Ministry of Home Security and not an Army responsibility. The developments referred to above in the Middle East and North Africa, and requirements for A.A. smoke screens which were foreseen in connection with an invasion of the continent of Europe, led to the War Office taking over responsibility from the Ministry of Home Security. It was partly pressure from below and from formations overseas which led to the setting up in the General Staff at the War Office of an organization adequate to handle the subject (*see* Section 1 of this Chapter). In May, 1943, smoke was officially incorporated in the duties of technical officers (C.W.).

Responsibility for flame was assumed in the first place by the C.W. staffs in the War Office (S.W.V. Directorate) but staff officers trained in chemical warfare and technical officers (C.W.) became increasingly involved in flame warfare as flame weapons started to come on the ground. Flame was added to the duties of G.S.Os. trained in chemical warfare in November, 1943, but it was not until April, 1944, that it was officially included in the duties of technical officers, by which time it had long been an accomplished fact. The duties of technical officers (C.W.), as they stood at the end of the war, are given in Appendix XI.

Future Requirements

As the operations of 21 Army Group on the continent progressed, the G.S.O. (Ops.) trained in C.W. at Corps and above functioned less and less in relation to gas, smoke, and flame warfare except in their purely operational aspects, whereas the technical officer (C.W.) came to function as both the staff officer and the technical adviser to his commander and to all branches of the staff, in particular in relation to smoke and flame, and provided expert advice to "I" branch as stocks of German gas weapons began to be captured.

Experience proved that it was of the greatest value to have a scientifically highly trained staff officer within the General Staff, and that it was anomalous to have a technical officer (C.W.) acting as the adviser only to a General Staff officer trained in C.W. What was required was a General Staff officer (C.W.) who had the scientific knowledge and background to apply to his duties, and the two appointments could then be rolled into one. In 21 Army Group this in effect happened but was not regularized up to the time hostilities ceased.

In other overseas theatres there had been a similar trend, as commanders became aware of the qualifications of these officers and the functions they could fulfil as technical staff officers, and the officers themselves gained experience in staff work. In Eastern Theatres they also became concerned with the General Staff problems connected with the use of anti-malarial and other insecticides.

It has been shown in Section 1 of this Chapter how, at the War Office, the gas warfare staffs of the early war days came to handle all "special weapons" in the latter part of the war, and how the staff executive and technical advisory functions became combined in one General Staff Arms Directorate (S.W.V. Directorate). The G.S.O. I handling "special weapons" (Lt.-Col. Wansbrough-Jones) was previously a technical officer (C.W.) and the majority of subordinate staff officers were officers with high scientific qualifications.

It has been shown above how the technical officers (C.W.) in formations gradually, in fact but not in title, became staff officers "special weapons" through being the only officers on the staff with the high professional scientific qualifications needed as a background to apply to the use of "special weapons" in the field.

The trend at the end of the war therefore was, both at the War Office and in the field, for a staff officer with high academic degrees and background to be the adviser on "special weapons" to his commander and all branches of the staff. To try to make the average military officer who lacks the basic scientific background reach the requisite high scientific standard by means of attendance at courses is impracticable. A scientist is required who is also a fully trained military officer and staff officer. This applies to corps and above, including the War Office. It is probably impractical and unnecessary to implement it in divisions, where a normal staff officer who has received practical training in "special weapons" should be adequate.

SECTION 3.—ORGANIZATION IN THE ARMY FOR THE OFFENSIVE USE OF GAS

The Inter-War Period

It was not until 1935, after the emergence of the Western Plan with Germany as the main opponent, that serious consideration was given to the offensive gas organization required by the Army. The findings of a committee (The Elles Committee) which had been set up in 1932 " to ascertain the extent of the threat presented to the Army by a hostile gas offensive in the event of a national war " and to recommend the policy to be pursued to meet such a contingency, had been shelved by the Secretary of State. He did this for political reasons connected with the Disarmament Conferences, at which proposals were being discussed that the manufacture or possession of all appliances or substances, exclusively suited for chemical or bacteriological warfare, should be prohibited, both in peace and in war.

In 1936 the Corps of Royal Engineers was made the parent arm for all offensive and defensive gas warfare in the Army, except gas shelling. The Anti-Gas Wing of the School of Military Engineering at Chatham was opened to train R.E. officers and the offensive use of gas was included in the curriculum ; chief engineers and Cs.R.E. became the gas advisers to their respective commanders (*see* Section 2 of this Chapter).

The political position at the time was delicate. In view of the Draft Disarmament Convention, we did not wish to be accused of setting up an organization in the Army for the offensive use of gas. Further, public opinion had been roused by the use of gas by the Italians in Abyssinia. Offensive gas training was greatly hampered by the requirements of security and exercises in commands were not allowed. This ban was not lifted until 1937.

By then, offensive weapon policy and the means to implement it had begun to take firmer shape. Production of mustard gas began in 1938, and this and liquid tear gas were the only gas chargings implemented by our pre-war policy (*see* Chapter III, Section 5). The Air Arm was recognized as the " big brother " of gas warfare, and measures were taken to equip it with mustard gas bombs and spray equipment (*see* Chapter IV, Section 2). For the Army, in addition to artillery gas shell (mustard gas and tear gas), chemical warfare companies at the scale of one for each division were visualized. These were to be armed with the Livens projector, bulk contamination vehicles and ground contamination bombs (all charged mustard gas)—the two latter being also for use by all field companies, R.E. The tactical implications of this policy are discussed in Chapter V.

No. 58 (Porton) Company, R.E., which had been the range and experimental company on the establishment of a Chemical Defence Experimental Station at Porton, was separated from that establishment in June, 1937, and re-organized as a regimental unit. It was re-designated No. 58 (Chemical Defence) Company, R.E. It was an offensive unit and this designation was a security cloak. The unit was used for tactical research. Initially, it was kept at a cadre strength, except during the collective training period when it was brought up to a minimum establishment essential for collective training.

Although the ban on offensive training had been lifted, undue publicity was to be avoided and offensive gas training continued to be hampered by the need for secrecy. The manual which formed the basis for the study of the offensive use of gas and for its inclusion in exercises was issued, down to brigade level, as a secret memorandum. In 1937 and in 1938 special gas exercises were included in the collective training of the 7th and 9th Infantry Brigades respectively, largely to determine the defensive requirements of the Army in the field. Advantage was also taken of this opportunity to clarify policy in regard to the offensive organization required.

Early in 1939, No. 58 (Chemical Defence) Company, R.E., was put on a proper footing with a normal peace establishment. This establishment was so arranged that expansion could be effected by means of training Supplementary Reservists to provide the four chemical companies then visualized as being required for the Field Force. Before this scheme could be implemented, compulsory military training came into force and the scheme was no longer adequate to meet the increased commitments of an expanded army. In June, 1939, the requirements of chemical companies were visualized as 22 companies by mobilization plus 11 months ; 12 of these companies were to be raised in peace from militia sources. A scheme for a Militia Chemical Training Centre was projected and was due to start functioning in November, 1939. Its initial location was to be at Figsbury Camp, Winterbourne Dauntsey, but it was to move to Barton Stacey in 1940 when accommodation for it had been completed there. The nucleus for this centre was attached to No. 58 (Chemical Defence) Company for training in August, 1939.

The Raising of Chemical Warfare Groups and Companies

On the outbreak of the war the long-term scheme to form a Chemical Warfare Training Centre at Barton Stacey was proceeded with. In the meantime the nucleus of the Militia Chemical Training Centre was diverted to form No. 11 Chemical Warfare Training Battalion, R.E., which began to function on 15th October, 1939, and from 15th December, 1939, turned out chemical warfare companies at the rate of one company a month.

The cadre for an enlarged training centre started to assemble at Barton Stacey in March, 1940. In April, 1940, Nos. 12 and 13 C.W. Training Battalions, R.E., were formed. In July, 1940, they were joined by No. 11 C.W. Training Battalion, R.E., from Winterbourne Dauntsey and by September, 1940, the Chemical Warfare Training Centre was running normally and turned out three chemical warfare companies a month until May, 1941. It was organized as follows :—

- H.Q., C.W.T.C.
  - Officer Cadet Training Wing
  - No. 24 Depot Battalion, R.E.
  - 11 C.W.T.B.
  - 12 C.W.T.B.
  - 13 C.W.T.B.

In all a total of 35 companies were turned out, of which one company (No. 63 C.W. Coy., R.E.) was " cannibalized ", after the return of No. 1 Chemical Warfare Group, R.E., from France after Dunkirk, to provide reinforcements for other companies, and another at a later date (No. 76 C.W. Coy., R.E.), on re-organization and re-armament with the rocket and the mortar. This left a total of 33 companies (11 groups). The order of battle of chemical warfare troops is given at Appendix XIII.

As every three companies were formed, a chemical warfare group headquarters was raised to take charge of them, a fourth company being added to the group as it became available.

As the completion of the target for chemical warfare companies came in sight the functions of the C.W. training battalions were completed. The disbandment of these training battalions and the depot battalion accordingly began in April, 1941, and was completed by July, 1941, with the exception of No. 11 C.W.T.B. which remained for the time being as a depot and reinforcement training unit.

In December, 1941, in order to avoid the manpower overheads resulting from a separate C.W. reinforcement training unit, No. 11 C.W.T.B. was disbanded. Future reinforcements for the C.W. companies received their normal recruit training at No. 2 L. of C. Training Battalion, R.E., their special training being carried out with No. 1 Chemical Warfare Training Company, R.E., which was attached to the Training Battalion for this purpose. This organization held good until the chemical warfare troops were abolished early in 1943.

In June, 1941, the Chemical Warfare Training and Experimental School was formed at South Molton, Devonshire. This School was formed to instruct officers and N.C.Os. of chemical warfare units in technical methods, particularly in connection with the 5-in. U.P. rocket, with which they were shortly to be re-equipped. It was also responsible for experiments and trials in connection with the development of chemical weapons and for the provision of technical instructions for their use. It provided the permanent staffs for the chemical warfare ranges in the Exmoor area which were used by the chemical warfare groups.

Organization of Chemical Warfare Troops

A chemical warfare group as it had evolved by the time chemical warfare troops were abolished in 1943, consisted of a headquarters, two rocket companies and one mortar company.(1) The detailed organization is shown in Appendix XIV. All the companies were administratively self-contained and the group headquarters was an operational H.Q. Broadly, the mortar company was visualized as a " light " company which worked well forward during mobile phases in close contact with the forward infantry in order to take advantage of opportunity targets, while the rocket companies were the " heavies " which remained well to the rear until a suitable target offered itself. Wireless communication was therefore essential as it would otherwise always have been a race against time to bring the rocket companies into action. A survey officer was included in the H.Q. of the group since the C.W. group brought fire to bear on the target by survey measures similar to those used in the artillery procedure for field regiments.

Note (1).—For the development of the armament of chemical warfare troops, *see* Chapter IV, Section 1.

In the earlier days, before 1942, when all companies were armed with the Livens projector, the group normally consisted of four companies, although the headquarters could have controlled up to six companies had this been necessary for any particular operation. Each Livens company was organized similarly to a rocket company as shown at Appendix XIV. There was no wireless communication within the group, and there was no survey officer in the group H.Q. These were not necessary owing to the shorter range of the weapon and its more static nature.

All chemical warfare companies, R.E. were also trained in engineering, other than wet bridging, and when not employed in their primary role could be used for general engineering duties. The proportion of tradesmen was smaller than in a field company, R.E. and they were unable to carry all technical tools and equipment for general engineering tasks. Their scope was therefore limited unless they were specially provided with the necessary equipment.

No. 1 C.W. Group R.E. (Nos. 58, 61 and 62 C.W. Companies, R.E.) proved a useful reserve of engineer troops in France in 1940. During the retreat they carried out many demolitions successfully and were constantly in action with various formations on the defensive. Had they been armed with mortars instead of with the Livens projector and been capable of adopting an H.E. role, they would have been even more useful.

At the end of 1942 and early in 1943, three C.W. (Mortar) Companies, R.E. (Nos. 68, 83 and 86) were withdrawn from their chemical warfare role and sent to North Africa in an infantry close support role using H.E. They were frequently in action until the formation of infantry brigade support companies, when they were disbanded. No. 86 Coy. received very heavy casualties since they were over-run by the enemy, and this company was disbanded first to provide reinforcements for the other two companies. In the Middle East No. 66 C.W. (Mortar) Coy. had also been in action with great success in an infantry support role at El Alamein.

While gas warfare was in abeyance the use of chemical warfare companies as flame troops was also visualized—particularly during the invasion period of 1940–41 in connection with the use of flame fougasses and hedge hoppers. They were never armed with flame throwers. A small Flame Warfare Experimental Section came under the command of the Commander C.W. Troops. A smoke role was always visualized as a function of C.W. troops. In the early days this was limited to the use of such small smoke generators as were then available ; it was not until the 4·2-in. mortar was introduced into their armament that smoke support became a normal role.

The total requirements of chemical warfare units were calculated broadly on the basis of one C.W. company to each division. Allocation in the field was planned on the basis of one C.W. group for each corps. As they were not divisional troops, co-operation with other arms was difficult and required particular attention.

The appointment of Commander C.W. Troops was implemented in Home Forces after Dunkirk. It was first held by Col. R. H. Maclaren, R.E., and after his death in an accident in 1941, by Brigadier W. H. H. Aitken, R.E.

Abolition of Chemical Warfare Troops

Chemical warfare troops were abolished in February, 1943. The chemical warfare companies, R.E. were converted to field companies, R.E. and the H.Q. chemical warfare groups to H.Q. Army, G.H.Q., or Corps Troops Engineers.

The factors leading up to their abolition were :—

(*a*) The change in the operational background in relation to gas warfare, particularly with regard to the role of ground gas weapons in mobile warfare and the reliance which came to be placed on the Air arm for putting down heavy concentrations of gas in the forward area. Experience on exercises showed the rarity with which suitable targets for heavy gas concentrations were encountered in the mobile phases of battle. (*See* Chapter V.)

(*b*) The overall inefficiency of the 5-in. U.P. rocket, in the stage of rocket development then reached, and the cumbrous nature and slowness into action of the chemical warfare companies armed with it, which led to the abandonment of this weapon. (*See* Chapter IV.)

(*c*) The incompatibility of the engineer role with the gas warfare offensive role of the chemical warfare companies. A high standard could not be maintained in both roles. The carrying of mortar equipment and the tools and equipment necessary for a field company was impracticable without an impossibly unwieldy transport "tail." If either was not carried, the companies were not immediately available in that role, and training in that role also suffered. If employed entirely as field companies with periodic refresher training in their chemical warfare role, they would have had no opportunity, while gas warfare was in abeyance, for using their main armament in battle and thereby gaining battle experience, or of working with other arms.

(*d*) With the formation of the infantry brigade support companies which included the 4·2-in. mortar in their normal armament, the retention of special units similarly armed for a C.W. role would have been a duplication of effort to carry out the same job, and this would not have been justified in the existing manpower situation.

(*e*) Manpower.—The compelling reason which precipitated the abolition of C.W. troops was the grave shortage of trained engineer manpower. 8,000 trained sappers were locked up in the chemical warfare groups for a contingency which might never arise. In June, 1942, a proposal had been made to use these sappers to complete R.E. requirements, and to form new C.W. groups from Infantry. This proposal had been held in abeyance until the effect could be seen of the formation (which was then being considered) of infantry brigade support companies on the requirements for C.W. units.

The "Existing Weapons" Policy

When the decision to abolish C.W. troops was taken, the threat of Germany using gas had not receded. Manpower and efficiency considerations overrode all others, but in one respect some concern was felt over the decision. This was that, should gas warfare start, we knew that the Germans might bring a heavy weight of attack to bear with ground weapons. In some quarters it was thought that the effect on the morale of the Army might be serious if they saw that, weight for weight, we had not got the same immediate ability to give them similar support with ground weapons.

The contrary view was that retaliation with analogous but *a priori* comparatively inefficient weapons was unnecessary. Air gas weapons, by their size and nature, were better suited to the putting down of heavy concentrations of gas than were ground gas weapons and a balance in the respective roles of air and ground gas weapons must be maintained. Whether or not special C.W. troops were available had little bearing on the ability to provide support from ground gas weapons in their existing state of development. There was no inherent reason why such support should not be forthcoming from the normal arms of the Service provided that they were adequately trained in the use of gas and that weapon development was directed to this end.

After the abolition of chemical warfare troops, the arms of the Service with a gas role were the Artillery and the infantry brigade support companies (later re-organized as divisional machine gun battalions), which included the 4·2-in. mortar in their armament. The divisional machine gun battalion had a heavy mortar company of four platoons, each of four 4·2-in. mortars, totalling sixteen mortars for each division. With this armament and organization, it was accepted that the main role of Army gas weapons was harassing, reliance for neutralizing concentrations in forward areas and in close support of troops being placed in the R.A.F.—a charge accepted by the Air Ministry. (*See* Chapter V, Tactical Doctrine.)

This organization continued until the end of the war and neither the discovery of nerve gases (*see* Chapter III, Section 4) nor any weapon development nor general war experience pointed to any modification of this policy or evidenced the necessity for special chemical warfare troops.

Meteorological Organization in the Field for Gas Warfare

The meteorological organization for the Army and Air Force in the field as developed in 21 Army Group by the end of the war is shown in Appendix XIV A. It fully covered all requirements for smoke ; although gas was never used, its requirements were always borne in mind and, had gas warfare started, they would have been met by this organization. In so far as it concerns gas warfare, the evolution of this organization is briefly outlined below.

Prior to and in the early days of the war the requirements of gas warfare had a considerable bearing on the development of the meteor organization in the Army below corps level. Although it had always been accepted that it was the responsibility of the R.A.F. to provide the Army in the field with the meteor information it required, this policy was departed from shortly before the war in so much as it was visualized that the Army would be responsible for taking its own local observations by means of field meteor equipment(1) to provide the weather data required for gas warfare purposes and for smoke—the latter being in those days mainly the concern of the Artillery.

Note (1).—Details of field meteor equipment will be found in the Chemical Warfare Pocket Book, 1942.

This organization consisted of a divisional chemical warfare section R.E. of six other ranks, working under the G.S.O. III (C.W.) of the division, and of two specially trained N.C.Os. in each chemical warfare company R.E. The organization was never actually implemented in so far as the divisional chemical warfare section was concerned. In July, 1940, before the divisional chemical warfare sections came on the ground, the R.A.F. decided that for the benefit of the meteor organization as a whole, they required a wider network of observational posts in the field manned by their own personnel, and divisional R.A.F. meteor sections (1 R.A.F. officer and 3 other ranks) were established in each division. The divisional chemical warfare sections became redundant and their implementation was abandoned. The meteor personnel of the C.W. companies were, however, retained until these units were abolished early in 1943.

During 1941, it was found from experience in exercises in the United Kingdom that this organization did not function well, largely by reason of inadequacies in the method and system of communications by which the divisional R.A.F. officers received the essential weather data and forecasts sent out from the Central Meteorological Office of the Air Ministry and without which they were not in a position to provide all the information required from them. It was decided that the interests of the organization in the field as a whole would be better served if the corps sections were enlarged to forecasting sections and the divisional sections abolished. Arrangements were made for meteor code messages to be enciphered and sent by wireless where necessary. This reorganization was effected in October, 1941, the divisional R.A.F. meteor units being abolished and enlarged corps sections established, including an officer whose primary function it was to maintain constant liaison with the user formations and units, *e.g.*, chemical warfare units.

Subsequent to this there was little fundamental change, the ultimate requirements of A.A. and large scale tactical smoke screening being met by the organization shown in Appendix XIV A, and by the attachment of meteor personnel to smoke companies. The organization for smoke is dealt with in more detail in Volume II, Part II A.

After the abolition of the chemical warfare troops, the inclusion of personnel trained in meteorology in the establishment of divisional machine-gun battalions for the purpose of taking local observations of weather data in connection with gas shoots from the 4·2-in. mortar was discontinued for the reasons given below. Forecasts supplied by means of the " Snuff Code " (*see* Chemical Warfare Pocket Book, 1942) would have been applied to munition tables or the gas concentration slide rule.

For shoots with base ejection shell the normal artillery meteor telegram was adequate.

Meteorological Information Required for Gas Operations

For gas operations information is necessary on the meteorological conditions near the ground, although a knowledge of conditions in the upper air may also be required for predicted fire with gas projectiles. Information is required regarding the values in the surface air layers of some or all of the following :—

(*a*) Turbulence.
(*b*) Wind velocity.
(*c*) Wind direction.
(*d*) Air temperature.
(*e*) Ground temperature.
(*f*) Relative humidity.

The effects of weather and ground on both non-persistent and persistent gases are described in the Chemical Warfare Pocket Book, 1942, and in greater detail in the Air Ministry publication "The Meteorology of Gas and Smoke (S.D. 216)." Attention is drawn here only to the outstanding development which affected both the method of obtaining the required data and its application to a gas operation.

In the early part of the war the weather data referred to above was to be determined by direct measurement on the spot by the divisional meteor section, and by the meteorological personnel of chemical warfare units. This included an assessment, using field meteor equipment, of the degree of atmospheric turbulence, which is the weather factor having a predominant effect on non-persistent gases. Experience showed that this attempt to determine turbulence by direct measurement was both unsatisfactory and unnecessary.

Turbulence is determined by two factors—the thermal turbulence caused by the difference in temperature of the lower and upper layers of the air, *i.e.*, thermal instability, and the mechanical turbulence arising from the gustiness of the wind and produced by the obstructions to the air flow by trees, buildings, ground surfaces and contours. Of these, thermal instability is of much greater significance.

There are three states of thermal stability—inversion, neutral conditions, and lapse. In inversion conditions, the air at ground level is colder that the air above and stable atmospheric conditions exist with very little mixing ; in lapse, the reverse is the case, and the warm air rises giving a continual mixing of the air layers which leads to unstable atmospheric conditions ; neutral conditions are those where there is no appreciable change in air temperature up to a height of about 50 ft. above the ground and the interchange of air between the layers arises from mechanical turbulence only. Mechanical turbulence also affects the degree of turbulence under inversion and lapse conditions.

In the Chemical Warfare Pocket Book, 1942, will be found the methods of calculating the turbulence factor ("R") by means of field meteor equipment for application to ammunition expenditures for any given weather conditions and ground surface. This factor included the effects of both thermal and mechanical turbulence. It assumed a large number of different values for minor variations in the degree of turbulence present, and the prevalent value was applied in calculating the munition expenditures required.

Later work, however, showed that the values of the turbulence factor as determined in the field were not truly representative of the degree of turbulence prevailing. Further, the conditions where instrumental readings were taken when calculating the "R" factor, namely our own lines, were not necessarily identical with those at the target area. The methods shown in the Chemical Warfare Pocket Book, using the field meteor equipment, were not only too meticulous in assessing very minor variations in turbulence, but did not in fact give a reliable assessment of the degree of turbulence. It was found that, for practical purposes in the field, turbulence could best be estimated from general meteorological conditions. Tables were drawn up giving average values of the turbulence factor based upon season, time of day, cloudiness and wind speed. These tables could be used to calculate turbulence from reports of existing conditions or from forecast data.

Minor variations in meteorological and topographical conditions can be ignored for practical purposes and munition expenditures calculated only for the major conditions (*i.e.*, strong lapse, moderate lapse, neutral conditions and inversion) in major types of terrain (*i.e.*, open or wooded country).

Normally attacks with non-persistent gas would only have been carried out in favourable meteor conditions, *i.e.*, low wind speed and neutral or inversion conditions. Against certain types of targets, *e.g.*, personnel without respirators, attacks during lapse periods with large aircraft bombs might have been worthwhile.(1) Expenditure tables for various munitions would have been drawn up accordingly for the major conditions of terrain, *i.e.*, open or wooded.

For attacks with persistent gas, tables would have been drawn up for the four conditions of turbulence defined above and for a wider range of wind speeds. In this case additional factors would have had to be included in the tables for the effects of ground temperatures and of types of surface, *e.g.*, open grassland, jungle, barren soil, or sand, since, in the case of liquid on the ground, these two factors, as well as wind speed, have a considerable effect on the rate of evaporation. Relative humidity has also to be taken into account owing to the increased vulnerability of the body surface to mustard gas in hot humid conditions.

Tables for all British and American munitions were in process of being drawn up when the war ended, some being included in an interim report issued by the Project Co-ordination Staff (*see* Section 9 of this Chapter) in P.C.S. Report No. 9). This Report was republished in book form in April, 1946, as " Technical Aspects of Chemical Warfare in the Field Part I (Confidential), Part II (Secret)."

Note (1).—Large scale field trials brought out the " pancaking " effect which occurs with non-persistent gases, particularly from the burst of large capacity bombs. After burst there is a momentary pillaring but, owing to the chilling caused by evaporation and sudden expansion, the cloud is very dense in relation to the air and sinks to the ground by force of gravity, spreading out in a circular shape from the point of burst. During this process, the cloud is unaffected by meteor conditions, other than the wind, by which it is bodily moved. For 30 seconds the cloud stays near to the ground and the concentration is very high. With large aircraft bombs it is therefore possible to use non-persistent gases effectively in lapse periods for surprise attacks even against troops in possession of respirators by exploiting the heavy concentrations which are effective for 30 seconds irrespective of the degree of turbulence prevailing. Another point of interest in relation to " pancaking " is that the concentrations at ground level are some five times greater than at the height of a man and a standing man has therefore a better chance of timely adjustment of the respirator than a man lying down.

SECTION 4.—DEFENSIVE ORGANIZATION IN BACK AREAS AND IN THE FIELD

Introductory

During the 1914–18 war the organization which developed for defence against gas was entirely a forward area organization within range of enemy artillery or attack with gas clouds. It consisted mainly of the introduction of a system of alarms and the extensive gas-proofing of dug-outs, etc., such measures being referred to as "collective protection." From July, 1917, onwards, when the Germans introduced mustard gas, certain simple decontamination measures were instituted and some reserves of clothing for reclothing purposes were held by formations, but no extensive organization for the decontamination of personnel or material was implemented. ("Official History of the War—Medical Services—Diseases of the War, Vol. II," pages 356–361.)

From 1932 onwards, as the gas threat from the air developed, and in particular the threat of air attack on back areas with mustard gas, the necessity for a more comprehensive organization became apparent, and the organization known as Passive Air Defence (P.A.D.) in the Services and Air Raid Precautions (A.R.P.) in the civil administration began to take shape. Anti-gas measures formed part of the responsibilities of both these organizations. To begin with, anti-gas defence was the senior partner and in both cases the organization for passive measures of defence against air attack as a whole was fostered in its early days by the gas experts. For a time the gas threat overshadowed the other dangers of air attack. By 1939 a proper perspective had been attained, and both P.A.D. and A.R.P. were well established to deal with the air threat as a whole ; a very large measure of attention was still, however, directed to anti-gas defence.

Principles underlying the Development of Anti-Gas Organization

In 1936 two fundamental principles governing the development of the defensive organization in the Army to meet the gas threat were established.

The first was that the threat must be met from within the existing organization of the Army, and that no special gas units should be created. This principle has always been adhered to with one unavoidable exception, namely the formation of anti-gas laboratories whose role and functions are referred to in Section 6 of this Chapter.

The second was that every unit and sub-unit of every arm and branch of the Service must be responsible for its own anti-gas defence, and that no special anti-gas sub-units would be incorporated in the establishment of units for this purpose. Reliance must be placed in having an adequate number of unit personnel trained in anti-gas duties to implement unit defence schemes. The unit was responsible for the protection and decontamination of its personnel, vehicles, and equipment other than clothing.

The fact that in 1936 the Royal Engineers was made the parent arm for anti-gas measures did not vitiate these principles. At that time the whole position regarding the organization and the measures required was nebulous and under examination. It was desirable to have some branch of the Service to study the problem and initiate action. The Royal Engineers had fathered gas in the 1914–18 war, and it was natural to turn to them. In 1939, shortly before the war, this Corps ceased to have this function since the need for it had disappeared. It had in fact become in every way undesirable, since as long as the Royal Engineers was designated the parent arm there was a tendency for other units and the services to shelve their responsibilities and regard anti-gas defence as something which would be largely done for them, rather than something they must essentially do for themselves. From 1939 onwards the relation between the Royal Engineers and the other Arms in anti-gas matters became in principle the same as for other tasks of that corps, *e.g.*, large scale decontamination where both decontamination and engineer work were involved.

A third principle was also enunciated in the early days of our gas rearmament, namely, the personal responsibility of the individual for preventing himself from becoming a casualty by the use of the personal equipment issued to him. To begin with, however, the onus of this responsibility was not placed fairly and squarely on the individual, since special gas cleansing and reclothing facilities were established in all unit lines in back areas and also visualized in forward areas to a degree admitted by the tactical situation. This is discussed in more detail below.

Organization current at the Outbreak of War

Within the framework of the above principles the organization as it had developed up to September, 1939, consisted of :—

(*a*) A system for detection and warning of attack. The equipment introduced for this purpose and its method of use is discussed in Chapter VI, Section 2. The posting of special gas sentries was not advocated, normal sentries combining gas duties with their usual functions.

(*b*) The allocation of responsibilities for reconnaissance after attack.

(*c*) A system for the gas cleansing of unwounded but contaminated personnel, and of contaminated wounded, for the delimitation or decontamination as necessary of contaminated areas or material, and for the decontamination of vehicles and of clothing. In back areas this involved respectively the organization of gas cleansing centres, of decontamination facilities at aid posts or M.I. rooms, of gas defence centres and decontamination squads, of vehicle decontamination centres, and of arrangements for decontamination of clothing.

It is not proposed to discuss this organization here except in a broad outline sufficient to trace development to the end of the war. Details of the organization to be implemented in back areas at the beginning of the war will be found in the handbook "Protection against Gas and Air Raids, Pamphlet No. 3—Passive Air Defence—1939." The organization current at the end of the war will be found in "Regulations for Passive Air Defence Organization, 1944."

The application of this organization to forward areas as visualized in 1939 will be found in " Protection against Gas and Air Raids. Pamphlet No. 1—Protection against Gas in the Field—1939." The modifications which occurred during the war were incorporated in " Gas Training, 1942 " and " Advanced Gas Training, 1943."

Organization in Back Areas

Gas Cleansing Centres

1939 instructions were that gas cleansing centres were to be provided on a scale " which would insure the possibility of treatment of all personnel within 10 minutes(1) of contamination." This in practice meant one in every unit area, although there was some grouping of small units for this purpose. The staffing and running of centres was a unit responsibility.

The centres themselves were over elaborate. Special splinter-proof buildings with undressing, bath, and dressing-rooms, and hot water systems were, in many cases, erected, and the most meticulous arrangements made for separate staff to supervise reception and segregation of lightly from heavily contaminated men, their undressing and the collection of their valuables, bathing or bleach treatment according to the degree of contamination, and the issue of fresh clothing and anti-gas equipment. Essentially all that is necessary for gas cleansing is the siting of undressing and dressing spaces in relation to an existing ablution room and this in due course became the standard to be provided.

Much of the viewpoint underlying the earlier elaborate organization had its roots in the over-appreciation of the efficacy of spray attacks referred to in Chapter V.

With the increased emphasis that came to be placed on the responsibility of the individual for avoidance of becoming a blister gas casualty in all circumstances, and with the inclusion in the drill for personal decontamination of improved methods of dealing with contaminated clothing, except for the worst cases of heavy contamination (*see* Chapter VII), there was less need for attendance at a gas cleansing centre. In fact the very provision of these centres went against the fundamental principle of personal responsibility, and, in any case, in the circumstances where they could be provided, *viz.*, camps, billets, barracks, etc., the optimum conditions existed both for avoidance of becoming contaminated, and for dealing with the contamination if it occurred.

By 1943 soldiers were forbidden to attend gas cleansing centres, but if a man was so grossly contaminated, *e.g.*, as by a near burst of a blister gas projectile, that he could not deal with the contamination himself, he could be sent to the gas defence centre (*see* below), but only on the orders of an officer. Gas cleansing centres, based on existing ablution sites, were only necessary for civilians not normally provided with personal equipment. They were authorized for War Department establishments where civilians worked regularly, or where special facilities were desirable, *e.g.*, in such places as ammunition depots where gas munitions were kept.

Note (1).—The time limit for penetration of serge clothing by liquid mustard gas, other than very small drops of spray.

Organization for Decontamination of Wounded Personnel

Such personnel were dealt with through the same channels as ordinary wounded. At the entrance of all first aid posts or hospitals receiving casualties, undressing and decontamination arrangements were organized through which contaminated casualties could be passed before receiving medical attention. As with gas cleansing centres, the arrangements visualized at the beginning of the war were rather more elaborate than at the end. The responsibilities of stretcher bearers in relation to decontamination are discussed later in this chapter (*see* Medical Services).

Gas Defence Centres and Decontamination Squads

The organization for decontamination of areas and materials centred on the gas defence centre which was established in each unit area. It was manned by unit personnel and was the headquarters for decontamination squads. Similarly to gas cleansing centres the fundamental requirement was simple, and special buildings, as initially allowed for, were an unnecessary frill. Bathing after a task was largely for hygienic reasons. The essentials for this centre were some bathing or shower facilities, or even water in buckets, in the vicinity of a storeroom (or tent) where the stores for decontamination squads could be kept, together with simple means for undressing after completion of tasks and the segregation of any clothing, protective types or otherwise, that might have become contaminated during work.

The duties of decontamination squads involved the marking of contaminated areas and safe routes, and the decontamination of ground and buildings, where considered essential, and of contaminated stores and equipment. The squads were provided under unit arrangements.

Vehicle Decontamination Depots

The 1939 organization visualized the establishment of vehicle decontamination depots on an area basis. Special wash-downs with the necessary hoses and decontamination materials were set aside and a storeman was put in charge. Although a driver was normally responsible for decontamination of his own vehicle, where work was considerable special squads were to be detailed. The extensiveness of the probable contamination of vehicles was over-estimated both in relation to the number of vehicles likely to be affected and to the degree of contamination of the vehicle itself. Latterly, methods of decontamination were simplified (*see* " Gas Training 1942 "), drivers were held responsible, and the special vehicle decontamination depots became redundant.

The Decontamination of Clothing

This has never been a unit responsibility. In the United Kingdom at the beginning of the war decontamination of clothing was to be carried out by civilian laundries under a national scheme. The organization in the field is discussed below, and, latterly, field formations in the United Kingdom would have utilized this organization.

Back Areas Overseas

In back areas overseas the organization outlined above was to be applied in as far as local conditions permitted. Overseas the decontamination of clothing was an R.A.O.C. responsibility.

Towards the end of the war, in view of the necessity for passive air defence in liberated Allied or occupied enemy territory in order to ensure that there should be no interference with military operations, the responsibilities of the Civil Affairs branch, in conjunction with P.A.D. staff officers, for P.A.D. measures (which included anti-gas measures) were laid down. Details will be found in " Regulations for P.A.D. Organization—1944 ".

Organization in other Services

An anti-gas organization as outlined above was set up by all other Services, including the civilian A.R.P. services, and adapted to their special requirements.

In the case of the Ministry of Home Security the organization for providing cleansing facilities for the general public presented a special problem. Civilian protective equipment consisted of a respirator only, although towards the end of the war some anti-gas ointment was on sale at chemist shops. Public gas cleansing centres based on any public, local or improvised washing facilities were provided on as wide a basis as possible, but latterly instruction was specially directed towards making people cleanse themselves.

Special measures were taken to meet the particular problems of contamination likely to be encountered by utility services, railways, docks, shipping, and food distribution, to ensure that such services continued to function with the minimum dislocation. Details will be found in the Ministry of Home Security Air Raid Precautions publications, a list of which is included in the bibliography at the end of this volume.

Liaison between the Army and other Services has been referred to in Chapter I.

Organization in Forward Areas

Before 1938 there was a strong tendency to apply an organization similar to that in back areas, to the maximum extent permitted by tactical conditions.

It was visualized that improvised gas cleansing centres would be opened up in forward areas as soon as possible after troops had been sprayed or otherwise contaminated, whether on the move or in the line. The early attendance of a contaminated individual at a cleansing point, and his re-clothing, was thought to be as essential in forward areas as in rear areas, and the system of personal responsibility of every officer and man for his own decontamination had not become established. The improvisation of a gas cleansing centre was to be effected by rapidly marking a space with a " dirty " side in which contaminated clothing could be removed before proceeding to a " clean " side for such bleach or ointment treatment as was practical, and for re-clothing.

In 1938, a brigade exercise was held which brought out the hampering effect of such measures and the impracticability of putting them into timely operation, since, to be of value, decontamination of clothing has to be carried out within ten minutes of contamination.

This exercise also proved a healthy corrective to the tendency which had existed of being too " gas-minded." There was a leaning on the part of specialists in gas warfare toward providing a degree of protection in the field that was higher than was practicable without severe loss of military efficiency, and of a standard above that which could be expected against other battle hazards. The principle was firmly enunciated that, in the forward area, the soldier must accept the risks of contamination as he must accept the risks of death by shot and shell if fighting efficiency was not to be impaired—and, in any case, contamination was curable. A higher degree of protection could be aimed at by units such as artillery and ancillary corps whose duties are less affected by the wearing of such items as the anti-gas cape. The report on this exercise was emphatic that normal risks must be taken if the use of gas was not to cause stagnation of all movement. As a result, a better perspective in the use of anti-gas equipment began to be evolved, *viz.*, that the use of such equipment must be governed by tactical requirements, and that the onus of responsibility for decontamination of his person must, in the first place, be put on the individual.(1) The procedure known as " Personal Decontamination " (*see* Chapter VII), which has since become the basis of defence against blister gas in the field, was born as a result of this exercise.

Although the 1938 exercise clarified procedure for troops in contact, as well as many anti-gas equipment matters, the implications of the re-clothing problem in relation to spray attacks on the L. of C. and in the forward area as a whole, particularly of troops on the move, required further examination. Also the whole problem of organization of decontamination in the field and the responsibilities of the ancillary services in this respect required more detailed consideration.

In 1939 a special re-clothing exercise was held and a committee—The Taylor Committee—was set up by the Army Council, under the chairmanship of the Deputy Master General of Ordnance, to consider and make recommendations on the decontamination organization required in the field. This Committee confirmed that the channels for evacuation of contaminated clothing, etc., and for its replacement, should be separated from the normal maintenance chain and should be combined with the petrol chain. Replacement was to be automatic as for ammunition. The reserves of clothing to be carried in various echelons in the field were also decided on (*see* Chapter VI, Section 4).

It was accepted that changing of clothing could only be carried out to an extent admitted by the tactical situation, and that for deployed units it was unlikely to be practicable to do more than decentralize the unit reserve of clothing to sub-units. Clothing changing, where necessary, would be effected as opportunity offered.

The responsibility for the decontamination of ordinary clothing, anti-gas clothing, blankets, web equipment and respirators was laid on the R.A.O.C. It was to be carried out by laundries at the base or advanced base. Responsibility for the decontamination of all other unit equipment and of vehicles remained that of the unit.

Note (1).—Extract from report on gas trials held by 9th Infantry Brigade (September, 1938) Commander, Field-Marshal (then Brigadier) Viscount Montgomery of Alamein.

> "The menace of gas must not be allowed to throw the Army off its balance. The tendency is to over-estimate the danger, whereas in fact the problem is quite simple. The onus of protecting himself should be placed initially on the individual, and he must be given the necessary means for the purpose as part of his personal equipment. Except in certain battle situations which might occur in the forward area, it may almost be said that, with the means now available, an unwounded man who develops blisters as a result of contamination from blister gas has only himself to blame."

During the war there was no fundamental change in organization or in the division of responsibilities as between units or services.

With the introduction of mobile laundry and bath units R.A.O.C., decontamination of clothing, including anti-gas clothing, of the formation they served became a responsibility of these units. Base laundries were responsible for base areas. In all theatres the Army accepted the responsibility for decontaminating R.A.F. clothing and protective clothing at applicable Army laundries.

Adjustments were made in the scales of reserves of clothing and the scales in force from 1940 onwards are indicated in Chapter VI, Section 4.

Considerable modification occurred in methods of training; more reliance was placed in the use of commonsense and more practical means of meeting various gas situations were adopted for all arms and services. These measures were all adopted to one end, namely getting on with the job in hand. Reasonable risks had to be taken if the delay factor caused by the use of gas was to be reduced to the minimum. Great insistence was placed on the personal responsibility of every officer and man for not becoming a gas casualty; this was the bedrock of all anti-gas action and is referred to in more detail in Chapter VII.

Neither experience gained during the war nor the advent of nerve gases indicated the necessity for any fundamental change in the organization in the Army to meet the threat of gas. It is open to question, however, whether it is still essential that the provision of clean clothing to replace that contaminated by gas should be on a special channel. Clothing has to be changed in the field for many reasons other than contamination and it is for consideration whether normal " Q " channels would not be adequate irrespective of the reason for which fresh clothing is required.

The Medical Services

No special channels of evacuation have ever been created nor have any special medical units been raised to deal with gas casualties or contaminated wounded. Stretcher bearers are responsible for giving first aid and such immediate decontamination as is practicable, followed by evacuation to the regimental aid post (R.A.P.). At all aid posts, or at any medical unit receiving casualties, arrangements for decontamination of casualties are necessary. These vary according to the location of the unit. In the case of a R.A.P. it merely consists of a " dirty " area where contaminated casualties are decontaminated.

The gas threat demands no special organization of the medical services except for decontamination arrangements where casualties are received, for the training of all medical personnel in the recognition and treatment of all types of gas casualties to a degree necessitated by their status, and for the provision of the necessary decontamination and medical stores. On occasions the deployment of a field ambulance, or a section of it, might be necessary to meet an abnormal gas situation, but this is in accordance with normal tactical principles of deployment.

All medical units are responsible for their own anti-gas defence in the same way as units of all other arms and services.

The Royal Army Service Corps

The R.A.S.C. are faced with particular problems in relation to protection and decontamination of foodstuffs. The main danger in the past has been contamination of food by blister gas. No special organization, however, was set up to deal with it. It was a regimental problem peculiar to that Corps.

The emphasis is on protection. As long as food is covered, it is safe. Food in transit is vulnerable when being loaded and unloaded and when bulk is broken. The danger is, in any case, relative to the type of packing in which the food is put up.

The essential factors were that supply officers should be adequately trained in the problems peculiar to their duties and that strict gas discipline should be maintained in connection with keeping food covered at all times.

Full details of the measures required, the effects of war gases (other than nerve gases) on food supplies, the decontamination of food, and the protective value of different types of food packs, will be found in Royal Army Service Corps Training Pamphlet No. 19 " Protection of Army Food Supplies Against Gas—1941 ".

The most difficult task for supply officers would have been to determine the extent to which vapour, arising from liquid contamination, had affected stocks of food and, in particular, sacked goods, although, in the majority of cases, a sound knowledge of gas together with common sense would have provided the answer. Technical officers (C.W.) were available for advice and assistance. Should large scale condemnations have been involved, the advice of an analyst would have been necessary, and one would have been available at the base.

In India a testing set was introduced which could be used by officers or warrant officers of the R.I.A.S.C. to carry out a qualitative test on samples of food suspected of contamination. It was provided down to field supply depots. It was so packed that it could be carried as part of a mule load if necessary. It did not, however, prove satisfactory, partly owing to its fragility and partly to the difficulty of keeping officers and W.O.s trained in its use. Latterly it was withdrawn except for a few sets retained at selected base and L. of C. depots for use by technical officers (C.W.) should the necessity have arisen.

The Royal Army Ordnance Corps

Reference has already been made to the special responsibilities of R.A.O.C. laundry units for the decontamination of clothing, etc. The decontamination capacity of the various types of base and mobile laundries and details of the processes adopted and time taken will be found in " Regulations for the Army Ordnance Services—Part 8 (now Vol. 3)—Pamphlets No. 32, 33 and 35 ".

No special ordnance units were ever created for storage, supply, maintenance or repair of any defensive or offensive gas stores. These duties were carried out by the applicable ordnance depots and units, and channels of supply were the same as for all other equipment and munitions.

Repair of anti-gas equipment was carried out at base ordnance depots. Returned stores companies at these depots contained certain specialist personnel to deal with repairs to respirators, oilskin clothing, etc. In some cases oilskin clothing was returned to factories for repair.

The Transportation Services

A number of trials and exercises were held during the war, in some of which real mustard gas was used, to determine more precisely the hazards likely to be encountered by these services, and the best means of combating them with the minimum of traffic dislocation. The results of these trials, as well as of the general developments in anti-gas procedure which occurred during the war, were incorporated in Military Training Pamphlet No. 66 " Anti-gas Measures—Railways and Ports "—1944.

SECTION 5.—ORGANIZATION FOR TRAINING

General

The main principle underlying anti-gas training for all arms is that it is not a specialist subject. It forms part of the normal training of all ranks, and all commanders from the infantry section or equivalent sub-unit commander upwards are responsible for the gas training of their commands. Junior leaders are trained at unit cadres by instructors previously trained at a central army gas school. Regimental instructors are designated unit gas officers and N.C.Os. and are maintained on a scale of one officer in each lieutenant-colonel's command, or in the case of the Royal Engineers one officer in each squadron or company, and one N.C.O. in each battery, squadron, company or equivalent sub-unit. This system has always been adhered to and it continued to be applied throughout the war in all formations including the A.T.S. and the Home Guard. In the Home Guard, since their scope was limited, representatives attended special short courses at army gas schools and were trained to a lower standard than the remainder of the Army, *i.e.*, a junior leader as opposed to a regimental instructor standard. Duties of gas officers and N.C.Os. and the standards to be attained by them, and by junior leaders and by other ranks, will be found in " Gas Training, 1942 ".

Senior officers (lieutenant-colonels and above) attended senior officers' courses at gas schools to enable them to be brought up-to-date in the latest developments and methods of gas training, the employment of gas, and the tactical and administrative problems arising from gas warfare ; courses for senior A.T.S. officers were also held.

Chemical warfare staff officers assisted by technical officers (C.W.) (*see* Section 2 of this Chapter) were charged with the maintenance of a high standard of training in their formations. They organized local courses and demonstrations within their formations and had special responsibilities in relation to training with actual war gases when this was introduced (*see* Chapter VII).

During the early part of the war local courses were also arranged by commands for training junior leaders. This training was in many cases implemented by the establishment of anti-gas wings at local weapon training schools. There was no de-centralization in the training of regimental instructors which was always carried out at central gas schools. Local courses, with the exception of special courses run by C.W. staff officers in their formations, were abolished in 1942 since they gave rise to a divergence in doctrine.

The Gas Schools

Up to 1936 there was only one gas school—the Anti-Gas Wing of the Small Arms School—for the training of regimental instructors of all arms and branches of the Service. In 1936, when the Royal Engineers were designated the parent arm for defensive and offensive gas warfare (less gas shelling), an anti-gas wing was established at the School of Military Engineering at Chatham, and became responsible for the training of all R.E. personnel in gas measures which affected that arm. The Anti-Gas Wing of the Small Arms School was responsible for the study of possible means of development in personal and unit anti-gas equipment and the training necessary for its use in war, while the Anti-Gas Wing of the School of Military Engineering was responsible for the study of offensive gas warfare and of defensive gas measures which affected the R.E. such as field engineering problems, organization for large scale decontamination of areas, and gas proofing, etc.

In 1937 the instruction of the R.A.M.C. in the medical aspects of gas casualties and the organization for their evacuation was started, and an anti-gas training officer (lieutenant-colonel) was added to the staff of the R.A.M.C. Depot. In addition to carrying out instructional duties he visited units to assist in training and acted as adviser for anti-gas training. This was a major step forward since the adequate training of all medical personnel in their special responsibilities cannot be overstressed. The system continued throughout the war.

Training of the R.A.S.C. and R.A.O.C. in problems peculiar to those Corps was also started. This was the responsibility of the corps concerned and was carried out at their own schools and training establishments. The Anti-Gas Wing of the Small Arms School was still responsible for the training of regimental instructors for all ancillary corps in those aspects of anti-gas defence common to all arms.

In 1938 the Territorial Army P.A.D. (and Anti-gas) School was opened at Fort Tregantle, Plymouth, to train instructors for the Territorial Army.

In 1939, when the Royal Engineers ceased to be the parent arm for anti-gas measures (*see* Section 4 of this Chapter), the Anti-Gas Wing of the School of Military Engineering was merged with the Anti-Gas Wing of the Small Arms School and became the Army Gas School. Thenceforward this School was responsible for the training of regimental instructors of all arms and services in the defensive aspects of gas warfare. At the same time the capacity of the School was increased, as it was the intention that every young officer should attend a course at this school, and young officer's courses were instituted. Consideration was in fact given to making attendance at the School obligatory as a qualification for promotion to captain. This was, however, never implemented since, although large numbers of young officers did attend, it was impracticable after the outbreak of war for every officer to attend a course.

Training for offensive gas warfare (other than gas shelling which was the concern of the Royal Artillery) remained the responsibility of the Royal Engineers, until the abolition of C.W. troops early in 1943. The training of C.W. units was a responsibility of that Corps, as was the setting up of establishments to carry out this training. This aspect of gas training is dealt with in Section 3 of this Chapter.

In October, 1939, the Territorial Army School at Tregantle was reformed as a wing of the Army Gas School and was no longer concerned with the T.A. only. The organization of the Army Gas School became a Headquarters with one wing at Winterbourne Gunner and a second wing at Tregantle. Both these wings expanded to keep pace with the expansion of the Army and maintain the requisite number of gas officers and N.C.Os. in units. In January, 1942, a third wing was opened at Glenridding. The capacity of the wings was : Winterbourne Gunner, 96 officers and 192 N.C.Os. ; Tregantle, 48 officers and

240 N.C.Os. ; Glenridding, 48 officers and 240 N.C.Os. The Winterbourne Gunner, Tregantle and Glenridding wings were self-supporting units under command of a lieutenant-colonel supervised by the Commandant, Army Gas School and his headquarter staff at Winterbourne Gunner.

In August, 1942, the Tregantle area was taken over by the U.S. Forces and, as it was found that the capacity required to maintain unit instructors could be met by the Winterbourne Gunner and Glenridding Wings, the Tregantle Wing was closed down. The Glenridding Wing was closed down in March, 1944, but a caretaking party and stores were kept ready there so that had gas warfare started this wing could have been reopened immediately. The capacity at Winterbourne Gunner remained 96 officers and 200 N.C.Os.

Early in 1943, when the War Office assumed responsibility for A.A. smoke screens in the United Kingdom from the Ministry of Home Security, the Army Gas School became responsible for the training of Pioneer Corps personnel in all aspects of A.A. smoke screens, including the operation of the equipment, lay-out of screens, meteorological aspects, etc. The Ministry of Home Security School at Compton Verney, for the training of other ranks of the Pioneer Corps in the operation of mechanical smoke equipment, came under the supervision of the Commandant, Army Gas School. It was closed in September, 1943.

In the middle of 1943 the Army Gas School also became responsible for the technical training of regimental instructors in the use of infantry flame throwers—the Wasp and the Lifebuoy—and this was undertaken by the Headquarters and the Winterbourne Gunner Wing.

The School was re-named the Army School of Chemical Warfare in December, 1943, the term " Chemical Warfare " embracing gas, smoke, and flame warfare.

The following courses were run at the School :—

(*a*) *Gas Courses to train Unit Gas Officers and N.C.Os.*

These courses were held at all three wings of the school. Up to 1940, when a separate P.A.D. School was established, instruction in passive air defence methods and organization were included in these courses. From July, 1942, smoke training was included, since it had been found that there was a general lack of knowledge in the Army, as a whole, on the behaviour of smoke and its tactical uses.

(*b*) *Courses for C.W. Staff Officers.*

These began in 1939 on the outbreak of war. They were held only at Winterbourne Gunner under the direct supervision of the Headquarters of the School. Initially they were concerned with gas warfare only, including the offensive aspects and, in particular, the planning of gas attacks and the connected administrative problems.

In 1942 tactical smoke, in 1943 A.A. smoke screening, and at the end of that year flame warfare, were also included.

The objects of the course were to train C.W. Staff Officers to fulfil the duties shown in Section 2 of this Chapter and Appendix XII.

(*c*) *Senior Officers' Courses.*

These were held at all three wings of the school under the supervision of the Headquarters staff until 1943, after which they were held at Winterbourne Gunner only. They were instituted before the war and were continued throughout the war, approximately two to three courses a year on an average being held at each wing of the school. It is to be regretted that circumstances did not permit courses being held more frequently, since it is unanimously agreed by all C.W. staff officers, from directors of chemical warfare downwards, that these courses did more than anything else to combat lack of interest and lethargy in gas training (*see* Chapter VII). It was frequently noticed by inspecting officers that formations and units whose commanders had attended the senior course at the school maintained a higher standard of gas training generally than those formations whose commanders had not.

The syllabus initially covered gas warfare only, but in 1942 tactical smoke was included and in 1943 the planning aspects of A.A. smoke screens. The final Senior Officers' Course to be held was a Chemical Warfare Course, which covered all aspects of gas, smoke, and flame. Chemical warfare courses for company and battery commanders were also held during 1944.

(*d*) *Home Guard and A.T.S. Courses.*

In 1942 gas courses were begun at the School for the Home Guard and for the A.T.S. Senior A.T.S. Officers' Courses were also held. All these courses were discontinued when the threat of invasion had passed.

(*e*) *Courses for Regimental Instructors in Infantry Flame Weapons.*

These courses were held separately from gas courses. They covered the technical handling only of the Wasp and the Lifebuoy, tactical handling being the responsibility of the Infantry Schools. The courses are discussed in more detail in Volume II, Part III.

(*f*) *Pioneer Corps Smoke Courses ; Senior Commanders' Smoke Courses for A.A. Command ; Training of A.A. Smoke Companies.*

These responsibilities of the Army School of Chemical Warfare are discussed in Volume II, Part IIA " Screening Smoke."

Responsibilities of the Commandant of the Army Gas School.

In April, 1939, an Inspector of Anti-Gas Training (brigadier) was appointed, who combined this appointment with that of Commandant, Army Gas School. In his capacity as an Inspector, the Commandant of the School was the adviser to the War Office over the whole range of gas warfare policy and doctrine, as well as being responsible for the inspection of gas training in the Army.

Shortly after the outbreak of war the Inspector was re-named Inspector of Chemical Warfare. In November, 1940, the duties of Inspector and Commandant of the School were divorced. The Commandant appointment was re-graded as colonel, and from then on the Commandant of the School had no advisory function to the War Office, this being assumed by the Inspector of Chemical Warfare and succeeding appointments (*see* Section 1 of this chapter).

In addition to being responsible for the efficiency of training at the School, the Commandant had special responsibilities for liaison with other Service schools, and also with the Chemical Defence Experimental Station, on the developmental aspect of anti-gas equipment. In 1943, on re-designation of the School as the Army School of Chemical Warfare, the Commandant's charter was extended to cover field trials, from the user's viewpoint, of gas and smoke weapons, and the co-ordination of flame training throughout the Army.

After the abolition of the chemical warfare troops in January, 1943 (*see* Section 3 of this chapter) the responsibility for the training of the Royal Artillery and of brigade support units (later divisional machine gun battalions) to fire gas munitions rested with the arms concerned. The training of senior and staff officers in the employment of gas, the planning of gas attacks, the administrative problems connected with the supply of gas munitions in the field, etc., remained the responsibility of the Commandant, Army School of Chemical Warfare.

SECTION 6.—DEFENSIVE UNITS—ANTI-GAS LABORATORIES

From the early days of chemical warfare in 1915(¹) provision for laboratory facilities in the field has always been included in our war establishments. During the inter-war period it was included in the War Establishment of G.H.Q. 1st Echelon. As the technical personnel of an anti-gas laboratory require an applied knowledge of the chemistry of chemical warfare, they were drawn in peace-time from the civilian staff of the Chemical Defence Research Department, who were enlisted into the Supplementary Reserve for the purpose.

The anti-gas laboratory proceeded with the 1st Echelon B.E.F. to France in September, 1939. It was evacuated on 21st May, 1940, after destroying as much of its equipment as time permitted. On return to the United Kingdom it was not immediately reconstituted, as there was no requirement for an anti-gas laboratory for Home Forces. Owing to the proximity of the Chemical Defence Experimental Station at Porton, where the necessary examination of equipment could be carried out, the provision of an anti-gas laboratory for Home Forces would have been a duplication of effort.

Anti-gas laboratories were subsequently raised for overseas theatres. No. 2 Anti-Gas Laboratory, R.E. was raised in August, 1940, for the Middle East, and No. 1 in January, 1941, for the Far East. This latter was *en route* to Malaya when Singapore fell, and was diverted to India. No. 3 Anti-Gas Laboratory, R.E. was originally raised in 1942, on the entry of Japan into the war, to be available for India if required, but on the diversion of No. 1 it remained in the United Kingdom as a general reserve and was eventually allotted to 21 Army Group. In February, 1945, it was re-designated No. 3 Chemical Defence Laboratory which was more descriptive of its proper function (*see* Appendix XV).

In addition to British anti-gas laboratories, No. 1 Canadian Chemical Warfare Laboratory, R.C.E. and No. 2/1 Australian Chemical Warfare Laboratory, R.A.E. were raised to serve their respective forces. No. 23 Anti-Gas Laboratory, South African Engineer Corps served in the Middle East.

Provision of Equipment for Anti-Gas Laboratories

Non-technical equipment which was a normal ordnance issue was obtained through normal channels. Provision of scientific apparatus presented difficulties. Ordnance neither stocked nor had the knowledge or experience of the technical equipment peculiar to this type of unit. Although the responsibility for supply was officially that of the War Office, it was in fact carried out from the Ministry of Supply by Lt.-Colonel W. A. Salt of the Chemical Defence Department (Liaison), who had commanded the Anti-Gas Laboratory in the B.E.F. and from long association with these units had an

---

Note (¹).—Chemical warfare was one of the major activities of the "Central Chemical Laboratory" in France.

intimate knowledge of their special requirements. He was in close technical liaison with them and favourably placed to purchase their needs direct from the trade, and to arrange shipping. Supplies were despatched to units by the quickest available channel—air mail where necessary. The arrangement worked well and was comparatively rapid.

An emergency reserve was accumulated for each laboratory and despatched overseas for storage at convenient ordnance depots, except in the case of No. 3 Chemical Defence Laboratory whose reserve remained in London.

Towards the end of the war a committee was set up to endeavour to standardize the types and sizes of specialist equipment in use by the many laboratories, in addition to anti-gas laboratories, then in being. As a result a special section was established at the Ordnance Depot at Didcot for the purchase of scientific equipment and its supply to Service laboratories. This channel was just about to function when the war ended.

Function of Anti-gas Laboratories

The function of a Chemical Defence Laboratory is the examination of captured enemy chemical warfare materials and also of our own chemical warfare materials in relation to their performance in the field. The term "chemical warfare materials" includes smoke weapons and appliances.

The examination of enemy materials which can be carried out by the laboratories is of a short term nature and, in the case of gas warfare materials, is related to the obtaining of early and accurate information on which staffs can act, and to the development, if necessary, of immediate counter-measures capable of being taken in the field. On the first use of gas by the enemy it is a particular function of the laboratory to identify the gas used (or confirm the technical officer's identification), and, should any of our protective equipment prove inadequate, to devise temporary means of protection. This especially applies in overseas theatres. In 21 Army Group, however, owing to the proximity of the theatre of operations to the Chemical Defence Experimental Station in the United Kingdom, arrangements for confirming the first use of gas centred on this establishment (*see* Section 7 of this Chapter). No. 3 Anti-Gas Laboratory, R.E., did not form up with 21 Army Group overseas until September, 1944, No. 1 Canadian C.W. Laboratory, R.C.E., arrived in Normandy in August, 1944.

The scope for research work in the laboratories is restricted by the type of equipment with which it is practical to equip them. More detailed technical examination of captured equipment which has a long term bearing on the enemy offensive and defensive potential, raw material and productive capacity, etc., and on our own research and countermeasures, is carried out by the Chemical Defence Experimental Station, Porton.

The examination of our own chemical warfare equipment is related to special investigations regarding the protective condition of anti-gas stores both in use in the field and awaiting issue, and to the behaviour of chemical chargings of weapons in the climatic conditions peculiar to the theatre. Work in this connection was directed by the technical officer (C.W.) at G.H.Q. Technical officers (C.W.) of formations submitted samples for examination as required.

The laboratories were in a sense out-stations of the Chemical Defence Experimental Station, Porton, for observation in relation to development requirements and the performance of our own and enemy equipment in the field. Although at first sight their functions in relation to our own equipment may appear to impinge on those of the ordnance inspection branches, in practice they did not. The closest liaison was maintained between inspecting ordnance officers and technical officers (C.W.).

Since laboratories were comprehensively equipped, they were in a position to carry out investigations of a scientific nature not connected with chemical warfare. Examples of such work during the war are the research work in the tropics on the insecticide D.D.T., carried out by No. 1 Anti-Gas Laboratory ; the analysis of salvaged metals from the Western Desert to enable them to be melted down and used to make new equipment (No. 2 Anti-Gas Laboratory) ; and the investigations by No. 3 Anti-Gas Laboratory into a number of cases of soldiers' deaths due to drinking poisoned wines and spirits.

Organization

Anti-gas laboratories were normally provided on the scale of one for each theatre of war ; although an Army unit the laboratory served all three Services in the theatre. They were base units for which frequent moves were not normally visualized and they were not, therefore, entirely self-contained in transport.

The laboratory with the B.E.F. in 1939/40 had been organized for semi-static conditions. It was deficient in transport, and certain of its equipment was bulky and not easily transportable. During our reverses in France it was moved from Arras to Boulogne. Transport was hard to find and much equipment had to be abandoned. Considerable improvements in design and portability of equipment were effected during the war.

Subsequent laboratories were all organized to be transportable so that they could open up in the best available accommodation. They could function in any building, Nissen hut, etc., where adequate water facilities could be provided. They carried their own electric generators and weapon stripping equipment for the examination of enemy weapons. The better the housing, water and lighting facilities provided, the greater was the scope of the work which could be carried out.

Canadian and South African laboratories were completely mobile, but owing to the lack of space in fitted vehicles they preferred to work off-loaded when they could.

Although the war establishments of all anti-gas laboratories were basically similar, each laboratory had its own W.E. That for No. 3 Chemical Defence Laboratory is given in Appendix XV. The W.Es. of No. 1 and No. 2 Anti-Gas Laboratories in India and the Middle East were considerably modified to suit the local requirements of those theatres. In all cases they were organized so that they could throw off detachments. In 21 Army Group these were particularly required to be sent forward to German chemical munition dumps as located. In the Middle East No. 2 Anti-Gas Laboratory had to serve several fronts—the East African and Greek campaigns, the Eighth Army in the Western Desert and the Tenth Army in Iraq, and they subsequently covered the Italian campaign by sending a detachment to Tripoli. In India the possibility of having to provide for widely separated land and sea invasions of Burma had to be catered for.

The organization of No. 3 Chemical Defence Laboratory is generally suitable as a basic organization for this type of unit and is sufficiently flexible to meet the strategic and tactical requirements of any theatre. It was somewhat lacking in workshop staff and facilities, but in 21 Army Group they were able to find suitable accommodation without much difficulty. It is desirable that laboratories should be independent of R.E. assistance in fitting up their buildings, the more so in less civilized theatres of operations.

In some cases it might be desirable to have a specialist medical officer on the establishment, as was the Canadian practice, and as was included in the original laboratory with the B.E.F. in France. This, however, would be dependent on the medical laboratory facilities available in the particular theatre of war, and the proximity or otherwise to such an establishment as the Chemical Defence Experimental Station at Porton.

SECTION 7.—ORGANIZATION—C.W. INTELLIGENCE

Introductory

Gas intelligence is the responsibility of the " I " branch of the Staff as is all other intelligence. In the case of gas warfare, however, in order to appreciate the significance of information received, it is essential that officers with a comprehensive scientific and technical knowledge of all aspects of the subject are either included within the " I " branch itself, or are immediately available for consultation and advice as necessary. The organization as it developed was a combination of these two alternatives—the former being adopted in higher formations such as the War Office, and the latter in formations in the field.

In the field, the general principles regarding all intelligence and responsibility of all ranks, all units and all staffs and formations apply equally to gas intelligence. Until gas is used and troops understand from experience the particular requirements of gas intelligence, effort has to be concentrated on maintaining as high a standard as possible of knowledge and alertness, in order to interpret the significance of all indications relative to the enemy's intentions to use gas thereby obviating to some degree the surprise effect of the first attack.

Retaliation, being a matter of high policy, was a War Cabinet decision co-ordinated with Allied Governments, and it was therefore of the utmost importance that the first use of gas should be immediately reported and reliably confirmed beyond doubt.

The early examination of captured gas equipment is essential, both in relation to the deductions which can be drawn from it about the enemy's methods and intentions, and also in relation to its bearing on our own research and development.

It was with these three aspects, *viz.*, maintenance of a high standard of intelligence training ; the reporting and confirmation of the first use of gas ; and the examination of captured equipment, that we were mainly concerned during the war.

Organization at the War Office and Ministry of Supply

Before the war the collection, collation and dissemination of all information regarding enemy gas warfare activity was the responsibility of the intelligence section of the Chemical Defence Research Department, who compiled an annual report which was amended quarterly. Such a procedure was obviously inadequate for war.

In October, 1939, a new branch of the Military Intelligence Directorate at the War Office (M.I.10) was formed for liaison with technical branches and departments regarding foreign arms, equipment, transportation and land warfare industrial potential. In this new branch one officer was primarily concerned with gas warfare and smoke intelligence. Major D. C. Evans, R.E., was commissioned from the Chemical Defence Research Department for this appointment. In 1943 he was transferred to C.D.R.5, the M.I.10 appointment being held thereafter by a selected technical officer (C.W.).

A central clearing house to deal with the subjects was essential, and one based on the existing facilities in the Chemical Research Department grew up, the intelligence section of that Department (C.D.R.5) becoming the nodal point in the organization. Information was collected by M.I. Directorate and C.D.R.5 from the respective sources peculiar to each, and this was pooled.

C.D.R.5 was responsible for the collation of all information irrespective of its source. They compiled summaries in great technical detail and included an assessment of the value of the information and their deductions from it. These reports were circulated to the War Office, Admiralty, Air Ministry, Ministry of Aircraft Production, Ministry of Home Security, C.W. Research Establishments and authorities at home, abroad and in the Dominions, Anti-Gas Laboratories and the U.S. Chemical Warfare Service Intelligence Section in London.

M.I.10 was responsible for its consideration in the light of purely military factors. They prepared and issued summaries of matter relevant to and suitable for distribution to formations and units, and kept other War Office branches informed of all detail affecting them. The War Office branch primarily concerned, *i.e.*, S.W.V.2, also received C.D.R.5 summaries direct. M.I.10 worked in close liaison with the G.S. branch controlling C.W. in the military assessment and evaluation of information, and also with other branches of the M.I. Directorate and the Joint Intelligence Committee in the preparation of appreciations. They also worked in close conjunction with U.S. Intelligence Staffs and to a lesser degree with the C.W. representative on the Soviet Military Mission.

Technical officers (C.W.), and through them the C.W. staffs of formations, were fed with much fuller technical information than that provided in normal M.I.10 summaries. This special channel emanated from S.W.V.2 and its predecessors.(1)

This organization continued throughout the war and proved successful. M.I.10 and C.D.R.5 worked in the closest collaboration. As an example, the Handbooks on German, Italian and Japanese Chemical Warfare were drafted by C.D.R.5 and edited and published by M.I.10.

Organization in the Field

The normal "I" channels were maintained from the unit upwards for submission of C.W. intelligence. An exception to this was the direct disposal of early samples of captured C.W. equipment, particularly of suspected new types, direct to the nearest formation having a technical officer (C.W.), *i.e.*, corps or above.

Note (1).—A monthly series of "Technical Notes" for technical officers C.W. was initiated by G. (C.W.), G.H.Q., Home Forces, in August, 1940. This series was taken over, and their scope and circulation extended, by the War Office (E.4) in November, 1942, and by the General Staff from April, 1943 onwards until after the end of the war. Their object was to keep technical officers up to date on all scientific and other developments in chemical warfare, both our own and enemy, which affected their duties. Latterly, a special C.W. Technical Bulletin on enemy intelligence was prepared by M.I.10 and issued with "Technical Notes." Although these notes carried authority on technical matters only they became in the end a valuable means of liaison between the War Office and C.W. staffs in formations, both at home and overseas, on all gas, smoke, and flame matters. In so far as these subjects were concerned, the notes were complementary to S.W.V. Liaison Letters which covered equipment policy on all subjects within D.S.W.V's. purview, and which were issued monthly from April, 1943 onwards to overseas theatres.

In a unit the submission of gas intelligence was the responsibility of the unit intelligence officer who had the unit gas officer and N.C.Os. to assist him with advice, *e.g.*, training of the battalion intelligence section. Unit gas officers and N.C.Os. however, were not responsible for the collection of information or submission of reports.

Brigade intelligence officers qualified on a course at the Army Gas School, or, if attendance at such a course was not practicable, they received instruction from C.W. staffs from a higher formation. All intelligence officers attending intelligence courses at Matlock received a short training in C.W.

From division upwards it was the particular responsibility of C.W. staffs to advise the " I " staff about matters on which gas information was particularly required and to assist them in assessing the value of information received.

At corps and above there were technical officers C.W. who had special responsibilities in connection with advice from a scientific and technical standpoint, and with the examination of gas and smoke weapons and equipment, samples of contaminated ground and materials, etc., and their submission, as required, to the Chemical Defence Experimental Station, Porton, to an overseas experimental station or to an anti-gas laboratory according to the facilities or orders prevailing in any particular theatre. (For detailed responsibilities of staff and technical officers, *see* Section 2 of this Chapter and Appendices XI and XII. The functions of anti-gas laboratories are given in Section 6 of this Chapter.)

In obtaining information on non-persistent gas, medical evidence is of a decisive nature. Assistant directors of hygiene and deputy assistant directors of hygiene, working with technical officers C.W., had a special responsibility in connection with obtaining and forwarding the necessary pathological specimens and arranging autopsies on non-persistent gas fatalities.

The division of responsibility between " I " and " C.W." staffs was normally defined in formation standing orders. No precise division was ever laid down by the War Office, nor does this seem practicable, as much depends on local conditions and the tactical situation. The closest liaison between " C.W." and " I " branches was essential. Normally it was the responsibility of " I " to include general information of the enemy use of gas in intelligence summaries and battle intelligence reports, and the responsibility of " C.W." to see that such reports were technically sound. Staff officers trained in C.W. were particularly responsible for collecting information about the enemy's methods of gas warfare.

The details of the information required are not dealt with here. They will be found in the current Gas Training manuals, the Field Service Pocket Book and the Chemical Warfare Pocket Book.

Organization in other Services

In order to understand the procedure for reporting the first use of gas by the enemy, had such attack taken place on the United Kingdom, a brief outline of the organization of other Services is necessary.

The Civil Defence Organization was controlled by the Ministry of Home Security. At the Ministry itself there was a Scientific Adviser to the Minister who controlled the subordinate technical officers within the organization. At Regional H.Q. there was a senior gas adviser, and gas identification officers (G.I.Os.) were provided in all sectors. The gas identification officers were provided with a small gas identification apparatus with which they could test for and recognize the standard war gases only, and both their training and functions were limited in scope. In the civil organization the senior gas advisers were broadly analogous to the technical officers C.W. in the military organization. Regional hospital officers investigated any suspected gas casualties. Naval shore establishments made use of the civil organization.

The R.A.F. had command gas advisers at the H.Q. of each R.A.F. command in the United Kingdom. These officers were chemists of similar standing to military technical officers (C.W.). They were controlled by the Ministry of Aircraft Production who were the technical advisers to the Air Ministry. There were no technical staffs on R.A.F. stations. Station commanders were to make use on demand of civil gas identification officers or military technical officers pending the arrival of the command gas adviser.

The U.S. Forces relied on the officers of the Chemical Warfare Service.

Procedure for Confirming the First Use of Gas

This presented a special problem and was an extension of normal intelligence problems.

The procedure for substantiating the first use of gas in the United Kingdom was based on the submission of technical and medical evidence and samples by technical officers (C.W.) and deputy assistant/assistant directors of hygiene, by R.A.F. command gas advisers, or by civil gas identification officers, gas advisers and regional hospital officers, or by all channels. The procedure was kept constantly under review and amended in detail as required by the circumstances of the phases of the war and was co-ordinated with the U.S. Forces.

In short, it involved :—

(*a*) The transmission of reports of the use of gas through civil and Service (including U.S. Service) channels to their respective higher headquarters, and arrangement for the lateral passage of information at various levels, *e.g.*, Army commands/corps and Civil Defence regions.

(*b*) The sending of gas and pathological samples to the Chemical Defence Experimental Station, Porton, followed by their examination and report. The U.S. Forces submitted samples to their own chemical laboratory companies and general medical laboratory companies.

(*c*) The preliminary examination of early reports on the use of gas by the technical experts of the ministries and services concerned.

(*d*) The final assessment of the reports.

The authority responsible for co-ordination of (*c*) and (*d*) varied in different phases of the war. Initially it was the Director of Chemical Warfare, G.H.Q. Home Forces ; latterly the Ministry of Home Security.

(*e*) The submission of agreed findings to the highest authority for subsequent action in relation to the employment of gas offensively and to the warning of the public.

At the outbreak of the war, there was no pre-arranged plan between the French and British Governments to get an immediate joint decision to retaliate. The matter was raised, but the retreat through Dunkirk occurred before any machinery to effect a rapid decision came into force.

The procedure to obtain a Cabinet decision should gas have been used against the United Kingdom had, initially, only to cater for speed in obtaining the decision for our own forces to retaliate. Speed was necessary for tactical reasons as, in the event of the enemy employing gas in an invasion, delay in retaliation might have been serious.

As the war progressed and Italy, Russia, Japan and America were drawn into the conflict, other factors as well as speed intervened. Strategic issues became involved. Retaliation in all theatres in all circumstances was a War Cabinet decision in consultation with Allied Governments (*see* Section 9 of this chapter for co-ordination of policy with the United States), and a system similar to that which had been established in the United Kingdom was set up in all theatres of war and co-ordinated with Empire and Allied Forces.

It was the common policy of the United States and British Governments only to use chemical warfare after its use had been begun by the enemy. It was implicit in this policy that the unplanned and unauthorized use of chemical warfare weapons on a small scale would not necessarily constitute adequate reason for starting gas warfare, although this point could obviously not be expressed in open international declarations.

The implementation of this agreed policy, for which co-ordination by the combined Chiefs of Staff had been arranged, required early and, if possible, simultaneous receipt in Washington and London of identical reports of all incidents which might prove to be the basis for the initiation of the use of gas, and the procedure agreed on was directed to this end.

The procedure for the United Kingdom and continent of Europe as it stood on " D " day is given in Appendix XVI. Procedure in other theatres was similar but adapted to the system of command and to local conditions.

It should be noted that at no time was any procedure co-ordinated with the Russians. Had the Germans used gas on their Eastern front, there were no facilities for independent confirmation by British or American observers and we should have had to take the Russian word for it.

The Final Uncovering of German Gas Warfare Activities

The channels for the examination of enemy C.W. equipment captured in the normal course of operations in all theatres has already been described.

As the war in Europe drew to its close a special organization came into being to uncover all German scientific discoveries, trade secrets, military, civil and commercial records, etc.

This organization consisted of a Committee under the Deputy Chiefs of Staff known as the Combined Intelligence Objectives Sub-Committee (C.I.O.S.) on which representatives of the Services and the Civil Ministries sat. This was linked through S.H.A.E.F. with the United States world-wide organization, Field Intelligence Army Technical (F.I.A.T.), which dealt with the U.S. formations in the theatre and had H.Q. in London and in North West Europe. In 21 Army Group, " T " (Target) Force was the C.I.O.S. agency for British formations.

The staff direction and command of " T " Force was vested in the Director of Chemical Warfare, 21 Army Group, whose branch was redesignated G(T) and C.W. Although a separate " T " Branch of the Staff would have been desirable, this was precluded by reasons of manpower. Superficially it might appear that the " I " Staff would have been the most suitable to direct such activities, but much of the intelligence to be gleaned was of no concern to Army Group G.S.(I) and, in many aspects, the activities of " T " Force were operational and administrative. The C.W. Branch was already actively concerned in uncovering German C.W. activities, munition dumps, etc., as operations progressed, and, although heavily involved in smoke and flame operations, could not be considered fully occupied owing to the continuance of gas peace.

The object of " T " Force (a force eventually comprising three battalions, two bomb disposal companies, six pioneer smoke companies whose transport([1]) made the force comparatively independent, together with document teams, interpreters, etc.) was the locating and securing intact of all Intelligence targets, *e.g.*, war factories, research and experimental establishments, Government ministries, static military H.Q., mines, banks, fuel dumps, at the first opportunity that operations permitted, and the guarding of them to prevent destruction or looting until examined by teams of experts, or until essential material or documents had been removed.

The planning of target lists was carried out in consultation with C.I.O.S., branches of Army Group H.Q., and the Inter-Service Topographical Department at Oxford.

As targets were overrun, they were occupied by " T " Force and then immediately investigated by teams of experts which were held ready in the United Kingdom and flown over. These teams were known as consolidated advance field teams (C.A.F.T.). The job of these teams, which were so organized that their members covered a number of related subjects([2]), was to make a quick preliminary evaluation of the target, and to recommend an appropriate team to investigate it in detail. These teams were administered by " T " Force.

The final uncovering of all German C.W. activities was therefore carried out by teams consisting of representatives of the Chemical Defence Department of the Ministry of Supply and technical officers (C.W.) under the aegis of the above organization.

Note ([1]).—This was subsequent to the Rhine Crossing after which smoke producing equipment was dumped because large tactical smoke screens and A.A. screens were no longer required.

Note ([2]).—C.W. was included in Group 3, Chemical Warfare, Miscellaneous Chemicals, Medical, Fuels and Lubricants.

SECTION 8.—ORGANIZATION, ALLIES

UNITED STATES ORGANIZATION TO HANDLE CHEMICAL WARFARE

NOTE.—THE TERM " CHEMICAL WARFARE " IN THE UNITED STATES EMBRACES ALL ASPECTS OF GAS, SMOKE, AND FLAME WARFARE (INCLUDING INCENDIARIES)

The Chemical Warfare Service

The organization to handle chemical warfare in the United States Forces centres on the Chemical Warfare Service. This service is a Service of Supply under the Commanding General, Army Services Forces, and serves the Army Ground Forces, the Army Air Forces and the Army Service Forces. The Chemical Warfare Service is responsible for gas, smoke, and flame research from every angle, and for design, production or procurement, provision and supply (including supply and maintenance in the field) of all offensive and defensive materials and equipment ; it is the parent corps for certain chemical units ; and it is responsible for the formulation of tactical and training doctrine in connection with these types of warfare, but, being a " G–4 " (or " Q ") organization, it does not exert any operational control.

This Service at the end of the war comprised :—

(1) The Chief of the United States Chemical Warfare Service and an Assistant Chief.

(2) A Headquarters organization. This combined within itself all functions in relation to gas, smoke, and flame warfare (other than those of operational control) which, in the case of the British Army, were performed by the War Office, the Air Ministry, the Ministry of Supply, the Ministry of Aircraft Production and the Petroleum Warfare Department. It had control of research establishments, proving grounds, and factories connected with chemical warfare.

(3) A corps of officers.

(4) The following units :—

(*a*) CHEMICAL MORTAR BATTALIONS

H.Q. and H.Q. company and 3 mortar companies (each twelve 4·2-in. rifled mortars) ; during non-gas warfare used in close support of Infantry, firing H.E. and smoke (W.P.) ; allotted to tactical formations as required. Total strength (including attached medical) : 38 officers, 1 W.O. and 633 other ranks.

(*b*) CHEMICAL WARFARE SERVICE UNITS—ARMY

(i) Chemical Depot Coy.—Handling and issue of C.W. equipment and munitions (*i.e.*, all gas warfare, smoke, flame and incendiary stores and munitions) ; normally allotted one for each army; generally self-sufficient, but in certain circumstances required temporary assistance from labour troops and additional motor transport ; could be divided into three depot sections to operate small depots close to front line. Strength : 5 officers and 173 other ranks.

(ii) Chemical Base Depot Coy.—Storage and issue of all classes C.W. materials ; operated a chemical base depot or chemical section of a general depot ; required extra labour and vehicles from labour and vehicle pools ; could handle chemical munitions and materials for 100,000 individuals ; did not maintain or repair chemical material. Strength : 7 officers and 100 other ranks.

(iii) Chemical Maintenance Coy.—Allotted to field armies for repair and salvage of chemical warfare equipment ; repair of all C.W. protective equipment and appliances, and repair and re-conditioning of weapons used exclusively by chemical combat troops ; required transport from pool sources for movement and for salvage operations. Strength : 4 officers and 115 other ranks.

(iv) Chemical Decontamination Coy.—Assigned to field armies; large scale decontamination of vital areas or installations ; equipped with twelve 400-gallon power-driven decontaminating apparatus ; also undertook decontamination of large quantities of materials ; completely self-contained. Strength : 5 officers and 158 other ranks.

(v) Chemical Processing Coy.—L. of C. unit for impregnation and re-impregnation of clothing ; allotted on the basis of one company for every 50,000 troops ; equipped with two M.1 or M.2 semi-fixed " Theatre of Operation " type impregnating plant ; not mobile. Strength : 5 officers and 135 other ranks.

(vi) Chemical Base Processing Coy.—Function as for (v) above ; normally allotted to base chemical depot or chemical section of base general depot on the basis of one for every 300,000 troops ; equipped with one Zone of Interior fixed impregnation plant (ten times the size of " Theatre of Operation " type in (v) above) ; not mobile and required certain additional labour. Strength : 4 officers and 62 other ranks.

(vii) Chemical Laboratory Coy.—One company normally allotted to the theatre of operations ; a base unit—not mobile ; analogous to British anti-gas laboratory. Strength : 8 officers and 50 other ranks.

(viii) Chemical Smoke Generator Coy.—Analogous to British Pioneer Corps smoke company (*see* Part II—Smoke). Strength : 4 officers and 129 other ranks.

(ix) H.Q. and H.Q. Detachment Chemical Smoke Generator Battalion.—For administration and supervision of three to eight smoke generator companies within an area of command.

(x) Chemical Service Coy.—A flexible organization normally allotted to a force or troop concentration of limited size ; combined on a reduced scale the functions of all Chemical Warfare Service units and had maintenance, supply, decontamination, impregnation, and laboratory sections, any of which could be omitted or duplicated according to requirements ; the laboratory facilities were mobile on a 2½-ton truck.

(xi) Chemical Warfare General Service Coy.—Normally allotted to a depot ; provided, on a reduced scale, supply, maintenance, laboratory and decontamination services for 30,000 troops.

(c) Chemical Warfare Service Units—Army Air Forces

(i) Chemical Depot Coy., Aviation.—Designed to set up and operate certain types of chemical munition depots or chemical supply warehouses ; responsible for storage and issue of all types of chemical munitions used by the Army Air Forces EXCEPT chemical filled bombs which were an Ordnance item.

(ii) Chemical Coy., Air Operations—A specialized organization to provide the necessary service in connection with the delivery of chemical munitions and filled chemical spray and smoke tanks to squadrons, and with the removal from aerodromes, maintenance and refilling of such tanks as were brought back by returning planes.

(iii) Chemical Maintenance Coy., Aviation.—Organized on the same establishment as a Chemical Maintenance Coy. ((b) (iii) above) ; provided corresponding services for the Army Air Forces.

With the exception of smoke generator companies and the chemical mortar battalions, Chemical Warfare Service units remained largely under the control of the Chemical Warfare Service on the " G–4 " (or " Q ") side of the staff in Army formations at home and in the field or under the equivalent branch of the staff in the Army Air Forces.

Chemical mortar battalions, when allotted to formations, were controlled operationally and administratively by the " G–3 " branch of the staff (Operations). The Chief of the Chemical Warfare Service was responsible for the basic training of these units, for the formulation of doctrine as to their employment, and for equipping them.

Smoke generator companies were controlled operationally by appropriate commanders, but the Chemical Warfare Service retained more administrative control than in the case of chemical mortar battalions.

C.W. Staffs—Unit and Formation

Unit gas officers and N.C.Os. were provided from regimental personnel. Their functions were mainly defensive and broadly analogous to those of British gas officers and N.C.Os. These officers and N.C.Os. were trained at schools run by the Chemical Warfare Service. Generally, each company or similar organization had a decontamination squad or detachment, and it was one of the functions of the gas N.C.O. to supervise and maintain its training.

Chemical staff sections, staffed by officers of the Chemical Warfare Service, were included in the establishment of theatre H.Q., armies, corps and divisions, and the equivalent formations of the Army Air Forces. In the British Army, staff officers trained for chemical warfare, with technical officers as advisers, were appointed in the General Staff and were mainly concerned with operational, training and intelligence aspects of chemical warfare, while also providing a source of specialist advice for other branches of the Staff, and Arms and Services as required. In the United States Forces the chemical staff organization was oriented more on the administrative side. Their chemical staff sections came under the " G–4 " (or " Q ") branch of the staff. Their main function was one of supply, including provision, storage, maintenance, repair and technical inspection of C.W. material and munitions, while acting in an advisory capacity on all chemical matters to the commander and " G–2 " (Intelligence) and " G–3 " (Operations) branches of the staff, and having certain supervisory and inspection functions in relation to chemical warfare training generally, decontamination, and captured enemy material. Their position in relation to Chemical Warfare Service units was broadly analogous to that of a British A.D.O.S. to R.A.O.C. units, except in certain circumstances outside the immediate theatre of operations where it amounted to command.

The United States Chemical Warfare Committee

Early in 1943, when the Combined Chiefs of Staff issued a directive for a joint Anglo-American Chemical Warfare policy and supply and procurement programme (*see* Section 9 of this Chapter), the United States Chemical Warfare Committee was established. The Chief of the U.S. Chemical Warfare Service was the chairman of this Committee, and it included, in addition to representatives of the Army Ground Forces, Naval and Air representatives. These latter reported to the Navy Department and Commanding General Army Air Forces respectively. This Committee was formed primarily to advise in matters connected with supply and procurement and to co-ordinate these with the British, and was responsible to the Commanding General Army Service Forces. It did not have direct access to the United States Chiefs of Staff in the same way as the British Inter-Service Chemical Warfare Committee, which was directly subordinate to and responsible to the British Chiefs of Staff.

Comparison between the British and United States Organization

It is difficult to draw useful deductions from the differences in the British and United States organizations to handle chemical warfare, since these differences were largely fundamental to the organization of the respective Armies and Air Forces as a whole. It would be a fallacy to conclude that either method would necessarily produce better results in all circumstances, *viz.*, a unified Chemical Warfare Service on the U.S. model, or, as in the British case, the selection of chemical warfare staff officers from all Arms, strongly reinforced by civilian chemists, and the assumption of applicable chemical warfare functions by the various Arms, Services and Ministries without resort, in the main, to special corps or units. There are, however, certain points which are worthy of attention.

A wholly centralized organization, such as the United States Chemical Warfare Service, tends to create a vested interest in the Army and Air Forces in the specific aspects of warfare with which it deals. This is likely to lead to lavish provisioning of materials, particularly when supported by unlimited manpower and productive capacity. Where resources are severely limited, as in the British case, such a tendency would be the more serious.

When, as in the last war, gas is not actually being used, it is the prime responsibility of officers dealing with chemical warfare to make the most careful assessment of the requirements of gas weapons and associated equipment, and of manpower requirements, and to keep them down to a minimum compatible with safety, in the interests of the war effort as a whole. A balanced appreciation of requirements is more likely to occur with an organization on the British model, where all interests are fully consulted in all stages of the formulation of policy and where no single interest is predominant. Further, it is easier to keep down the total requirements of men and materials where the " overheads," so to speak, are shared amongst the different arms and services.

The very existence of a separate Chemical Warfare Service establishes a separate loyalty and somewhat militates against the proper co-operation between the chemical warfare experts and the General Staff branches concerned. This in turn leads to a lack of a true appraisement of the worth of chemical warfare *vis-à-vis* other weapons in the individual circumstances of particular theatres of war or campaigns. In the field, the fact that the U.S. chemical warfare staff sections were part of the "G.4," (or " Q ") organization, and were therefore normally located to the rear with the services rather than forward with the " G.3 " (operations) staff, tended to cause a lack of intimate and immediate liaison with the " I " and " O " staffs, which gave rise on occasions to an unbalanced evaluation of information.

The U.S. Chemical Warfare Service considered gas to be a conclusive weapon of warfare in itself, and were prepared to subordinate other methods of attack. British authorities responsible for gas warfare considered it to be a means to an end, only to be employed if its use would achieve that end better and more cheaply than would other means. This divergence in doctrine (which is referred to in Chapter I, Section 7 and in Section 9 of this chapter) fundamentally had its roots in the differences in organization, which reflected the different conceptions of the functions of gas warfare.

Had gas warfare started there can be little doubt that in the British case, with our acute manpower problems, many difficulties would have arisen during the transition period from non-gas warfare to an " all-out " gas effort. The comparatively small staffs and establishment, both in the Services and the Ministries, would soon have been overwhelmed by the work entailed and would have had to be diluted with imperfectly trained personnel.

Although fundamentally a problem of manpower rather than organization, the fact that the United States possessed a large regular chemical warfare service would have eased the problems of transition.

SECTION 9.—ANGLO-AMERICAN ORGANIZATION FOR THE FURTHERANCE OF A JOINT CHEMICAL WARFARE POLICY

Before the entry of the United States into the war there had been some exchange of information on chemical warfare matters between ourselves and the United States both through the U.S. Assistant Military Attaché in London and through the representative of the Ministry of Supply (Mr. Kingan) with the British Purchasing Mission in the United States.

Direct liaison with the U.S. Army and Air Forces was established on their arrival in the United Kingdom in 1942. United States personnel visited British training establishments, studied British defensive methods, and attended Service trials, demonstrations, and exercises, connected with chemical warfare.

In the case of the Air Forces, British anti-gas equipment was retained on R.A.F. stations taken over by United States Air Forces, and N.C.O. instructors were loaned to them for short periods to instruct them in our organization, methods, and procedure in co-operation with other Services and the civil authorities. An R.A.F. officer was appointed to the 8th U.S. Air Force as a gas defence/fire officer to co-ordinate the training of American personnel in our methods. The result of the information and instruction given to the chemical warfare personnel of the U.S. Air Forces in the United Kingdom was that, to some extent, they were re-organized on the same lines as anti-gas personnel in the R.A.F.

Co-ordination of Policy

In May and June, 1942, the Prime Minister and the President of the United States made public announcements regarding the policy of their respective countries should the Germans or the Japanese initiate gas warfare.

On May 10th, 1942 the Prime Minister declared :—

"I wish now to make it plain that we shall treat the unprovoked use of poison gas against our Russian ally exactly as if it were used against ourselves, and if we are satisfied that this new outrage has been committed by Hitler we will use our great and growing air superiority in the West to carry gas warfare on the largest possible scale far and wide upon the towns and cities of Germany." The full text of the Prime Minister's declaration is at Appendix I.

On June 6th, 1942, the President declared :—

"Authoritative reports are reaching this Government of the use by Japanese armed forces in various localities of China of poisonous or noxious gases.

I desire to make it unmistakably clear that if Japan persists in this inhuman form of warfare against China or against any other of the United Nations, such action will be regarded by this government as though taken against the United States and retaliation in kind and in full measure will be meted out. We shall be prepared to enforce complete retribution. Upon Japan will rest the responsibility."

These declarations had been made unilaterally, and initially there was no co-ordinated policy for joint means to give effect to them or to avoid the delay in last minute consultations when retaliation was required ; nor had the implications of the effects of gas warfare on the joint Anglo-American war effort been examined.

It must be remembered that the British Commonwealth was bound by the Geneva Protocol, whereas the United States were not so bound and had consistently refused to undertake any international obligations to restrict herself from gas warfare.

Both nations had to be prepared for the initiation of gas warfare by the Axis powers—the time, place and purpose of which could be chosen by the enemy. Each of the United Nations faced the possibility of instituting retaliation on the spot and it was realized that an individual decision by any one power might be a matter of the utmost gravity for the others.

In August, 1942, the United States Chiefs of Staff initiated conversations with the British Chiefs of Staff with a view to the formulation of a joint chemical warfare policy for the two nations and of a joint supply and procurement programme. Initially there was some divergence of views between the two staffs on whether the responsibility for the decision to retaliate in the event of the enemy using gas could be delegated in any theatre to any individual, *e.g.*, Supreme Commander. In British theatres the decision to retaliate was to be taken only by the War Cabinet on the evidence produced. Owing to the fact that once gas warfare was being waged in any one theatre it could not be restricted to that theatre only, the British Chiefs considered that the decision to retaliate should be taken only by the Governments concerned and not by any representative.

The U.S. Chiefs of Staff accepted this point of view and the following policy directive was issued by the Combined Chiefs of Staff in November, 1942 :

" (*a*) Gas warfare will be undertaken by both the United States and British Commonwealth Forces on the order of the Combined Chiefs of Staff after approval by the United States and United Kingdom Governments, or independently in retaliation on the decision of the Government concerned.

(*b*) The United States and British Commonwealth Forces will provide the evidence of the enemy's use of gas warfare and make prompt confirmed transmittal of the information to the Combined Chiefs of Staff."

The circumstances leading up to the public announcements by the Prime Minister and the President in 1942 and their re-affirmation in 1943 have been dealt with in Chapter I. The 1942 statement of the President referred to the use of gas by Japan only, but his statement of June, 1943 was to the effect that the use of gas by any of the Axis powers against any of the United Nations would be taken as its use against the United States itself, and would be met with full and instant retaliation. The co-ordination of world-wide arrangements to confirm the use of gas by the enemy is referred to in Section 7 of this Chapter.

At the end of 1943 the question of the association of Commonwealth Governments with the decision to retaliate was raised by the Canadian Government who asked to be so associated ; they also raised the question that the orders to use gas by any Allied Supreme Commander, after his own Government had decided to retaliate, should be binding on all troops under his command irrespective of nationality.

This led to a re-examination of policy between the British and United States Chiefs of Staff. The situation had changed since the policy announcement of 1942. There was complete integration in almost all theatres of U.S. and Empire Forces, coupled with the appointment of Supreme Allied Commanders of differing nationalities. It made the idea of independent action by any one Government inappropriate and largely inoperable. The effect of a unilateral decision would have been widespread. For example, if the Japanese had used gas and the Americans had retaliated, this would have involved British Empire troops in that theatre and might have had serious repercussions in the European theatre where the outbreak of gas warfare before invasion of the continent would have been to our disadvantage (*see* Chapter I).

Further, British and Commonwealth territories were geographically more exposed to gas attack than the United States, and our civil populations and bases of production would have had to bear the brunt of the use of gas against them as a result of unilateral action by the United States.

The British Chiefs of Staff did not consider it appropriate that any one Government should be in a position to embroil the whole world in gas warfare.

The British Chiefs of Staff, therefore, would have liked to delete the provision for independent action by any one Government as laid down in the 1942 policy announcement, although from a purely military standpoint they recognized the impossibility of swift retaliation if a number of Governments had to be consulted, and were sympathetic of the U.S. Chiefs of Staffs position in the light of the President's public announcement of June, 1943.

The U.S. Chiefs of Staff considered that the dangers inherent in a policy which would bind the individual Governments to obtain approval of U.S., British and Dominion Governments before taking retaliatory action were far more serious than those attendant on the existing policy. They considered that any inability for prompt retaliation might have the gravest military consequences, and they felt that the disapproval of any Government, no matter how little concerned, could prevent effective action. The U.S. wished to be entirely free to implement the President's public announcement.

After some discussion a compromise was effected. If a decision to retaliate was made independently by any Government that Government undertook to inform the Combined Chiefs of Staff before taking any action.

In May, 1944, a revised directive was issued by the Combined Chiefs of Staff as follows :—

"(A) Gas warfare will be undertaken by both U.S. and British Commonwealth Forces on order of Combined Chiefs of Staff after approval by the U.S. and U.K. and other Commonwealth Governments concerned, or independently in retaliation, including the fulfilment of commitments([1]), on the decision of the Government concerned. In the latter case, the decision of the Government concerned will apply with full force to all troops operating under Commanders of that Government, which will assume full responsibility for their actions.

(B) U.S. and British Commonwealth Forces will provide evidence of the enemy's use of gas warfare and make prompt confirmed transmittal of the information to the Combined Chiefs of Staff. When the decision to retaliate is made independently by any such nation it will give immediate confirmed advice to the Combined Chiefs of Staff of its intention before taking any such action."

Combined Chiefs of Staff Directive for a Co-ordinated Supply and Procurement Programme

The directive for a co-ordinated Anglo-American Procurement and Supply programme was issued in November, 1942, and embodied the following :—

(*a*) The responsibility for carrying out the programme was charged, on the United States side, on the Commanding General, Army Service Forces who was to designate the Chief of the Chemical Warfare Service, in co-ordination with Naval representatives, to contact the appropriate British agencies. On the United Kingdom side the British Chiefs of Staff designated the Inter-Service Committee on Chemical Warfare as their agent to implement the programme, except in respect of flame and incendiary weapons which were not a responsibility of this Committee, and which were handled by the War Office, the Petroleum Warfare Department and the Ministry of Supply through departmental channels, liaison with the United States being effected through the British Army Staff in Washington.

(*b*) Information was to be exchanged and composite appreciations made, and kept under review, of the Axis powers' capabilities to wage gas warfare. On the British side this was carried out by the Joint Intelligence Sub-Committee ; on the U.S. side it was a responsibility of the "G–2" (Intelligence) branch of the staff. Combined appreciations were handled by the Combined Intelligence Committee in Washington.

Note ([1]).—The phrase "including the fulfilment of commitments" referred, for example, to the Prime Minister's declaration that if the Germans used gas against the Russians we should immediately retaliate.

(*c*) The agencies designated in (*a*) above were responsible for the following :—

(i) The establishment of a potential production capacity capable of rapid expansion to meet the needs of gas warfare. Current production was to be kept on a minimum level compatible with the policy to expand and with the building up and maintenance of initial stocks. Initial stocks were to be maintained at levels which would enable gas warfare to be sustained pending expansion of production.

(ii) Uniform initial stock levels of all types of equipment were to be established in combined theatres.

(iii) Uniform logistic factors for anti-gas equipment, gas weapons, and munitions were to be maintained as practicable.

(iv) Uniform minimum levels and ultimate standards of individual and collective anti-gas protection were to be established.

(v) All types of C.W. material in use by the United States and British Commonwealth were to be interchangeable and, as far as possible, standardized. This included screening smokes and incendiary weapons.

(vi) The extent of the measures adopted to effect the above were to be limited to those compatible with a balanced overall munition programme.

As regards (i), British policy was already based on these principles. In connection with (ii), British policy took into account the difficulties of supply and the distance from sources of production in the case of each theatre ; (iii) presented little difficulty, and close liaison had already been initiated in the case of (iv). Item (v) was one of great consequence and difficult to attain in full. It involved much detailed discussion with U.S. experts and the extent to which it was implemented is shown on page 128.

Item (vi) had been inserted on strong representations from the British Chiefs of Staff who realized the necessity for ensuring that gas enthusiasts should not be allowed to embark on the excessive production that might result from an intensive programme of joint gas warfare preparations. A balanced overall munition programme was already a major factor in British policy.

Initial Steps in Implementing the Directive

An initial step to assist in giving effect to the above policies was the appointment, in February, 1943, of a War Office chemical warfare representative, Lieutenant-Colonel H. Paget, R.E., to the British Army Staff in Washington, whom U.S. C.W. agencies could contact on matters connected with supply and procurement. Since the Royal Navy and the R.A.F. had no C. W. experts on their staffs in Washington, Colonel Paget acted on their behalf in relation to policy matters, in particular in respect of anti-gas stores which were common to all three Services. At a later date the position was formalized when Colonel Paget became the agent of the British Inter-Service Committee on Chemical Warfare to the United States Chemical Warfare Committee. He was also authorized to represent the Australian Army.

The Ministry of Supply already had a chemical warfare expert with their Mission in Washington, and the Ministry of Aircraft production later (at the end of 1944) had a representative.

Concurrently with Lieutenant-Colonel Paget's appointment H.Q., E.T.O U.S.A. were approached to nominate a U.S. representative to the Inter-Service Committee on Chemical Warfare. Colonel (later Brigadier-General) Hugh W. Rowan was nominated and he later became the agent of the U.S. Chemical Warfare Committee with the British Inter-Services Committee on Chemical Warfare, Colonel J. E. Zanetti being nominated as his alternate.

In March, 1943, General William N. Porter, Chief of the United States Chemical Warfare Service and Chairman of the U.S. Chemical Warfare Committee, visited the United Kingdom, and Air and Army problems affecting Anglo–American participation in gas warfare were discussed. As a result the need became apparent for more detailed examination of the doctrines and policies of the two countries in relation to supply and production. Two sub-committees of the British Inter-Service Committee on Chemical Warfare were formed in May, 1943 (the Army Technical Sub-Committee and the Air Ministry Technical Sub-Committee). These two Sub-Committees consisted of United States and British members and were set up to study and make recommendations regarding supply and production problems in relation to the furtherance of a joint Anglo–American C.W. policy.

The United States Chemical Warfare Committee had been set up consequent on the issue of the Directive. This paralleled our Inter-Service Committee on Chemical Warfare, although it did not have direct access to the U.S. Chiefs of Staff (*see* Section 8 of this Chapter).

There were suggestions in the United States that their Chemical Warfare Committee should be a combined (U.S. and British) chemical warfare committee to report direct to the Combined Chiefs of Staff. In the British view, with which the United States eventually agreed, this was not considered justified. Combined committees already existed which handled, as they arose, chemical warfare questions within their purview, *i.e.* Combined Planners (Plans), Combined Intelligence Committee, the C.P.R.B. (Procurement), and the Munitions Assignment Board (Assignment). A Combined Chemical Warfare Committee would have overlapped with all these, and this would have tended to throw C.W. problems out of their true perspective and to the direction of an undue proportion of effort to chemical warfare. Further, increased commitments would have arisen for expert representation on such a committee of all three British Services and of the British Ministries concerned (*i.e.* Ministry of Supply and Ministry of Aircraft Production). Finally, there was no authority for a combined committee in the Directive issued by the Combined Chiefs of Staff.

It was thought that the formation of the two Sub-Committees of the Inter-Service Committee on Chemical Warfare with joint British and U.S. membership was sufficient. These Sub-Committees could submit reports on which comprehensive proposals could be made to the U.S. Chemical Warfare Committee. If agreement could not be reached it might be necessary for a delegation to proceed to America to discuss the whole question with the U.S. Chemical Warfare Committee before submission to the Combined Chiefs of Staff.

Measures Implemented under the Directive

The Sub-Committees referred to above met throughout the summer and produced their first interim reports early in September, 1943. These reports were discussed with Brigadier-General Alden Waitt, Assistant Chief of the United States Chemical Warfare Service and with Lieutenant-Colonel Javits, Secretary to the United States Chemical Warfare Committee, who visited London during that month.

The object of Brigadier-General Waitt's visit was to discuss with the British authorities the proper allocation of the chemical warfare effort of both countries, the furtherance of the inter-changeability and the standardization of munitions and equipment, and the co-ordination of tactical questions in relation to the supply and procurement programme. As a result of meetings with the War Office, the Air Ministry, Ministry of Supply and the Ministry of Aircraft Production, Brigadier-General Waitt produced his report before his departure from the United Kingdom.

This report was discussed by the British Inter-Service Committee on Chemical Warfare in October, 1943, and it was agreed that it represented to a great extent the combined views of the two countries. The report was endorsed by the United States Chemical Warfare Committee on the return of Brigadier-General Waitt to America.

The principal points on which a measure of agreement had been reached or on which discussion had been initiated are outlined below.

(i) Agreement had been reached on :—

(*a*) Tasks for which gas would be used in the European Theatre and concentrations required to effect them.

(*b*) A basis for computing requirements of A.A. smoke screens for overseas landing beaches and ports.

(ii) Comparative data of U.S. and British gas weapons and protective equipment were under publication.

(iii) *Equipment*

(*a*) British and U.S. scales of protection against gas had been found to be the same in principle, although there were some differences in respirator containers and protective clothing.

(*b*) Agreement had been reached for calculating bleach requirements for Western Europe and the United Kingdom and for a joint Anglo-American reserve. The object of this was to overcome the over-provisioning which would have resulted if the British and U.S. Services and the British Civil Defence Forces each provisioned separately for concentrated attacks on their own installations without reckoning on the lull which would occur during attacks on installations of other authorities.

(*c*) Supply of stores in common use, especially bleach and smoke materials, to United States and British Forces in combined theatres had been examined with a view to their provision from the most economical source in so far as shipping and production were concerned.

(iv) *Training.*—Training in individual protection had been found to be substantially similar in both British and U.S. Forces with one major exception. British instruction was (and still is) based solely on *categories of gas* whereas United States instruction involves *details of individual gases.* This difference in instruction led to difficulties in obtaining a common policy on divulgence of information on new chemical warfare agents including those of the enemy.

Instructions on the use of British and United States protective equipment had been drawn up and issued to enable U.S. Forces to understand and use British equipment, and *vice versa,* should the necessity arise.

(v) *Policy on the divulgence of information* on new chemical warfare agents had been agreed. This policy was that, before the outbreak of gas warfare, the British and the United States would exchange notice of the proposed release by either of them of such information for training or for public discussion. If gas warfare started the practicability of continuing this procedure was to be reviewed. We should have preferred a combined release of information in an agreed form, but the difficulties connected with the basic differences in training referred to above precluded this.

(vi) *Measures for the co-ordination of research and development* had been established including the necessary routine for exchange of information on both sides of the Atlantic.

(vii) *Standardization and Interchangeability of Material.*—The basic policy was recognized that in the development of new items and in the revision of existing items interchangeability should be attained wherever practical. The extent to which this was fulfilled is shown on page 128.

(viii) *Gas Output.*—Had gas warfare started we should have required the whole output of our gas factories to implement operational plans. Only a small surplus would have been likely to accrue through gas plant output being at a steadier rate than the operational air effort. This amount could have been stored in U.S. forward filling depots for charging into spray tanks and was to be supplemented by importation of mustard gas from U.S.A. or Canada. Charged mustard gas bombs were to be imported from the U.S.A.

(ix) *Air Gas Weapons*—

(*a*) The United States Air Force's requirements of British 400-lb. SL3 mustard gas spray bombs had been completed, and 500-lb. phosgene bombs were also being supplied to them pending the arrival of American 1,000-lb. bombs in the European theatre. These latter could also have been used by British aircraft.

(*b*) The U.S. M.10 spray tanks were to be provided from America for British " Typhoons " of the Tactical Air Force.

(*c*) An " aimable " cluster of 65-lb. mustard gas bombs interchangeable between British and United States aircraft was a development requirement to increase operational flexibility.

Divergence of Views about the Method and Extent of Gas Retaliation

The principal point on which a measure of agreement had NOT been reached was the method and extent of retaliation by gas.

The U.S. Chemical Warfare Service's view was that the gas weapon could of itself be decisive in character if used in great enough quantities and at the right points. They placed a higher premium on this form of attack on the assumption that huge personnel casualties would have a very damaging effect on the enemy. They also assumed that enemy protection among large masses of people and troop concentrations would not catch up quickly and that the surprise effect of non-persistent gas would continue to get heavy casualties for some time and would always be substantial. With this in view they visualized a most intensive air gas effort during the first few weeks of gas warfare, during which the scale of bombing would be stepped up and 75 per cent. of missions flown would be gas missions, about half of which would be with non-persistent gas. Subsequent gas effort was visualized as up to 50 per cent. of the bombing effort, one-third being with non-persistent gas. They considered that definite gas plans ought to be drawn up for attack on special targets and alternative plans prepared, any of which could be put into operation according to the conditions obtaining at the time retaliation became necessary. The above views were based on the absence of any limitations on the capacity to supply and to lift the gas munitions required.

The British view was that gas was essentially a weapon which should be combined with and supplementary to attacks by H.E. and incendiaries. This did not mean that " all out " gas attacks on the heaviest scale would not be made on individual targets, but that such attacks for maximum effect must be co-ordinated with H.E. and incendiary attacks. It might happen that at the outbreak of gas warfare bomber forces would be achieving success with a particular bombing policy which it might be unwise to abandon : gas would then be used to increase the effectiveness of the attack. Furthermore, gas targets might, for operational reasons, not be accessible at the time, and it was to be anticipated that when the enemy first used gas he would be in the strongest position to receive retaliation.

In addition to our strongly held views on the role of gas in air warfare we could not have afforded, in the light of our manpower and production capacity, to switch an unduly large proportion of our air effort from weapons of proved worth to one of which the effects were in large degree conjectural.

The supply problems connected with a scale of air gas attack of the magnitude visualized by the Americans would have unbalanced our production effort as a whole. British policy has been dealt with in detail in Chapter I, Sections 5 and 7.

Visit of Chairman of the Inter-Services Chemical Warfare Committee to United States.—In January, 1944, the United States Chief of Staff, General Marshall, invited Air Marshal N. H. Bottomley, Chairman of the Committee, to visit the United States for discussions with the Chairman of the United States Chemical Warfare Committee, and interested War Departments, on the states of preparation for chemical warfare and the procurement and supply arrangements connected with them, including an appraisal of the use of smoke munitions.

Air Marshal Bottomley was accompanied by Group Captain K. D. F. Collier, Deputy Director of Explosives, Air Ministry. Visits of a somewhat longer duration were arranged simultaneously by Major-General Brunskill, Director of Special Weapons and Vehicles at the War Office, accompanied by Lieutenant-Colonel Wansbrough-Jones (G.S.O.I.), and by Mr. Davidson Pratt of the Ministry of Supply (Controller, Chemical Defence).

For a short period it was, therefore, possible for these representatives to act as a nucleus of the British Inter-Service Committee on Chemical Warfare and they were able to discuss the various aspects of policy, operations, designs, supply of equipment, research and development with all U.S. departments interested.

Amongst the major subjects discussed were :—

(i) A comparison of American and British reviews on preparedness for gas warfare.

It had for some time been British practice to submit periodical reviews to the British Chiefs of Staff. A similar practice of submitting reviews to their own chiefs was now started by the Americans, and these reviews formed a useful complement to our own.

(ii) An appreciation of the strategic gas effort on the outbreak of gas warfare.

The difference in viewpoint on the method and extent of retaliation by gas, already referred to, always remained fundamental between ourselves and the Americans. A sufficient measure of agreement was, however, reached to enable a basis for the production planning of the two nations to be agreed.

(iii) The co-ordination of supplies of gas and smoke chargings for weapons where there was any limitation in these supplies, or where general economy of effort could be achieved by the sharing of inter-Allied resources.

(iv) Standardization of equipment. This is discussed below.

(v) The co-ordination of research, whether controlled by the Ministry of Supply, by the United States Chemical Warfare Service, or conducted extra-murally.

(vi) The production of a concerted plan for the examination of all aspects of chemical warfare in the tropics. This proved a major matter for discussion. While it was recognized that adequate data were by then available on which to appreciate and plan for chemical warfare in Europe the same was by no means true of the Pacific with its hot, humid conditions. There were grave doubts about the adequacy of certain defensive equipment in these circumstances.

Arrangements were therefore made for the co-ordination of work to elucidate these matters, and to evaluate the results obtained at the experimental stations in Florida, Panama, Australia and India. This is dealt with in detail below.

Standardization of Equipment

By the time the United States entered the war it was too late to standardize equipment to any great extent owing to the effects on production schemes that had already been started. An important item was in relation to aircraft bombs. British bombs are attached to aircraft by one lug ; American bombs were attached by two. Arrangements were made for three lugs to be fitted to all bombs in such a manner as to make them usable by British or American aircraft.

Considerable uniformity of policy on the use of anti-gas equipment was attained, but uniformity of equipment itself was largely unattainable. In the light respirator the screw thread to receive the container was made a standard thread which could take either the British or the American container. As it was found impossible to produce a wide range of standardized equipment, the major effort was directed to the preparation of tables of British and American equipment so that each side could understand the other's equipment and use it if required.

In the case of coloured smoke some standardization of colours used in the field was achieved to facilitate operational aspects, but there was no standardization of the shells and generators themselves.

There was practically no standardization of flame throwers. There was a difference in theoretical outlook—notably in relation to the rate of flow—, performance was different and fuels were different. Thickened fuels for flame throwers were developed by the Americans, ourselves and the Australians, largely according to the availability of materials. However, agreement was obtained on standard methods of testing all new types.

If the standardization of equipment or interchangeability of components is to be effected between allies, exchange of information from the very inception of a project is essential and samples and prototypes must be provided at early stages before the final sealing of designs. Even if fundamental differences in theory or policy persist, interchangeability of some components may still be possible.

Co-ordination of Experimental Work in relation to Chemical Warfare in the Tropics

In April, 1944, the Advisory Committee on the Effectiveness of Chemical Warfare Material in the Tropics was set up in Washington. The Committee consisted of representatives from the United States Chemical Warfare Service, the United States Government, the Canadian Field Experimental Station, Suffield, and the United Kingdom (Colonel Paget of the British Army Staff and Mr. Kingan, Ministry of Supply, C.W. representative). The Chairman was the Chief of the United States Chemical Warfare Service. The function of the Committee was to provide adequate operational data for the planning of chemical warfare in tropical theatres of war.

The Committee was served by a full-time team—the Project Co-ordination Staff—whose function was the co-ordination and evaluation of United States, British, Canadian, Australian and Indian trials on standard British and

United States C.W. agents, munitions and protective devices. The Project Co-ordination Staff consisted of United States representatives from the Chemical Warfare Service and National Defence Research Council, a British member from the Ministry of Supply, and a Canadian member.

The Committee and the Project Co-ordination Staff were advisory and co-ordinating bodies only. In effect, they were a clearing house for all reports. They did not exert direct control over any American or British experimental stations or proving grounds which continued to function under the direction and administration of their respective controlling authorities.

Before the Committee was set up the United States had established two proving grounds for the special study of tropical chemical warfare—one at Bushnell in Florida and the other on the island of San José off Panama. At Bushnell, under the direction of the Chief of the Technical Division of the Chemical Warfare Service, the more fundamental work on concentrations obtainable under sub-tropical conditions was carried out. At San José weapon trials were carried out under control of the Operations branch, with the object of obtaining quick results to enable recommended expenditure of weapons to be given to users.

On the British side much valuable work had already been carried out by a mixed team of Australian, Porton and United States workers under Lieutenant-Colonel F. S. Gorrill in Australia, and this pioneer work was of great value to the American stations.

In India the small Chemical Warfare Experimental Station which had been located in the hot, dry climate of the Punjab was in process of expansion and of being moved to a hot, humid area (Cannanore in S. India) more representative of conditions under which we had to fight against Japan.

Since no one experimental establishment could cover more than a few of the various types of terrain likely to be met with in the war against Japan, the British Chiefs of Staff agreed to strengthening the Indian and Australian establishments as an Imperial commitment. The strengthening of these establishments was implemented by the Ministry of Supply during 1944.

The overall Anglo-U.S. organization during 1944 and 1945 for the furtherance of a joint chemical warfare policy is shown diagramatically at Appendix XVII. This organization continued until the end of the war with Japan.

By May, 1945, as a result of intensive work at all field experimental stations, the Advisory Committee and the Project Co-ordination Staff had largely completed the primary task for which it had been set up, *viz.*, provision of adequate data for the planning of chemical warfare in the tropics.

An Interim Report (Project Co-ordination Staff Report No. 9 " Résumé of Recent Knowledge of Technical Aspects of Chemical Warfare in the Field ") was issued. This covered all factors affecting the use of chemical warfare, *viz.*, the effects of meteorological conditions and terrain ; available U.S., British, and enemy protection ; available British and U.S. war gases and munitions ; the choice of agents and munitions for given tasks ; and munition expenditures needed for these tasks. The expenditure tables covering all available air and ground weapons were in less complete form for British than for American weapons. The final edition of the report, embodying full details for British weapons and the results of further trials still in progress, was in preparation when the war ended. It was published in April, 1946.

The report was drawn up mainly from the standpoint of the war against Japan. In view of the fact that the more nearly the war approached the Japanese mainland the more would temperate and open conditions prevail, the report was not restricted to so-called tropical and jungle conditions only but covered a wide range of types of terrain and climate.

The title of the Advisory Committee was amended in May, 1945, the words " in the Tropics " being deleted.

Further Liaison Tours

In January, 1945, Major-General Brunskill, Lieutenant-Colonel Wansbrough-Jones and Mr. Davidson Pratt again visited America on completion of a tour of the Indian Experimental Station and the Burma front, the South West Pacific front and the Australian Experimental Station. All outstanding problems in relation to chemical warfare in the war against Japan were discussed with the Americans in the light of personal knowledge of prevailing conditions.

The continued exchange of visits on a high level between British and American authorities responsible for C.W. and smoke policy and development was of the utmost value. It was these personal contacts which contributed as much as anything to a mutual understanding of each other's viewpoint and the carrying out of a joint basic policy in a friendly and understanding spirit.

CHAPTER III

THE WAR GASES

Introductory

Detailed descriptions of the physical and physiological properties of the war gases are not given here. These will be found in the relevant training manuals and in greater technical detail in the Ministry of Supply histories. The object of this chapter is to record the outstanding characteristics as they affected policy and employment of gas.

The war gases are discussed under the following groups :—

Nose Gases.
Tear Gases.
Lethal Gases—(as available 1939–1945).
Blister Gases.
Nerve Gases—(which became known to us by capture from the Germans in April, 1945).

SECTION 1.—NOSE GASES AND TEAR GASES

Nose Gases

Nose gases are all solid arsenical compounds. There are a number of different compounds but their effect on the body is similar. The most effective means of dispersing them is by heating them in a generator which produces a non-persistent cloud of minute particles. This cloud of particles is effective at distances greater than that in which the smoke caused in generation is visible. The generator can be inserted in projectiles in a similar manner to the smoke containers in a base ejection smoke shell. Nose gases can also be dispersed by the use of a high bursting charge in a projectile of H.E./chemical type (Chapter IV, Section I), but this method is less effective than the heat method as the particle size is larger and the concentration lower.

It was by this latter method that the Germans first used nose gases—largely ineffectively—in the 1914–1918 war. The heat method of discharge from generators was a British development which would have been used in 1919, had the war not ended before, with the object of penetrating the German respirator container which was not proof against discharge by this method. A description of German and British development of nose gases during the 1914–1918 war will be found in " Gas—The Story of the Special Brigade " by Major-General C. H. Foulkes, (pp. 250 to 253).

During the period between the wars, although improved nose gases were developed which offered a greater possibility of penetrating the filter of the container, filter development by all nations matched the risk and there was little scope for the exploitation of nose gases to " break " protection.

The Germans possessed a nose gas shell of the H.E./chemical type assault shell indistinguishable from H.E. when mixed with an H.E. bombardment, but they subsequently developed an assault gas shell with a more potent charging (nerve gases). They recognized that the heat method was, *per se*, the most effective means of discharging nose gases.

The outstanding characteristics of nose gases are the difficulty of their detection and the delayed onset of the symptoms. Adjustment of the respirator after symptoms appear provides no immediate relief. This tends to make troops remove the respirator and thereby become vulnerable to lethal gas attack. The symptoms are coughing and sneezing, pain in the head, teeth and breathing tract, and nausea. Recovery is comparatively quick and treatment in hospital is not necessary. Nose gases are therefore not killers.

Since they are non-persistent gases their effective discharge, as also any follow-up attack with lethal gas, is dependent on favourable weather conditions. This is in some degree offset by the fact that nose gases are effective in very low concentrations and are therefore economical in use from weapons of any calibre. A simple hand generator can be used by any troops without special training. They can be mixed with smoke so that it is impossible to tell in advance whether a smoke cloud is toxic or not.

At first sight their characteristics indicate considerable potentialities for harassing. The Anglo-French conversations of 1939 revealed that the French were even more impressed with these potentialities than we were ourselves, but our laboratory work and results of field trials subsequently converted them to our view (*see* Chapter V).

The disablement caused is temporary and the effects of the symptoms, although distressing, are as much psychological as real. Any verbal description is prone to overpaint the picture, and it is for this reason that our training policy has been that every officer and man must experience nose gas. Determined men who know what they are up against and who realize that the effects are transitory can carry on[1]. It follows therefore that the effects of nose gases will be proportionate to the state of enemy morale and training.

Both before and shortly after the beginning of the war consideration was given to the use of nose gas from base ejection shell as an assault weapon. Overall, persistent tear gas was preferred as an " all-round harasser " and for use preparatory to an assault (*See* Chapter IV, Section 1.) Inclusion of both nose and tear gas in our gas armoury as alternative chargings was not considered justified owing to the complications arising in supply in the field of a variety of gas chargings. In general the results likely to be obtained from the use of nose gases were not sufficiently certain in normal circumstances to justify their inclusion. Production was established and a considerable number of hand generators produced. They were only used for training, and production was adjusted to cover maintenance for this purpose.

Note (1).—In 1942 a party of 36 officers and N.C.Os. at the end of a march of 35 miles were subjected without warning to a heavy concentration of nose gas. They were encouraged not to put on their respirators and were exposed to the gas for two minutes. Two men only fell out and the remainder, although severely affected, carried on until a standard halt was called 150 yards further on. After 10 minutes they re-formed and resumed the march and by the next day had marched a further 30 miles. Their ability to fight would have been markedly reduced for ½ to 1 hour after exposure, but less so during the first 10 to 15 minutes.

It is of interest to record that a remote possibility of the Germans projecting massive clouds of nose gas from the French coast against this country was examined in January, 1942. There are a few occasions in any winter in which peculiarly favourable meteorological conditions for such attack prevail for a day or so at a time over a wide portion of the British Isles and adjacent seas and the North West Continental area, combined with steady south-easterly winds. The amount of nose gas required (in the nature of 3,700 tons) to project a massive cloud from along a wide stretch (*e.g.* 200 miles) of the French-Belgian coast, which would have been effective over a large part of England, was within practicable limits, but the quantities required for an effective cloud of lethal gas were not. Such an attack with nose gas would only have been harassing, but the effect on civilian morale might have been considerable. Existing warning arrangements were considered adequate and no special measures were taken as it was unlikely that preparations for attack on such a scale could have escaped notice. The mounting of such an attack in relation to a major operation such as invasion would have presented great difficulty owing to the niceties of timing involved : the requisite conditions rarely occurred and it was difficult to forecast them more than a few hours ahead. There is no evidence that the Germans ever contemplated such attack.

Tear Gases

Tear Gases have no casualty effect against troops who have the means of protection : they only cause the wearing of the respirator. With persistent tear gas this may be enforced for long periods in circumstances where troops are unable to take avoiding action by leaving the affected area. Tear gases are economical to use in the field because they are effective in low concentrations and do not have to rely on a heavy weight of attack for their effect. This characteristic makes them suitable for discharge from weapons of almost any calibre.

In the static conditions of 1914–18 war they were a most effective means of causing harassment. Targets were thickly populated and of a nature where avoiding action was difficult ; approaches to the forward area were restricted and well defined ; in the static conditions prevailing, attacks could be maintained continuously for very long periods, or periodically at stated intervals over long periods, and the target was always there. In the early days protection was inferior, and later, when protection was perfected, the wearing of the respirator, as then developed, caused far greater harassment than does the wearing of the modern respirator. Tear gases, under the above conditions, were a valuable bonus as a weapon of attrition, particularly as the question of supply in the field did not present a major problem.

The type of liquid (persistent) tear gas used in the 1914–18 war was K.S.K. ; B.B.C. was introduced shortly before the 1939–1945 war. B.B.C. was more persistent, more stable, superior in its physiological effect, and, in respect of the raw materials required, more economical to produce. Thickened B.B.C.—B.B.C.V. (Viscous)—was used for shell chargings in order to obtain the optimum size of droplet on burst (*see* Chapter IV, Section 1).

When our gas re-armament occurred in the middle 1930s, persistent tear gas was considered superior to nose gas for harassing purposes, since it was less dependent on favourable weather conditions for its discharge. With "sweetening," continued harassment, by compelling the wearing of the respirator, could be kept up over long periods where tactical conditions of the target permitted. It was for these reasons that it was adopted before the war as a standard war gas in preference to nose gas.

In trials during the war it was found that some degree of tolerance to steady or falling concentrations of tear gas could be acquired by troops.

Tactically, the conditions of 1939–1945 would not have lent themselves to the use of tear gas as did those of 1914–1918. In general, targets were more sparsely populated and more fleeting, approaches to the forward area were less well defined, opportunities for avoiding action were greater, and although of itself tear gas is an economical gas weapon, its supply in the field, had gas warfare started, would have had to be maintained in addition to supplies of H.E. and other gas weapons. Its bonus value as a weapon of attrition would have been far less than under the static conditions of the first war.

The inclusion of a gas which, although capable of harassing, has no casualty producing value, at the expense of H.E. or casualty producing gas, is not justified where considerations of supply in the field are involved. Tear gas is a puny weapon of minor nuisance value when compared with other modern weapons. In the static slogging matches of this war, had gas been in use, tear gas would not have had an appreciable effect on the result.

It can be argued that it would make an economical harassing charging for air gas weapons, but such a role would not justify the diversion of aircraft from more worth-while tasks[1]. It can also be argued that in certain particular instances, *e.g.*, attacks on isolated garrisons—the Germans in the Channel Ports, ourselves at Arnhem—it would be of similar value as in the 1914–18 war.

In no case, however, could the use of such a gas be decisive, and the retention in our gas armoury of a weapon which has only a small bonus in a few specified instances is not justified.

Certain non-persistent tear gases were used in the 1914–18 war but since then a non-persistent tear gas has only been used for training. The standard training tear gas generator contains C.A.P.

At the end of the war a requirement arose for tear gas grenades for riot control in conquered territories. The existing generator was unsuitable and a special grenade was developed for the purpose.

Note (1).—The possibility of the use of tear gas from the air as an assault weapon was also examined in 1943, but abandoned. Airburst and ground burst aircraft bombs gave a contamination which lasted for too long and thus would have affected attacking as well as defending troops. From S.C.I. functioned at heights up to 1,000 feet satisfactory result could be obtained, but as such a weapon would have been unlikely to be used except against heavily defended localities, aircraft casualties were likely to have been out of all proportion to the benefits derived.

SECTION 2.—LETHAL GASES

Introductory

Lethal gases are those whose primary purpose is to kill rather than to incapacitate. The fact that some deaths may occur through inhalation of blister gas vapour or through massive contamination with the liquid does not justify blister gases being termed lethal gases. Nerve gases are lethal and of far greater potency than any lethal gas known to us before April, 1945. They are not, however, dealt with here under the heading of "lethal gases," but separately under the grouping "nerve gases." Although they will have a marked bearing on future development, they did not affect either policy or doctrine up to the end of the war since we did not know of them until the closing stages of the war in Europe. Under this heading of "lethal gases" only those gases known to us before April, 1945, are discussed. Symbols used, *e.g.*, CG for phosgene, AC for prussic acid, are those which had been agreed on with the Americans and were in common use at the end of the war.

General

A number of substances were tried as lethal gases in the 1914–18 war. The most potent and insidious lethal gas found for use in the field was the choking gas, phosgene. The most extensive research by us between the wars and during this war revealed no better all-round lethal gas. The use of prussic acid and cyanogen chloride in large aircraft bombs or in special weapons was extensively tried out, but their general superiority to phosgene was never proved and they were not introduced into standard British weapons. The possibility of the use of arsine by the Germans led to special protective measures being taken, but subsequent research on the value of this gas showed that it was not a suitable gas for use in the field. These gases are discussed below and compared with phosgene.

The lethal gases known to us up to 1945 were all non-persistent gases. To be effective they required high concentrations. This meant that they were only suitable for use in weapons which had a high gas content, and even so a heavy scale of attack or the use of special weapons was necessary. In the case of ground weapons this ruled out their use from artillery shell, and special weapons were developed for their discharge, *e.g.* Livens projector and 5-in. rocket for phosgene, and the experimental jet anti-tank for prussic acid (*see* Chapter IV, Section 1). The most efficient medium for their discharge was the large aircraft bomb.

Modern respirators—both our own and enemy—gave protection against all lethal gases. The probability of breaking German protection was remote in normal circumstances in the field. The possibility of penetrating the Japanese respirator with cyanogen chloride or prussic acid was considered and is discussed later in this chapter.

The protection afforded by the respirator, combined with the fact that the lethal gases were detectable by smell or their irritancy, dictated the methods of attack with these gases in the field. Surprise was an essential element in order to catch the greatest numbers without their respirators adjusted. This meant that the gas had to be delivered on the target in the shortest possible time, *i.e.* the " crash " shoot or bombardment. Prolonged bombardment would have been a waste of munitions and unlikely to be effective as after the initial surprise the enemy would have adjusted their respirators. In some quarters it was argued that, with an efficient warning system, a lethal gas attack was unlikely to be effective against well organized alert troops of good morale and in possession of modern respirators. Such an argument postulated a 100 per cent standard of gas discipline, warning, and morale. This was unlikely to be achieved on all occasions even by the best troops. A number would always be caught out either through gas masks not being properly fitted, or becoming damaged or not being adjusted in time—particularly if the attack were made at night against tired troops and in the earlier stages of gas warfare before gas discipline improved with experience. The lower the morale of the enemy the lower would have been his standard of gas discipline and the more effective were attacks with lethal gas likely to be. The harassing value of lethal gas was considerable owing to the threat of lethal effect if the respirator was not adjusted in time.

If used against civilians, whose standard of gas discipline and protection could be anticipated to be lower than that of troops in the field, the critical time factor in delivering the attack was of less importance.

Except from special weapons against a special target (*e.g.* prussic acid in the jet anti-tank against the tank or pillbox), the lethal gases, being non-persistent gases, were dependent on favourable weather conditions for their discharge. In general such conditions mainly occur during the night and at dusk and dawn, when air turbulence is least, providing that the wind speed does not exceed 6 to 8 miles an hour, since high winds are very disruptive to non-persistent gas. It was at night that gas discipline could be expected to be lowest. To some extent, therefore, opportune tactical and weather conditions coincided. In the field lethal gases were accepted as being in the nature of a bonus only since their use could not be planned with certainty by a commander.

Had lethal gases been used for bombing built-up areas, the limitations imposed by the dependance on favourable weather conditions would have been less restrictive. Built-up areas are protected from the wind and wind speeds are lower than in the open. Since in most cases tactical consideration would not have been involved, the mounting of an attack could have awaited suitable conditions, which would have occurred with greater frequency than in the open.

In tropical jungle with thick overhead leaf cover, favourable conditions occur by day since air turbulence is absent and wind speeds are nil or low. In the war against Japan, therefore, the use of lethal gas would have been possible by day under such conditions, provided worthwhile targets could be located, but the achievement of surprise would have been less probable.

Phosgene (CG)

Phosgene was the only standard lethal charging for British gas weapons. We have always considered it superior to prussic acid, with whose potentialities our American allies were for a time so impressed. Its salient characteristics are the delay in appearance of casualty-producing symptoms and the fact that sub-lethal dosages cause severe casualties. This latter characteristic is noteworthy in comparing phosgene with prussic acid or cyanogen chloride.

There may be some initial coughing and choking particularly when the concentration inhaled is high. A period of from two to six hours in which there are no signs or symptoms may follow this initial irritation. Before symptoms appear men can fight or work with little or no discomfort. The onset of the symptoms may be either slow with painful coughing and rapid shallow breathing, or rapid with collapse. Total disablement occurs in about 12 hours. A considerable portion of those so affected would die and the remainder would be in hospital for from one to three weeks.

To obtain these effects, however, a certain minimum dosage is required (*see* Appendix XVIII). This would occur through breathing low concentrations for long periods or high concentrations for short periods. The former can be largely discounted in the field among well-trained troops equipped with respirators. In built-up areas, particularly among civilians whose protection and gas discipline may be less perfect, the inhalation of lethal or incapacitating dosages may frequently occur in this way—especially where a target has been " opened up " and " ventilated " with H.E. before phosgene attack, so that the gas will percolate into buildings and hang about for considerable periods in effective concentrations. This does not mean that such a target need not be saturated but that the chances both of the effectiveness of the attack being prolonged and of the number of casualties produced are greater against such targets than in the field. The closing of windows and doors affords considerable protection, and it would therefore have been normal bombing technique when using phosgene to include a proportion of H.E. to " ventilate " buildings and of incendiaries to bring out firefighters. The strategic bombing policy to exploit the characteristics of phosgene in combination with H.E. and incendiaries has been discussed in Chapter I, Section 5.

In the field the required dosage must be achieved in the shortest possible time (within one or two minutes) if the protection afforded by the timely adjustment of the respirator is to be defeated. This, combined with the delay in appearance of incapacitating symptoms, dictated the technique for the use of phosgene preparatory to an assault, *viz.*, heavy surprise " crash " concentrations some two hours or more before attack, but not long enough before attack to allow the enemy to reinforce.

Phosgene, therefore, although the most toxic of the lethal gases when the casualty-producing non-lethal dosages are included, was a good, but by no means ideal, assault gas owing to its lack of immediate effect. On the other hand, no better gas was available before 1945 for general use in the field.

A phosgene bombardment creates around the point of impact a cloud of gas which travels down wind at the pace of the wind. Under favourable conditions this cloud will flow with the wind in effective concentrations for considerable distances. It would therefore have been possible to put down a cloud on the upwind edge of the target area only and allow the gas to travel over the target, but owing to the short time limit available if protection was to be forestalled it came to be accepted that the whole target area had to be covered by the attack and that although some casualties would have been likely down wind of the target area this effect could only be regarded, in the field, as a bonus addition to the attack. Against a civilian target the bonus addition would have been greater.

Prussic Acid (AC)

Prussic acid (or hydrocyanic acid) is not a new gas. It was used by the French from artillery shell in the 1914–18 war but it was ineffective as lethal concentrations were not produced in the open by this means (*see* " Gas—The Story of the Special Brigade, by Major-General C. A. Foulkes). It can be described as " hit or miss " gas, since if a lethal dosage is not attained sub-lethal dosages have little effect, recovery being rapid and complete.

It is a paralysant gas. On inhalation it causes death by paralysing the nerves controlling the breathing muscles. It is not effective in very low concentrations since the body can absorb and neutralize small amounts of the gas continuously. In lethal dosages its action is rapid, unconsciousness occurring in a few minutes and death in 15–20 minutes. It is not irritant to breathe.

It is a highly volatile liquid—so volatile that it is non-persistent. Absorption of the liquid through the skin can cause death, but for a number of reasons, including its rapid evaporation, this is a negligible factor in assessing its value under field conditions.

Its toxicity is less than phosgene. Higher dosages are required for lethal effect and sub-lethal dosages are not casualty-producing.

It is more difficult to find protection against AC than against phosgene. The charcoal in the respirator container requires special treatment (*see* Chapter VI), and the protective life of containers is shorter against high concentrations of prussic acid than against phosgene. Even against the British or German containers, concentrations high enough to cause penetration in a short time could be effected in enclosed spaces, *e.g.* bunkers or caves, when used from special high capacity close range weapons. Very high concentrations are required and these could not be attained in the open. Both the British and Germans were well protected against dosages likely to be met with in the field. The Japanese respirator was less efficient against AC, but even so container penetration would only have been probable under particularly favourable circumstances.

During the war we reconsidered its use owing to its potentialities in high capacity close range weapons against enclosed spaces (*e.g.* its use in the jet anti-tank—*see* Chapter IV, Section 1).

The Americans introduced it as a charging for large aircraft bombs ; the Japanese had a glass grenade charged with it for use against tanks and covered defences. A single grenade of this type was ineffective against a tank.

A difficulty in the use of AC is its liability to ignite and thereby lose its effectiveness, particularly when used from burster type munitions unless special means to prevent this are taken. United States aircraft bombs charged with this gas, available up to the end of the war, had this failing.

In high concentrations prussic acid is superior to phosgene in one respect—it is immediately lethal. This characteristic would make it a good assault gas. Large aircraft bombs gave a means of putting down high concentrations such as were not available in the 1914–18 war. The Americans were greatly impressed with the lethality of prussic acid, and considered its use from large aircraft bombs to be practicable provided inflaming could be overcome. Their trial grounds where initial trials were carried out were so situated that abnormally favourable conditions existed for the discharge of gas. Further extensive trials ultimately showed that sufficiently high concentrations could not be reliably achieved in most circumstances against an open target. Against a built-up area, particularly against civilians not equipped with respirators, the probability of attaining the requisite high concentrations was greater, but even so, the higher toxicity of phosgene, including its ability to produce casualties in sub-lethal dosages, made it generally preferable to AC.

AC would be effective on the target area only. Owing to the tendency of the cloud to rise and the high concentrations required there would be no bonus value downwind of the target area.

It was against enclosed spaces from short range bursterless munitions, *e.g.*, the jet anti-tank or grenades, that this gas could have been best exploited. Thickened AC–ACV—would have been the best charging for the jet anti-tank.

Jungle with heavy leaf cover is, in a sense, an enclosed space, in that wind speeds are nil, or very low, and air turbulence is small. Such a target would have lent itself to the use of the American prussic acid air bomb provided a means was introduced of preventing ignition of the charging on burst.

Cyanogen Chloride (CK)

Cyanogen chloride (CK) is less toxic than either phosgene (CG) or prussic acid (AC) so that a higher dosage is needed to cause casualties. It causes death in a similar manner to AC and sub-lethal dosages are not casualty producing. Unlike CG and AC, it is very irritant to breathe, even in low concentrations, and causes severe lachrymation. It has even been claimed that it is physically impossible to inhale a lethal dose. It is also irritant to the skin, notably round the edge of the respirator, and in hot, humid conditions it causes severe irritation of the scrotum. Skin and eye effects pass off quickly in gas-free air, and there are no lasting effects. Owing to its irritancy there is therefore little likelihood of dangerous concentrations being undetected, even among other war smells. When breathed in high concentrations it is sufficiently irritant to cause an involuntary stoppage of breathing, which would tend to prevent a lethal dosage being breathed before the respirator is adjusted. Its intense irritancy is a decided disadvantage in its use as a lethal gas, although not so disadvantageous if harassment alone were aimed at.

Its use was considered owing to its potentialities for penetrating respirator containers, and extensive trials from large aircraft bombs were held in America. It is much less readily absorbed by charcoal than are the other lethal gases, unless charcoal is specially treated as was that of modern British and American respirators. Recent German containers also contained specially treated charcoal. Neither the older German nor the Japanese containers had this extra protection. The Germans took measures to treat the older containers in the field to improve their protection. Unless charcoal is specially treated it does not hold CK firmly as it does other gases, and as a result CK may be given off when the respirator is subsequently worn and, being very irritating, this would render the wearing of the respirator intolerable. If the German field process had been used for treating Japanese respirators the tendency to give off this gas would have been reduced.

Charcoal is more readily penetrated by CK when it is moist. Any action which will make the enemy wear his respirator in humid conditions before an attack with CK increases the chance of success in penetrating the container. In the case of the Japanese respirator, which was kept plugged when not in use, wearing for about 12 hours in the tropics in a humidity of 80 per cent. or more, would have so affected the charcoal as to reduce protection against CK to a low level.

To penetrate the container in sufficient amount to be certain of killing the wearer requires a very high dosage, and this would only have been feasible in situations where high dosages could be achieved, *e.g.*, in enclosed spaces, in jungle, or in the bottom of ravines.

The British jet anti-tank would have been a suitable weapon for its use against enclosed spaces, but CK would not have been a preferable charging to AC. Large aircraft bombs would have been the best vehicle for general use. The only ground weapon which lent itself to such use would have been the United States 4·2-in. mortar. The low charge weight of the British 4·2-in. mortar bomb made it unsuitable.

Although there was a possibility of penetrating the Japanese respirator with CK, very considerable munition expenditure would have been necessary to achieve the required high concentrations. On the whole the results obtained would not have been worth the effort.

Other Lethal Gases

Certain other lethal gases, *i.e.*, diphosgene, chloropicrin and chlorine are referred to in "Advanced Gas Training—1943". This is a defensive gas training pamphlet and the use of these gases by the enemy, although improbable, could not be entirely excluded. No useful purpose would be served in discussing them in detail here.

ARSINE (arseniuretted hydrogen—British code name "Arthur") is also mentioned in the pamphlet. In October, 1939, there was some evidence(1) that the Germans might possibly use this gas against us. The current British respirator container gave very inferior protection and immediate action was taken to correct this. Arsine as a gas was well known and although we had considered it unsuitable for use in the field, it was felt unsafe to assume that the Germans might not have discovered a means of using it.

Unfortunately, before all aspects had been exhaustively re-examined and adequate trials staged, the possibility of its use was broadcast to the Services, and emergency protection (Type EA container) and a special detector (Detector Paper Type A) was issued (*see* Chapter VI). These emergency measures gave

Note (1).—The main points of the evidence were (*a*) Captured German containers were found to give protection against arsine 10 times as great as that afforded by British and French containers. German containers had been changed on mobilization and the new container afforded this extra protection. (*b*) It became known that the Germans, before the war broke out, practically cleared the metal market of arsenic—the main constituent of this gas. (*c*) Persistent reports that the Germans possessed a new gas. (*d*) Pre-war reports that the Germans were experimenting with arsine. From information now available it is known that the Germans never had any intention of using arsine. The container which was provided on mobilization provided better all-round protection against gas and it was a coincidence that this included superior protection against arsine. Stocks of arsenic were obtained for general purposes and for the preparation of Arsenol which they used as a diluent for mustard gas.

rise to a scare and to a temporary loss of faith in protection afforded by the respirator. It was thought by some at the time that the action was untimely but political considerations outweighed all others. The repercussions both in the field and at home, had attack materialized without our being protected against it, would have been serious.

The lesson was, however, well learnt, and when the possibility arose in 1942 of the use of another substance—Substance " S "—by the Germans the matter was better handled (*see* **Blister Gases,** page 147). Subsequent exhaustive trials with arsine showed the technical difficulties in projection to be so great as to render its use unlikely. Low concentrations of arsine breathed over long periods are fatal. One method by which it was thought projection might be achieved would be by the dispersal of calcium arsenide (a black solid substance) on the ground which, when in contact with moisture, gives off arsine. It was proved, however, that the quantities required for effective attack by this means were so large as to preclude its use in this manner.

Lethal Gases—Conclusion

The above is a revue of the salient points connected with the possible uses of the lethal gases, as known to us up to April, 1945. Discussion of the gases included under this title is now academic. Their use in the future is improbable owing to the discovery of nerve gases with far greater potency as killers.

SECTION 3.—BLISTER GASES

General

Blister gases are of three main types :—

Mustard Gas (H) and (HT).
Nitrogen Mustard Gas (HN).
Lewisite (L).

Of the three, mustard gas was the only blister gas charged into British or American munitions. The Germans had some HN shell. The Japanese had shell with a mixture of H and L.

Mustard Gas

Types of Mustard Gas

Mustard gas can be manufactured in a number of ways and its purity and the nature of the impurities vary according to the method of manufacture. To manufacture a completely pure compound increases the difficulties of production and is in general uneconomical and unnecessary. A number of different types of mustard gas were manufactured in Great Britain, South Africa and Canada (HT, HS, HM, HB, DESA, to quote examples) but with one exception, HT, these can be conveniently grouped under the general symbol H. The various manufacturing processes give rise to slightly different characteristics but for practical purposes the blistering powers, persistence and method of employment of H are the same.

HT is different from other mustard gas chargings. It is a mixture of H with another blistering liquid T which is many times more persistent than H. The H evaporates more quickly than the T and creates a vapour danger approximately equal to other H chargings. A residue of T is left which by reason of its very slow rate of evaporation causes little vapour danger but gives a very long danger from contact of the liquid on the skin. It is more difficult to decontaminate articles contaminated with HT than with H.

In order to obtain a larger size of droplet when mustard gas is discharged from certain munitions it is thickened by the addition of certain agents. The thickening agent introduced in 1943—Perspex (MM)—was a more effective thickener, particularly under conditions of high temperature met with in the tropics, than were the previous agents used for the purpose. There is no major difference in the total yields of vapour from thickened chargings and unthickened chargings, although the rate of evaporation of the former is slower than that of the latter.

Broadly therefore, mustard gas chargings are of four main types :—

H.
Thickened H.
HT.
Thickened HT (HTV).

Thickened H chargings are not referred to generally as HV ; it is only individual types when thickened that have the symbol V added (*e.g.*, HMV, HBV).

Each type of charging is given a code symbol and number. For reasons of space these are not given here. They will be found in the Chemical Warfare Pocket Book, and technical data in detail will be found in secret documents issued on a limited distribution only to certain technical personnel for whom such detail is essential.

The policy underlying the allocation of a particular type of charging to particular weapons was not dictated by the General Staff for tactical reasons but rather according to the technical requirements, *i.e.* :—

(*a*) *The type of weapon and its purpose.*—Airburst base ejection shell and air spray weapons for direct anti-personnel contamination with liquid droplets required thickened chargings (*see* Chapter IV). The shattering effect on the liquid caused by the functioning of the shell or by impact with the air stream on release of spray from aircraft is reduced if the liquid is thickened, and a larger size of droplet is therefore obtained. In all other weapons unthickened chargings were used.

(*b*) *Supply and production factors* ; *factors connected with stability in storage* ; *and cost.*—HT is more difficult to make and requires a greater expenditure of money, time and effort than the H chargings. It is, however, the most stable of all mustard gas chargings and therefore has a longer storage life in weapons. Until gas warfare started the question of storage life was obviously an important one, and, although not strictly adhered to, our policy was to charge HT into weapons before the outbreak of gas warfare, particularly into the weapons which were expensive both from the point of view of money and manufacturing effort and for which a long storage life was economically and financially desirable. A policy of charging HT only into weapons in the pre-gas period could not be fully implemented since established sources of production of H had to be utilized. Had gas warfare started, long storage life would have been a secondary factor, but cheapness and ease of production of primary importance ; H chargings would then have been used in many cases instead of HT. The source of supply in relation to the theatre of war would also have been a determining factor, but regard would have been paid to the short life of H in some weapons when stored in tropical climates.

As stated above, the H portion of HT creates a vapour danger similar to the H chargings. The proportion of H and T in HT is 60 : 40. There is therefore weight for weight a smaller quantity of H in a shell charged HT than with the H chargings. The H chargings are therefore preferable where vapour effects solely are required.

The major tactical difference between the H chargings and HT is that where the maximum duration of liquid contact hazard from ground contamination is required, HT is preferable to H. Where residual contamination is not desirable, *e.g.*, when access to a contaminated area by our own troops is wanted at the earliest opportunity, H is preferable.

This tactical difference was not a factor affecting policy for charging various munitions. It would not have been possible beforehand to assess the proportion required for each tactical use. Had gas warfare started, owing to supply and production factors both types would have been available, but it is doubtful whether the tactical difference would normally have been taken into account in the field.

American mustard gas, known as H (Levinstein), has one peculiarity as compared with British mustard gas chargings. The proportion of inactive impurities, although no greater than in any British charging, imparts a stronger and more persistent smell which resembles H in odour. The impurities evaporate more slowly and leave a residual smell resembling H long after the danger has passed. This would have been liable to involve friendly troops in unnecessary gas precautions, and would have necessitated the use of detector equipment, or a knowledge of the time the contamination was put down, to assess when precautions could be relaxed. The formation of a solid deposit on storage also made this charging unsuitable for use in a rotated projectile. Thickened chargings were not used in any American munitions although methods of thickening H (Levinstein) were available.

German mustard gas although utilizing different stabilizers and diluents was fundamentally similar to British H in its characteristics.

Effects of Mustard Gas

It is presumed that the general effects of mustard gas in its liquid and vapour form are known to the reader. They are given in the Gas Training Manuals.

Outline of Lessons Learned during the War Period

During the war facilities became available both in the Empire and in America for more exhaustive laboratory experiments than had been possible before the war, and also for field trials on a scale comparable with actual war conditions over a wide range of climates and terrain. Large numbers of human observers were used and as knowledge increased and experience was gained the safety margin in submitting the observers to mustard gas, both in its liquid and vapour form, was narrowed, and more critical tests could be carried out. It is worth mentioning that although severe symptoms were induced in some observers no deaths occurred, nor were there any permanent after effects. Details of this work are covered in histories written by the Ministry of Supply.

The fact that a significant period of exposure to mustard gas vapour was required for casualty effects under temperate climatic conditions had led us to exploit the use of mustard gas by means of direct anti-personnel attack with the liquid (*see* Chapter V, Tactical Doctrine).

Trials and experiments carried out in the war period showed that we had underestimated the effect of mustard gas vapour in hot, humid climates. Our pre-war knowledge had been based on extrapolation from laboratory work only which had been carried out in India under hot dry conditions. Although under such conditions an increase in the vulnerability of the body surface had been noted and it was realized that higher concentrations of vapour would occur involving shorter periods of exposure for casualty effect, the full significance of these factors had not been appreciated ; we were not led to believe that our doctrine for the use of H under European conditions would be fundamentally changed in the tropics. Light tropical clothing offered a good opportunity for exploitation of direct liquid attack.

In 1943, field work, carried out by Lieutenant-Colonel F. S. Gorrill, R.A.M.C. and a team of workers sent to Australia, emphasized the aggressiveness and offensive potential of mustard gas vapour under hot, humid conditions. It also brought to light certain deficiencies in our protection—notably deficiencies in impregnated clothing and anti-gas ointments (*see* Chapter VI).

A hot and profusely sweating skin is highly susceptible to mustard gas vapour. It was found that under hot humid, as opposed to cool, conditions, much lower vapour dosages produced incapacitation through generalized skin blistering in men whose eyes and breathing tract were protected by the respirator, and this happened much more rapidly than had been supposed. If the respirator were not adjusted, very short exposures would ultimately result in temporary blindness from dosages insufficient to cause other bodily damage. In general, mustard gas vapour was found to be three to four times more aggressive in hot humid than in cool conditions. Not only was production of severe casualties more certain, but this could be effected with considerably lower munition expenditures.

Systemic effects, not previously associated with mustard gas vapour, were noticed in some cases. Nausea, malaise, weakness and, later, vomiting occurred —and these symptoms appeared earlier than generalized skin effects. Their incidence and severity were not, however, sufficiently well-known by the time the war ended fully to establish their military significance, but they aptly demonstrated the increased effectiveness of mustard vapour under hot humid conditions.

All this new knowledge led to a revised concept of the use of mustard gas in the tropics, *viz.*, the exploitation of the vapour rather than the direct action of the liquid. In the general run of circumstances the vapour hazard outweighed the liquid hazard ; the liquid became a valuable adjunct of the vapour rather than *vice versa.*

It was necessary, however, to determine the precise dosages required to achieve casualty effect under the numerous variables in conditions of discharge, *viz.*, type of country, nature of surface, ground surface temperatures, wind speed and air turbulence—all of which govern the vapour concentrations arising from liquid on the ground. The required dosages, under any given combination of the variables governing evaporation and vapour concentrations, had then to be converted into terms of munition expenditures for all British and American munitions.

Intensive trials and experiments in Australia, India and America were instituted and sufficient data were available by the middle of 1945 to draw up tables of expenditure for all American air and army contamination weapons and to revise or confirm existing British ones.

In general, the increased facilities for trial and experiment which became available during the war did not bring to light any new characteristics of mustard gas, but afforded a better perspective of the relative value of mustard gas in its liquid and vapour form in varying circumstances. Our pre-war knowledge may be said to have been mainly qualitative. Our knowledge at the end of the war was quantitative and precise. The exact dosages required to achieve certain casualty effect under any given circumstances were known as also the munition expenditures necessary to achieve them.

The enhanced aggressiveness of mustard gas under tropical conditions opened up possibilities for its exploitation against the Japanese, particularly as they were not equipped with impregnated clothing. These possibilities were examined in great detail both by ourselves and by the United States. This aspect is discussed in Chapter I, Section 7.

Nitrogen Mustard Gas (HN)

There is a range of substances collectively referred to as " Nitrogen Vesicants " and to which the symbols, HN1, HN2 and HN3 were ultimately given.

Although we had earlier rejected French pleas for greater interest in this range, considerable attention was drawn to these substances early in 1942 owing to reports that the Germans were intending to use one of them. The substance in question was originally given the code name of " Substance S " and latterly designated HN2.

These substances were known to us. HN3, in particular, had been considered of great potential value by the French. Our view had been that they were generally of less value than mustard gas. Exhaustive research and field trials, subsequent to reports on their possible use by the Germans, did not change our views, in particular about the more volatile type (HN2 or Substance S) which was insufficiently stable for use in weapons. In general, the HN group present considerable production and storage difficulties.

The blister-producing power of these substances is considerably less than that of mustard gas. They have a more rapid and severe effect on the eyes, but the lung effects are of a similar order to those produced by mustard gas.

Their salient characteristic is that they have little or no smell ; they are, therefore, difficult to detect, particularly when the smell is overlaid by other battlefield smells. The respirator affords complete protection to the eyes and breathing tract against both liquid droplets and vapour. The eyeshield protects against liquid droplets, but not against the action of the vapour. They are difficult to decontaminate by the standard method of using bleach.

At the time when the possibility of the Germans using these substances was envisaged, we had no adequate detector. Our existing detector paper for mustard gas did not react to the vapour, the main danger.

The capture of the German handbook " Gasscheissen Aller Werfen " in 1942 indicated that the Germans had an odourless charging, referred to as Green Ring 1, for their HE/chemical shell for use in surprise attacks mixed with H.E. bombardments. Captured notebooks, defensive manuals, and gas detection instructions indicated that they had an odourless blister gas " Stickstofflost ". This ultimately led to the conclusion that "Stickstofflost " and "Green Ring 1 " were one and the same thing, HN3. Information at the end of the war showed that this deduction was correct. The German Green Ring 1 shell did contain HN3, but the odourless shell charging of real significance was Green Ring 3—Nerve gas—which was not referred to in the captured handbook. The Germans had also considered HN2 unstable and confined their attention to HN3.

Since we had had experience, in the case of Arsine (*see* page 140), of the ill effects of early and wide dissemination of information based on incomplete knowledge, and in view of the fact that the detector in possession of troops was ineffective, dissemination of information on HN2, when its possible use by the Germans was first mooted, was made on a very limited Top Secret basis. Exhaustive experiments and trials with HN2 from weapons with a high bursting charge were put in hand at high priority. A test which could be carried out by technical officers was rapidly developed, and the development of a detector for issue to troops was proceeded with urgently.

Since it was likely that the Germans would withhold this gas until some vital operation was started and that they might then employ it against key points on a large scale for surprise effect and disguise its presence by the simultaneous use of high explosive, instructions were issued that, after the notification of the invasion alert, troops down wind of a heavy bombardment should wear their respirators.

In May, 1942, as more precise knowledge became available, it was desirable that the Medical Services should receive more detailed instructions to enable them to handle HN casualties adequately. Substance S (HN2) was therefore downgraded from Top Secret to Secret. A general statement was issued to the Services stating that it was possible that the enemy might use colourless and almost odourless blister gases which would make detection difficult and that respirators should accordingly be adjusted during an intense bombardment until the absence of gas had been established. This formed the basis for inclusion of instructions in Gas Training Manuals and Army Training Memoranda. Although a number of officers came to know that there was something called Substance S, no " scare " occurred, as had happened in the case of Arsine, and interest soon waned.

Our policy of including nitrogen mustard gas generically under blister gases was consistent with our normal training policy of basing instruction, for sake of simplicity, on groups of gases rather than individual types within the group. When the Pocket Vapour Detector became available, there was, therefore, little advantage to be gained from resuscitating interest in nitrogen mustard gases by starting to give detailed instruction about them. This would have tended to put them out of their proper perspective and would have been likely to have had the reverse effect to that desired. Mustard gas still remained the most important of the blister gases, and the one most likely to be used in quantity. Any high-lighting of the nitrogen mustard gases at this stage would have tended to deflect attention from the real danger.

American policy, however, had always been to base training on individual gases within a group. Consonant with this policy they had publicized HN2 to their troops to the extent of including it in a pocket reference card listing the types of war gases, which they issued to all ranks. This forced our hand ; in December, 1943, in order to avoid any difficulties arising in Anglo-American theatres it was decided that in lectures on blister gases nitrogen mustard gas would be referred to as an example of an odourless type of blister gas, and that HN2 or Substance S would be quoted as an example of a gas of this type of low persistence. Training manuals were amended accordingly.

Lewisite (L)

Lewisite is an arsenical liquid. It was first prepared by an American, Dr. Lewis, in 1917, in an effort to produce a compound which would combine the blistering effects of mustard gas with the systemic poisoning of arsenic. It has never been used in war. During the inter-war period it attracted considerable public attention and its properties were grossly exaggerated by " scare " writers and in the Press, and in some cases exploited for political purposes. The facts are that, though symptoms of arsenical poisoning can be produced through contact of the liquid on the bare skin (but not on clothing), the quantity of liquid required to induce severe symptoms in a human being or for lethal effect are large, and would not normally be achieved by dispersion from weapons, although if massive contamination with the liquid occurred severe symptoms could be induced.

Lewisite is therefore a blistering gas with no appreciable additional value in producing systemic arsenic poisoning. Food and water contaminated with it are less easily rendered usable than when contaminated with mustard gas.

Compared with mustard gas it is generally inferior. It is less persistent, has a distinctive smell and is irritant. The latter characteristic enforces an immediate adjustment of the respirator ; lung and eye casualties from the vapour are therefore less probable. The liquid penetrates the skin more rapidly and blisters occur earlier than with mustard gas, but the blisters are less severe. The vapour is relatively ineffective on the skin. An important characteristic is that lewisite, except for its arsenic content, is very easily destroyed by water. This has been a deciding factor against its adoption as a blister gas. It is more difficult to produce than mustard gas. The Americans produced considerable quantities, while we made smaller amounts for field trials, but it was never charged into weapons.

Lewisite has a lower melting point than mustard gas. Mustard gas normally has a diluent mixed with it to lower its melting point. Lewisite could be mixed with mustard gas for this purpose, but other types of diluent are more easily manufactured. A Russian Mission which visited this country in March, 1942, wished to obtain lewisite from us, largely for this purpose. The Japanese adopted a mixture of 50/50 H and L for shell and aircraft bombs for similar reasons. The Germans used an arsenical compound—Arsenol—as a diluent for mustard gas to reduce the melting point, but this would not normally have induced any arsenical symptoms.

Fake Gases

During the 1914–18 war malodorous but harmless substances, such as bone oil or balloon dope (amyl acetate), were used to simulate gas on occasions when, for safety reasons, mustard gas could not be employed. The use of these fakes induced the enemy to adjust the respirator, thereby meeting our infantry assaults on unequal terms.

In 1943, the question whether there was a requirement for fake gases under modern conditions was re-examined. Developments in the methods of detection of gas since the 1914–18 war were a major factor in assessing the requirements of a fake gas. Any innocuous substance used would have to defeat the enemy's unit detector equipment. A fake gas would therefore have had to be not only indistinguishable by sight and smell from a standard persistent war gas, such as mustard gas, but would also have had to react on the enemy's gas detectors and powders in a similar manner to the gas it was intended to simulate.

A substance was developed by the Chemical Defence Research Department known as " Fake H No. 2 " which fulfilled these requirements. A series of trials were held in which a high proportion of experienced gas officers, in addition to large numbers of inexperienced personnel, were deceived by the fake.

The advantages to be expected from the use of a fake gas would be to force the enemy either to modify his existing detectors or introduce and issue additional detectors to the troops. There would also be some effect on morale, owing to the doubt raised in the troops' minds regarding the efficiency of their detectors. Until the enemy's detectors had been redesigned or modified, the fake would be indistinguishable to units in the field from real gas, and would compel the taking of full protective measures. Tactically, it could be used before an assault to mislead the enemy about our intentions in relation to place and time of attack. Areas contaminated with the fake could be safely crossed by our own troops. The fact that German teaching was that mustard gas would never be put down on areas it was proposed to cross or occupy would have given the use of fake gas an enhanced value. There were possibilities in the use of such deception in relation to the forcing of obstacles or in pre-landing operations.

On the other hand, although the possibility of making the enemy react to gas without ourselves having to contend with its difficulties was attractive, the objection of adding yet another type of munitions to our already complicated supply in the field was a weighty one. Such ammunition would probably not be available when required. In any case, the effects of the fake gas would always be somewhat problematical, and commanders would hesitate to commit their troops on this uncertain value. It was therefore ultimately decided that there was no requirement for fake gases, and production was not proceeded with. Had gas been used, however, it is probable that an operational trial with fake gas would have been made.

(C29058) G

SECTION 4.—NERVE GASES

Nerve gases consist of a number of related substances, some of which are markedly more toxic than others of the group—although all are the most toxic war gases yet discovered.

" Tabun " was first discovered by a German chemist, Dr. Gebhardt Schrader, about 1936, during work on insecticides. " Sarin " was discovered as a result of further research on this group in 1938. The German War Ministry were impressed with their toxicity and put a large team of chemists on to study them. Plans for large scale production of " Tabun " were made in September, 1939. Initial production difficulties of " Tabun " were considerable ; it was not until 1942 that the Germans got into large scale production at a specially erected factory at Dyhernfurth, with an output of 1,000 tons a month. In all, some 12,000 tons were manufactured. " Sarin " presented formidable production difficulties ; only a pilot plant had been brought into operation by the end of hostilities. " Saman " was discovered in 1944. The Germans had only attempted laboratory preparation of this compound.

The existence of these gases was entirely unknown to us until the capture and examination of German 105-mm. Green Ring 3 H.E./chemical shells in April, 1945. Subsequently 250 kg. aircraft bombs with the same charging were found. That the Russians also know of the nerve gases must be assumed as certain. Although it appears that we captured the majority of stocks of German gas weapons, as well as the research establishment at Raubkammer (*see* Appendix XXIV), the " Tabun " factory at Dyhernfurth in Galacia is in the Russian Zone.

" Tabun " is ten times more toxic than phosgene, hitherto generally the most toxic war gas known, and " Saman " and " Sarin " are still more toxic (lethal dosage of phosgene is 3,200—of the nerve gases 300 or less).

Nerve gases are liquids of varying volatility ; in the field they can be regarded as colourless. Both the liquid and vapour are lethal, although unless actually inhaled in droplets the danger from liquid is probably not great. The liquid has no blistering or irritant effect. The respirator affords complete protection to the eyes, throat and lungs, but the problem of detection has not yet been satisfactorily solved.

As our own investigation of this group of war gases only began after the cessation of hostilities, our knowledge at the time of writing is based largely on German information, which has been completely supplied to us, of the types of nerve gases which they had developed. Further exhaustive research of this group may produce even more toxic types. Owing to their lethal characteristics experiments on human beings are circumscribed. The only evidence on which to base lethal effects on humans is by extrapolation from animal results, and from the records of a few factory accidents which occurred in Germany.

The action on the body is on the nervous system. They give no warning of their presence by sensory irritation.

Exposure to low concentrations gives rise in a few minutes to a slight tightness of the chest and some difficulty in breathing, gradual constriction of the pupil of the eye to pin point size (myosis), running at the nose, and excessive salivation. Myosis is accompanied by dimness and blurring of vision and gives rise to distressing frontal headache which reaches its maximum severity in 15 to 30 minutes and which persists for several days unless relieved by treatment. Nausea and vomiting frequently accompany the headache, particularly if the victim is carrying out any exacting duty.

If a lethal dosage is inhaled the following further symptoms occur progressively : loss of muscular power and co-ordination affecting balance and gait, a twitching of the muscles leading to convulsive movements, involuntary emptying of the bladder and bowels, respiratory distress, slowing of the heart, unconsciousness, and death which in general occurs during or within an hour of exposure to a lethal concentration. Initially it is impossible to say whether a lethal dose has been inhaled or not.

With massive dosages (several times the fatal) death may result in a very few minutes with convulsions as the main feature.

Effects are cumulative within a period of seven to ten days, *i.e.* an individual is more susceptible if he has had a small dose within the previous seven to ten days.

As far as is known the vapour is not significantly absorbed through the skin. Wearing the respirator therefore affords protection against the major hazard.

The liquid can be absorbed through the skin. A very large splash on the bare skin, or on clothing in close contact with the skin, would probably rapidly produce fatal results if it were not quickly removed. Symptoms follow a similar course to those from inhalation of the vapour. The lethal dose on the skin is from 1 to 3 cc. Absorption is rapid and with a high dose takes place in a matter of minutes.

Where the amount of liquid absorbed is only just sufficient to be lethal, death may not occur for hours or even days. If, however, the amount is many times the lethal dose death may occur in minutes. The resultant symptoms are especially rapid if absorbed through the eye or through an open wound or broken skin. A drop in the eye, if large enough, can be lethal in a matter of minutes, and there is no accompanying pain or irritation to the eye. Clothing has a certain protective value against initial cloud droplets ; large splashes or drops will soak through and require to be dealt with rapidly.

The German shell with a nerve gas charging—Green Ring 3—was a high explosive chemical shell. It was for use as an assault shell and would have gained its effect from a highly concentrated initial cloud of vapour and droplets, inhalation of which, before the respirator was adjusted, would have led to rapid death. The residual after effect from contamination on the ground would not have lasted very long. Although the respirator would have provided a protection from the major hazard, difficulties in timely recognition would have presented a most serious problem. If bombs containing Green Ring 3 had been used, contamination of ground would have given rise to dangerous concentrations of vapour and a contact danger would also have existed. Since the respirator gives protection against the vapour, surprise would have been necessary to take advantage of the difficulty of detection and the insidious nature of the gas.

Detection and recognition are difficult. The liquid reacts with detector paint and ground detectors, but such detectors cannot be relied on to detect the fine droplets of the initial cloud of an H.E./chemical shell. The papers in the Pocket Vapour Detector give no reaction, and there seems to be little chance up to the present of developing a detector which would be sufficiently quick and reliable to give the rapid warning necessary in the case of bombardment with H.E./chemical shell. A detector for use when the time factor is not so critical, *e.g.* to determine whether respirators can be removed, is a possibility, but is not available up to the time of writing (1947). Detection by smell cannot be relied on. Detection would therefore often be dependent on the earliest recognition of symptoms.

To date, the only answer would appear to be the wearing of the respirator for very long periods from the beginning of any heavy bombardment. There is need for the application of personal decontamination on the lines already taught for mustard gas, but existing anti-gas ointments are ineffective. A combined anti-nerve gas/anti-mustard gas ointment is a possibility. The early medical treatment of men showing initial symptoms would be necessary. Myosis can be relieved in about an hour by drops of atropine or hyocine in the eye. A man could not, however, be returned to duty immediately, since his sight might still be unreliable, and several hours must pass before it is known whether he has received a lethal dose or not. If the more severe symptoms then begin to evidence themselves, medical treatment is essential for a good chance of recovery. In our present state of knowledge, however, medical treatment is palliative rather than curative.

For decontamination, hosing, swabbing and aqueous bleach paste are efficacious, and boiling will decontaminate clothing.

The above is the résumé of the general effects of nerve gases. Had the Germans initiated gas warfare and used these gases against us there can be little doubt that the initial results would have been serious. Although the respirator gives complete protection from vapour to the eyes and lungs, the difficulties of recognition would have been considerable, and the danger from the liquid remained. Only battle experience would have shown the degree of effectiveness of these gases, but from laboratory experiment and extrapolation from animal results they obviously possess great potentialities for the future.

Hitherto, the known lethal gases, amongst which phosgene was pre-eminent, have required high concentrations to obtain lethal effect. This made them relatively difficult to use in the open under field conditions and necessitated a large munitions effort for their effective use. They acted on the body in their vapour form only, and were comparatively easy to recognize and easy to protect against. " Tabun " is a far more efficient killer ; both the liquid and the vapour are lethal, and other agents in this group are even more toxic. Munition expenditures for these are proportionately less as toxicity rises. Added to this is the tactical value of their speed of action, their greater insidiousness, and the fact that the problem of timely recognition has not yet been solved. On the other hand, although on existing knowledge they appear to have great potentialities as a weapon of war owing to the greater ease with which lethal casualties can be inflicted as compared with any other gas yet known, their weight for weight efficiency as compared to H.E. has yet to be proved.

At first sight it might appear that all other gases fade into the background. This is not necessarily so and, although it appears that phosgene will become a " back number," blister gases should continue to hold a place amongst the war gases for their harassing value arising from their long persistence, their insidious nature, their high rate of casualty production, and the fact that in both liquid and vapour form they affect the whole of the body surface.

An examination of Intelligence reports since before the war shows that although there were a few reports in 1943–44 which can now be interpreted to refer to nerve gases, there was nothing which gave a tangible clue to their composition or to their development by the Germans.

SECTION 5.—LESSONS LEARNT DURING THE WAR PERIOD

Two major lessons about war gases were learnt during the period of the war. Firstly, the improbable can happen, *i.e.*, the development of a new war gas ; secondly, that failing battle experience, facilities are essential for large scale trials over a wide range of climatic conditions, if war gases are to be evaluated in their true perspective.

The nerve gases were discovered by the Germans not through deliberate search for a war gas but during research incidental to other work. Such a possibility can never be excluded. The number of substances with a potential value as war gases which have been investigated and evaluated by the Allies over the last 30 years runs into thousands. Details of Russian research are not known but it is highly improbable that the Russians knew of nerve gases.

Facilities for trials on a large enough scale cannot be provided in a small closely-populated country such as England. Furthermore, the British Army may be called upon to fight in many climates, and substances which are suitable for use in war under temperate conditions may prove unsuitable or require modification for use in tropical conditions, or alternatively their value may be enhanced. The same applies to the means of defence against them. The continuance of Commonwealth facilities for large scale field trials such as became available during the war is, therefore, indicated, although this does not mean that such facilities need be kept permanently staffed at full strength. It has been proved that extrapolation from laboratory experiments and small scale field trials is inadequate.

SECTION 6.—GAS PRODUCTION POLICY

There will always be a school of thought which will desire to exploit the harassing and morale-weakening effects of gas by using the maximum number of different types of gas in the field to surprise and to confuse the enemy. Owing to the complications which would arise not only in supply in the field but in production, with its attendant reactions on man-power, such a doctrine for the employment of gas must be resolutely resisted.

Pre-war Policy

Before the war the principles of simplicity of production and simplicity of supply in the field were paramount. Only two major types of war gas were produced—mustard gas and tear gas (B.B.C.). Although phosgene was produced for the manufacture of carbamite (an intermediary of cordite), there was no capacity available for diversion to its use as a war gas. It had been left to be seen whether there was a post-mobilization requirement for it (for pre-war doctrine in relation to the use of phosgene *see* Chapter V).

Post 1939

A similar policy was adhered to. Phosgene was added to mustard gas and tear gas as a standard charging, and only these three gases were charged into standard weapons. Small scale production of certain other gases was established for trials or for training. Our production policy as affected by man-power and other factors has been discussed in Chapter I.

CHAPTER IV

GAS WEAPONS

SECTION 1.—ARMY GAS WEAPONS

Artillery Gas Shell—Standard Types

General Development

At the outbreak of war, the standard gas shell was a mustard gas shell of the " bursting chemical " type as developed during the 1914–18 war. This was a converted H.E. type with a percussion fuze and a low-bursting charge sufficient to break open the shell. It was a ground contamination weapon, the main hazard being the occupational danger arising from the effects of the vapour. It was uneconomical in use owing to the inefficient distribution of the charging on the ground around the point of burst and the loss of charging in the crater. Direct contamination of personnel with the liquid was incidental in the immediate vicinity of the point of impact, but it was not an efficient weapon for this purpose.

A " base ejection " shell was in the development stage when the war broke out, and in November, 1939, it was decided to adopt this type as the standard gas shell for the 25-pdr., the 6-in. howitzer and the 5·5-in. gun howitzer. The 25-pdr. did not get into production until September, 1940, and the larger calibres at a later date. Stock figures are given in Appendices II and III. It was not until early 1943 that adequate stocks were available in all theatres.

Owing to their small capacity, we had always regarded artillery shell as an uneconomical vehicle for choking gas, and no Army shell was so charged.(1) A few naval shell (converted H.E. type) were charged with this gas in 1941 for a specific purpose (*see* Chapter I, Section 5). Consideration was given at various times to nose gas shell—either of a base ejection type with nose gas generators similar to the base ejection smoke shell, or to a H.E./chemical type. Neither of these were introduced for varying reasons. For policy in relation to nose gas and tear gas *see* Chapter III, Section 1. H.E./chemical shell with other chargings were also experimented with but not adopted.

Base Ejection Shell

These shell were designed primarily as an anti-personnel weapon. They were airburst shell which distributed their liquid content directly on to personnel and equipment and on to the surface of the ground, and thus obviated the crater effect of the "bursting chemical" type. They were fitted with a time fuze (No. 221).

Note (1).—Compare this with the French 1939 policy which is referred to in Chapter V. 30 per cent. of French 75-mm. and 40 per cent. of medium calibre gas shell had this charging.

The base of the burster container acted as a piston and forced the liquid against the baseplate, which, being lightly threaded, sheared off, the liquid being ejected from the base of the shell in a spray of droplets of varying size. After burst, the spray drifted downwind to a distance determined by the strength of the wind and the height of burst, the larger droplets falling to the ground first. A smoke box provided the means of observing the height of burst or of giving an indication of the strength and direction of the wind on the target. The fuze was not very reliable and resulted in bursts at heights of from 300 ft. down to a small percentage on graze. If the fuze was set to percussion, the shell would penetrate buildings and burst inside effectively.

These shell were charged with either mustard gas or liquid tear gas. Initially they were provisioned for on a fifty-fifty basis, but latterly the policy was three mustard gas to one tear gas. No firm policy was laid down for the proportion of gas shell to be carried by artillery in the field. At one time a figure of 10 per cent. was visualized, but ultimately it was decided that such shell would be held centrally and brought forward as required.

The thickened types of charging, as described in Chapter III, Section 3, were used in order to reduce the shatter of the liquid on burst and obtain the optimum size of droplets. The thickener used in the latter part of the war (MM) improved the performance of these shell. This was of particular importance under tropical conditions where, before the introduction of this thickener, the direct anti-personnel performance had been much inferior to that in temperate climates.

There were certain limitations in the use of these shell. In the case of the 25-pdr., there were a large variety of marks—Mark I to Mark XI. The Mark VIII was the principal mark and was adopted partly for reasons of economy in production since the shell body was the same as that used for the H.E. shell. It was also the most ballistically stable of all marks. It was heavier shell than the other marks (25 lb. as against 20 lb.) and the gas content was less ($1\frac{3}{8}$ lb. as opposed to 2 lb.). The Mark XI was the Canadian production which varied from the Mark VIII in certain minor constructional detail. The Marks I, II and III developed ballistic instability at temperatures below 40° F. and above 100° F. All marks had to be fired from first and second quarter guns, *i.e.*, those which had fired less than 5,000 equivalent full charged rounds, and there were limitations on the charge to be used at certain elevations. Charge I was not used with these shell.

In the case of the 5·5-in., Marks I to IV, and the 6-in. howitzer, Marks I and II, there were fewer limitations. All marks could be fired at temperatures from 25° F. to 140° F. and at all elevations and charges, but from first and second quarter guns. In each case the Mark II design was an H.E. converted type which underwent trials and was sealed to supplement production if required. The 5·5-in. Mark I started in production but few were made, as a structural weakness leading to ballistic instability became evident under proof pressures. This was the cause of the delay in the provision of this shell. Rectification of this fault, in the case of the Marks III and IV, was only accomplished with a loss of maximum range which had to be accepted (15,000 yards as compared to 16,500). The liquid content of 5·5-in. and 6-in. shell varied in different marks from $7\frac{1}{8}$ lb. to $8\frac{3}{4}$ lb.

In use the difference between 25-pdr. and 5·5-in. or 6-in. was that munition expenditures were reduced in the case of the two latter owing to the additional chemical content. With the larger shell, contamination was heavier in patches and, therefore, in places was more persistent. The 25-pdr. was to be preferred when early traversal or occupation of the target by our own troops was likely to occur.

To allow for wind conditions on the target and the unreliability of the fuze, it was necessary to engage an area consisting of the target itself plus a " frame " of 100 yards all round it, *e.g.*, to engage a target area of one artillery square (100 yards by 100 yards), a total area of nine artillery squares had to be engaged. The target to frame proportion decreased with the larger target areas. Point targets were therefore less economical to engage than a larger area. The addition of the frame fundamentally increased munition expenditures.

B.E. Shell Charged Mustard Gas

When first introduced in the early days of the war, the munition expenditures for an anti-personnel role were assessed at a quarter of those found necessary at a later stage.(1) This was due partly to an underestimate of the density of droplets of spray (whether from shell or from aircraft) required to produce a casualty, and partly (*see* Chapter V) to the experience gained during extensive field trials throughout the Empire (including trials where shell were fired over human observers) about the actual charging distribution over an area when engaged by the normal method of searching and sweeping.

Any head cover—even leaf—gave protection against falling drops, and troops in the open could counter this hazard by the use of anti-gas equipment. On the other hand, the necessity for taking anti-gas measures imposed a very considerable degree of harassment with consequent loss of fighting efficiency.

In comparing these shell with H.E., on the credit side of gas must be remembered its searching effect into slit trenches, and the degree of harassment imposed by the enforcing of anti-gas measures. By their nature, artillery shell are not suited to the putting down of heavy concentrations of gas, and the gas role of artillery is, in general, harassing. The B.E. shell charged mustard gas, therefore, provided the artillery with a means with which they could effectively harass the enemy while also providing the threat of a high casualty rate as the penalty of inadequate counter-measures. They would have been of considerable value in counter-battery work. On the other hand, it is questionable whether, against an open target, a weight of high explosive, equivalent to that required in the case of gas shell to engage the target plus the " frame," would not, when fired into the target area itself, have been more effective. Since gas was not used, there is no battle experience on which to determine which of the above factors would have predominated. In drawing this comparison it must be remembered that, with the munition expenditures required for direct contamination of personnel, the vapour effects are small ; the persistence factor which normally operates when comparing mustard gas with H.E. is therefore of little significance here.

Note (1).—The munition expenditure required for 25-pdr. was initially laid down as 10 rounds for an artillery square but later increased to 40 rounds for each square. In the case of the 5·5-in. gun howitzer it increased from 2½ to 10.

To cause an occupational hazard in temperate climates from the vapour and from contact danger would have needed at least four times the munition expenditure for each artillery square required for direct anti-personnel effect. Ground contamination of areas was therefore not regarded as a practical role for these shell in mobile warfare. This object could have been more easily and more economically achieved with air gas weapons, or, if carried out by ground weapons, by the 4·2-in. mortar. Only in certain specific circumstances, such as enemy positions in caves, where a precision weapon was required, could mustard gas shell have been used more economically in this role than air gas weapons or mortars.

Under tropical conditions, when used in their normal role against open targets, the inferior performance of mustard gas shell before the introduction of the MM thickener has already been referred to. In jungle country, they would have been of little use in this role ; when the shells were airburst the leaf canopy provided protection, and, if the fuze were set at percussion, there would have been a few tree bursts with a very large proportion of ground bursts—and the performance on soft ground was poor. Significant vapour effects would, however, have been attained in hot humid conditions, but the munition expenditures would have been greater than would have been the case had the performance of the fuze been more reliable.

Although by no means an ideal shell for tropical gas warfare, any limitations on their use in their normal role for direct contamination of personnel would have been largely offset by their increased effectiveness as a source of vapour, particularly against small Japanese defended positions, and they would have been of value in complicating enemy defence and supply arrangements in circumstances where acute supply problems on attenuated L. of C. did not militate against their use.

B.E. Shell Charged Tear Gas

In support of an assault, mustard gas is an unsuitable weapon by reason of its delay in causing casualties. It was partly with the object of providing a gas shell to support an assault that base ejection shell were charged tear gas. Their other role was harassing. Their value for harassing depended on the nature of the gas rather than the method of its discharge, and this has been discussed in Chapter 3, Section 1. They would have been of value in counter-battery work either in circumstances where the use of mustard gas was tactically undesirable or to cloak the use of mustard gas.

For tear gas shell used in an assault role, it was the method of discharge which was important. In this case it was envisaged that a minimum shell expenditure would have been used, which, although insufficient to affect our own troops through evaporation off the ground when they arrived in the target area, would have directly contaminated the individual enemy's clothing with liquid and caused him to adjust the respirator. In some cases drops would have fallen on the face and, being difficult to remove, would have caused weeping after the facepiece had been put on, but an alert enemy would have been quick to learn the necessity for rapid adjustment of the respirator on the first signs of such attack. With the expenditures necessary to ensure that the area was adequately covered, the residual vapour effects would have affected our own troops in warm weather even in temperate climates. Under tropical conditions, therefore, the use of such shell in an assault role would have been impracticable.

Gas Shell Development Types

During the war consideration was given to two other types of gas shell—H.E./Chemical and the base ejection double ejection types.

H.E./Chemical Shell

This is a ground burst shell with a high bursting charge and a liquid content. On burst the liquid content is shattered into an (invisible) " mist "—commonly referred to as the initial cloud. Residual contamination on the ground is negligible. The Germans developed a shell of this nature as an assault shell and it was their standard type (*see* Appendix XXIV).

In 1943 a General Staff requirement was stated for such a shell for the 25-pdr., the 95-mm.(1) and the 5·5-in. gun howitzer. The shell was to be indistinguishable from normal H.E. both by sight and sound. Neither irritant action nor smell was to be sufficiently strong to make its recognition certain before harmful effects might be caused. It was to be capable of causing sufficiently marked effects in a significant proportion of personnel to render the wearing of the respirator for long periods preferable to the disadvantages resulting from the effects of the shell on unprotected personnel.

The object of the shell was the production of unexplained casualties which would undoubtedly have lowered morale and enforced the wearing of respirators in any shell fire, because the shell would never be distinguishable from H.E. and therefore all H.E. would have had to be regarded with suspicion at all times.

The full G.S. requirement could not be met. A shell was developed which could have been used on a neutralizing shoot to enforce the wearing of respirators, but it was most unlikely that a satisfactory chemical effect could be obtained without a very large ammunition expenditure, and this, in conjunction with the fact that it would have been difficult to find capacity to put H.E./chemical shell into production in very large quantity in addition to H.E. requirements, led to the matter being dropped.

Base Ejection Double Ejection Shell

This type of shell was designed so that on the functioning of the fuze a container was blown out from the base. The container fell to the ground and ejected its contents either above or on the ground. It was experimented with before the war but dropped in favour of the base ejection type. Consideration was given to it again for use in tropical jungle in the war against Japan. The object was to burst the shell above or on the leaf canopy so that the containers fell and ejected their contents close to or on the ground. Considerable technical difficulties were encountered in its development and the proposal was shelved. Full details of the development of this shell and of the H.E./chemical shell will be found in histories written by the Ministry of Supply.

Note (1).—Being a close support gun no chemical shell had been allotted for the 95-mm., but as the H.E./chemical type did not effectively contaminate ground, and also obviated the complications of a time fuze, a shell of this type was a possible requirement for this weapon. A design of base ejection tear gas shell was sealed. In 21 Army Group the 4·2-in. mortar was eventually adopted as the close support weapon in lieu of the 95-mm. Infantry gun and the requirement for a gas shell for the latter lapsed.

Possible Lines of Future Development

The base ejection shell remains a potentially valuable and economical way of applying mustard gas. Should a fuze be perfected with which a standard height of burst near the ground can be obtained, the disabilities under which these shell laboured during the war should disappear. The " frame " could probably be dispensed with and ammunition expenditure consequently reduced. A more effective dispersal of the liquid over the target would also be effected.

The introduction of nerve gas (*see* Chapter III, Section 4), however, points to the most probable line of development for the exploitation of gas by artillery, namely in an assault role. The employment of nerve gases in shell would have to be considered *vis-à-vis* their employment in artillery rockets or in mortar bombs.

The Armament of Chemical Warfare Troops

Offensive Weapons

The Livens Projector.—At the outbreak of the war the main armament of chemical warfare companies was the Livens projector. Briefly the projector consisted of a steel tube 3 feet in length and 8 inches in diameter. It fired a large cylindrical drum by means of a charge box placed in the bottom of the projector, variation in range being obtained by altering the quantity of the charge. The projectors were dug in or leant against a simple stand at an angle of 45° and aimed for line. The charge boxes in the assembled projectors were connected by cable to exploders. On being fired, the force of the explosion actuated a delay fuze in the bomb, and the bomb exploded a short interval after reaching the ground in the target area. The projectors themselves were thrown about the firing area by the force of the explosion or embedded in the ground.

They were one shot weapons which were always fired simultaneously in large numbers. A chemical warfare company could fire a salvo of 270 bombs. It was an inherently inaccurate weapon and had been developed for use against area targets only.

A " crash " concentration of lethal gas, *e.g.*, phosgene, could be put down on an area, or a heavy contamination with mustard gas effected. To bring a company armed with this weapon into action was a ponderous procedure, and it took some time to lay on a shoot. The above brief description combined with the photographs at Appendix XIX should be sufficient to convey its undesirable characteristics for modern mobile warfare. (Further details will be found in Military Training Pamphlet No. 32, Part 2, 1941). It was essentially a static warfare weapon.(1) It is now obsolete.

Before the war it had been visualized that in a future war on the Continent of Europe considerable periods of static warfare would intervene, which would admit of the exploitation of the Livens projector within its range, and that

---

Note (1).—Our development of the Livens Projector in the 1914–18 war was one of the major gas developments of that war. It was greatly feared by the Germans. Details will be found in " Gas—The Story of the Special Brigade " by Major-General C. H. Foulkes.

during mobile phases the use of gas would be for harassing only, which could be effected by artillery gas shell. It was largely as the result of this doctrine, combined with the outstanding gas efficiency of the Livens projector as demonstrated in the 1914–18 war, that it was re-adopted shortly before the last war as the main armament of chemical warfare troops in preference to the mortar.

The Livens projector as issued to chemical warfare troops on the outbreak of the war was of the same design as that used in 1914–18, with the exception that it was strengthened to give an increased range. The 1914–18 projector had a range of 1,500 yards—the 1939 projector, 1,800 yards. Further increase in range could not be obtained without modification in design which would have prejudiced its outstanding characteristic—the high gas content to weight ratio of the bomb.

The 5-in. U.P. Rocket.—The need was felt for a weapon of greater range and flexibility than the Livens projector whilst retaining its " crash " shoot effect. This led to the development in 1940 of the 5-in. U.P. rocket. This consisted of a 30-lb. aircraft gas bomb casing, with an impact fuze and a head burster, fitted with a rocket tail. It was fired electrically from a stand. Both the rocket and the stand as finally developed are illustrated at Appendix XIX. Some of its defects undoubtedly arose because, instead of being designed *ab initio*, it was made up from components in use for other purposes.

Similarly to the Livens the 5-in. U.P. rockets were discharged simultaneously in large numbers. A chemical warfare company fired a salvo of 432 rockets as opposed to 270 Livens drums. This compensated for the smaller gas content of the rocket. It was a much more flexible weapon and had a maximum range of 3,500 yards and a minimum safe range of 1,950 yards. At maximum range the 50 per cent. zone foı length was 120 yards ; and for line, about 340 yards. At 2,500 yards the zones were 370 and 180 yards respectively. Variation in range was obtained by elevation of the stand.

When the adoption of the 5-in. rocket was originally proposed it was visualized that it would supplement the Livens projector to engage targets beyond the range of the latter, but it ultimately superseded the Livens projector entirely.

The rocket was slow in getting into production. Owing to faulty design of the tail, which resulted in short bursts near the firing point and inaccuracy in flight, not one rocket fit for operational use was received in 1941. The spread of a salvo of rockets and the number of wides and shorts was greater than had been expected. It was an effective weapon for putting down a " crash " concentration of phosgene on an area, but when charged mustard gas both the distribution of the salvo and the unsatisfactory dispersion of the charging from each individual round made it an inferior weapon for ground contamination tasks. A lack was felt for a suitable weapon for such tasks, as well as for one of greater precision and flexibility for use with phosgene.

There had always been a school of thought in favour of the mortar for the armament of chemical warfare troops, and at one time consideration had been given to the purchase of foreign mortars, owing to the lack of a suitable and immediately available mortar of our own, but nothing came of this.

There was considerable divergence of opinion in 1940 which led to strong opposition to the adoption of a mortar for chemical warfare troops. For some time the view prevailed that, since the personnel of chemical warfare companies were trained engineers, they ought to be employed on field company work in the front line when not employed in their chemical warfare role ; to put a mortar into their hands would lead to the diversion of trained sappers to an infantry or artillery role. The opposing view was that, although the personnel were trained engineers, they were not fully trained, since wet bridging was excluded ; the C.W. companies could not carry all technical R.E. equipment in addition to the chemical warfare equipment ; as sappers they were, therefore, neither fish, fowl, nor good red herring, and there was a tendency to husband them in relation to the tasks on which they might be employed in order to conserve them for their primary role. Their basic training did not well suit them for handling a weapon of the type and tactical characteristics of the rocket, and their existing weapon (the Livens projector) was solely a gas weapon. The mortar was a more flexible weapon and could be used with H.E. and smoke chargings. So armed, the C.W. troops could have gained battle experience before the outbreak of gas warfare ; the fact that this could not occur with their existing armament was bad for their morale and for their fighting efficiency in a C.W. role.

By 1941 the latter views, strengthened by the disappointing performance of the 5-in. rocket in its then stage of development, gained ascendancy, and it was decided to introduce the 4·2-in. mortar into the armament of the C.W. Groups.

The 4·2-in. Mortar.—The mortar provided a weapon which could carry out sustained fire. With non-persistent chargings " crash " surprise shoots were impracticable ; a short period was necessary for the building up of the required incapacitating concentration with rapid fire. This period of build-up might have enabled the enemy to recognize such attack in time to gain protection, and the degree of casualty production would have been inversely in proportion to the enemy's standard of gas discipline. Tired troops were a particularly suitable target for such attacks. In all cases, however, the threat of lethal casualties as the penalty of bad gas discipline made it a powerful harassing weapon with this charging. The size of the area which could have been engaged depended on the concentration of mortars. The mortar was a more flexible weapon and better suited than the 5-in. rocket for engaging opportunity targets.

Using mustard gas it could carry out ground contamination of limited areas without waste and without a large concentration of mortars, owing to its greater precision and its capacity for sustained fire.

Initially the proportion of the different chargings decided on was 50 per cent. mustard gas, 25 per cent. phosgene, 20 per cent. smoke and 5 per cent. H.E. for ranging. In November, 1942, this was changed to 28 per cent. each of mustard gas, phosgene, and smoke, plus 16 per cent. H.E. A total of 100 rounds for each mortar (50 with each mortar and 50 in reserve) was carried by a C.W. (mortar) company, R.E.

Chemical Warfare Groups Armed Rocket and Mortar.—It was still desired, however, to retain the capacity to put down the " crash " surprise shoot with non-persistent gas from ground weapons. This led to the organization of C.W. groups of two companies armed 5-in. rocket and one company armed 4·2-in. mortar. Re-armament with these weapons occurred as they became available during 1942. This organization (for details *see* Chapter II, Section 3 and Appendix XIV) held good, with certain exceptions, overseas (*see* Chapter I, Section 5), until the 5-in. rocket was abandoned shortly before the decision was taken in January, 1943, to abolish the chemical warfare groups.

The Abandonment of the 5-in. U.P. Rocket.—Although the rocket was a more flexible weapon than the Livens projector, the companies armed with it were slow into action. The original set-up for firing was to fire a single rocket from a single trough and stand. Later six troughs were mounted on one stand (the sextuple mounting) which made it easier and quicker to deploy at the firing point. Even so it took a minimum of four hours to bring up a rocket company and get it into action from the time the target area had been decided on. The company was a cumbrous unit and a large number of vehicles had to be brought to the firing point. Development projects put in hand to improve mobility and speed into action included the mounting of sextuple stands in pairs on a 15-cwt. truck,(1) and the development of a mobile trailer to fire automatically more than the one salvo to which the existing equipment was restricted. Neither of these projects had borne fruit by the time the decision to abandon the 5-in. rocket was taken. In any case they would have been, at best, palliatives, not cures.

Our gas doctrine in the early years of the war had been influenced by the doctrine of the 1914–18 war particularly in relation to the putting down of heavy gas concentrations in the forward area by means of ground weapons. By the end of 1942 it was realized that, with the existing ground weapon development, the " crash " shoot was only possible in mobile phases during temporary pauses in operations. Further, the Arm which lent itself to the putting down of heavy gas concentrations was the Air Arm (*see* Chapter V—Tactical Doctrine).

As far as the rocket itself was concerned its development had started on wrong lines ; it was inaccurate and incapable of improvement worth the cost ; the units armed with it were unwieldy, and their general usefulness in an offensive campaign on the Continent of Europe was problematical. They had already proved unacceptable in overseas theatres (Middle East and India). It was consequently decided in December, 1942, to abandon this weapon.

The only other weapon with which the chemical warfare rocket companies could be re-armed was the 4·2-in. mortar. At this time the decision had recently been taken to form infantry brigade support groups which included the 4·2-in. mortar in their armament. For reasons given in Chapter II, Section 3, it was decided that the retention of special chemical warfare troops, also armed with this weapon, was unjustified, and the chemical warfare groups, R.E., were consequently abolished.

Defensive Chemical Armament of C.W. Troops

In addition to the main armament described above, the following defensive weapons were adopted for C.W. troops :—

Bulk Contamination Vehicles (B.C.V.s) (or Installation Type A/LP). Chemical Mines. Bombs, Ground, 6-lb.	All charged mustard gas.

Note (1).—It was during trials with this that Colonel R. H. MacLaren, R.E., Commander C. W. Troops, was killed. He was a severe loss to the C.W. organization as he was an officer of outstanding ability and long experience with the C.W. Groups.

Detailed descriptions of these weapons will be found in Military Training Pamphlet No. 23, Parts IV, V and VI.

In brief, the bulk contamination vehicle was a pressure "water" cart, containing 126 gallons of mustard gas. It was unarmoured and vulnerable to small arms fire. The object of B.C.V. was the gross contamination of ground in withdrawal or defence. It was designed so that, under ideal conditions at a speed of 5 to 6 miles an hour, one vehicle could effect contamination to a degree of one gallon of blister gas each 25 square yards over a continuous belt 14 yards wide and 200 yards long.

B.C.Vs. were not a permanent part of the armament of a C.W. group. They were to be held centrally and brought up as required. A C.W. company could operate up to 50 B.C.Vs.

As the nature of the war changed and we took the offensive, the need for purely defensive weapons of this nature disappeared (the tactical background to their use is discussed in Chapter V). The B.C.V. was diverted for use as a smoke producer, although never used operationally. In this role it was re-christened the "Apparatus Smoke Discharge No. 2".

The chemical mine was a commercial pattern cylindrical steel drum with a burster in the neck and two handles for carrying it. Its charged weight was 60 lb. It was intended for use in terrain to which a B.C.V. would not have access, and for the contamination of demolitions. It could also have been used by any field engineer unit. It heavily contaminated (50 gm/m^2) approximately 170 square yards and lightly contaminated (10 gm/m^2) about 1,000 square yards.

The bomb, ground, 6-lb. was for small scale contamination of trenches, rooms, etc., or small areas in circumstances where the use of mines or B.C.Vs. would have been uneconomical. It could be used by all arms. It heavily contaminated an area of 10 square yards, an area of 60 to 100 square yards being lightly contaminated.

Neither the mine nor the ground bomb were permanently carried by C.W. companies. If these companies had dumped their main armament, they could have carried 864 mines or 7,920 ground bombs in their own transport.

At the end of the war both the B.C.V. and the chemical mine became obsolete. The bomb, ground, 6-lb. was latterly only used for training (*see* Chapter VII).

The "Existing Weapons" Policy

After the abandonment of the 5-in. rocket and the abolition of C.W. groups, it was accepted that the Army gas effort would comprise the firing of gas munitions from existing weapons. The weapons available were Artillery, and the 4·2-in. mortar of the infantry brigade support companies—later re-organized as divisional machine-gun battalions—which have been discussed in Chapter II, Section 3. The tactical doctrine underlying this policy will be found in Chapter V.

No firm policy on the proportion of gas bombs to be carried by divisional machine-gun battalions was laid down, but they were provisioned for in the proportion of three mustard gas to one phosgene. As with artillery gas shell these munitions would not have been carried as a fixed proportion but would have been supplied by central planning as required, at any rate until gas warfare experience indicated the desirability of carrying a fixed quantity.

Lacking battle experience, it was difficult to assess the proportion of gas bombs which was likely to be required under conditions of gas warfare. A figure of 20 per cent. of the total of H.E. and smoke was visualized, although it was recognized that the assessing of probable expenditure in terms of a proportion of H.E. and smoke was not a very realistic approach to the problem since the tasks were so distinct.

The chemical bomb was a cylindrical type(1) which had been developed for use of the chemical warfare troops ; large stocks became available early in 1943. The Mk. I(2) had a range of 3,350 yards and had a gas content of 3½ lb. The H.E. bomb subsequently developed was a streamlined bomb(1) with a range of 4,100 yards. An increase in range of the chemical bomb could only be achieved at the expense of chemical effect. An H.E. streamlined bomb charged with gas had a gas content of 2¼ lb. Bombs so charged were experimented with in case it became necessary to charge this type of bomb.

Where ground contamination with mustard gas is carried out with ground gas weapons, the 4·2-in. mortar, within its range, is the weapon of choice, as opposed to artillery ; there is, therefore, a possible future requirement for a mustard gas bomb for the 4·2-in. mortar. With nerve gas chargings, it may be a valuable neutralizing weapon ; in this role it will need consideration *vis-à-vis* artillery shell and rockets, in relation to the best methods of discharging nerve gases.

Experimental Army Gas Weapons

Details of Army gas weapons which were experimented with during the war, but not introduced, will be found in the Ministry of Supply histories. A brief résumé is given below of the lines on which development occurred, to show that there was no sterility of outlook. Experimental types of Artillery shell have already been discussed (*see* Gas Shell—Development Types).

Rockets

After the abandonment of the 5-in. U.P. rocket in December, 1942, the provision of a gas rocket to increase the weight of our ground gas attack continued to be borne in mind in relation to experiments with rotated and unrotated rockets of various calibres for use as H.E. rockets, and the possibility of using a chemical head was kept in view. The gas rocket was never a high priority requirement, owing to the greater ease with which the effects of a gas rocket shoot, with any of the chargings available during the war, could be paralleled with air gas weapons.

The introduction of nerve gases, however, may lead to a requirement for an Artillery H.E./chemical rocket so charged, as an adjunct to a crash shoot with H.E. rockets.

Note (1).—The difference between a cylindrical and a streamlined bomb can be seen from the photographs at Appendix XIX. The H.E. bomb issued for ranging purposes to C.W. troops was a cylindrical type.

Note (2).—A Mk. II bomb with a range of 3,500 yards was developed which was more snub nosed and heavier in the nose than the Mk. I. It was superior ballistically and preferable from the production aspect and would have become the Service weapon when stocks of the Mk. I had become exhausted.

Anti-tank Chemical Weapons

The use of gas against a confined space, such as the interior of a tank, led to the consideration, in 1941, of the use from small arms and armour-piercing ammunition of those gases which were effective in low concentrations, such as tear gases and nose gases, and which, therefore, lent themselves to use in small quantities. A charging with immediate effect was necessary.

The use of such gas munitions against tanks was visualized in two ways, either by bursting them on the outside of the tank so that the gas would get into the interior through interstices in the armour or with ventilation air, or by armour piercing shot which would distribute the gas inside the tank. In the former category, ·303 tracer ammunition filled tear gas (CAP) were tried out but abandoned as being insufficiently effective. The 20-mm. H.E./I with a SAP nose-piece to close the round, instead of the H.E./I fuse, and charged with a mixture of nose gas and tear gas, or tear gas (B.B.C.) alone, was found suitable and a trial order was in production when the war ended. In the latter category, 20-mm. AP/T were also being put into trial production. A 40-mm. SAP shell was also developed but was little more effective than the 20-mm. and was therefore not put into production.

A CAP tear gas hand grenade (No. 78, Mark I) was also developed as an ambush weapon or for use from slit trenches. It detonated on impact and the H.E. charge was sufficiently large to break up the solid gas into fine particles and disperse them, driving the particles through the crevices in the A.F.V. It proved an effective weapon and was sealed, but there was no production.

A line of approach which arose early in the war was the development of a weapon to exploit the lethal properties of high concentrations of prussic (or hydrocyanic) acid against an enclosed space, such as a tank, with a special weapon developed for the purpose (*see* Chapter III, Section 2). This led to the development of the "Jet Anti-Tank Portable, Mark I". This was a cylindrical liquid gas projector with a flexible hose. An electric flash ignited a cordite charge which built up the necessary pressure to expel the liquid. It contained two gallons of prussic acid and weighed 34 lb. fully charged. It had a range of 25 yards. It is illustrated at Appendix XIX. A direct hit on a tank was lethal and would have killed all personnel inside. It was also effective against other enclosed spaces, such as a pillbox, if accurately aimed at a weapon slit.

It was developed as an infantry anti-tank weapon and the design was sealed. A small order of 500 only was placed for use in troop trials and no large-scale production was undertaken. It was simple to produce and could have been introduced rapidly, if required, had gas warfare started. Tactically it suffered from lack of range and from being a one-shot weapon. It was inaccurate in high winds.

The future of chemical weapons with an anti-tank role is linked with the exploitation of nerve gases and points to the development of projectiles with this charging for the principal artillery and infantry anti-tank weapons. They would have to be superior to H.E. or H.E./I types. Gas chargings may knock out the crew but not the tank itself. It is not improbable that such chargings will be more applicable to infantry anti-tank weapons than artillery.

Chemical Weapons for the War against Japan

Considerable attention was given from the end of 1943 onwards to the development of Army chemical weapons for use in the war against Japan to exploit the enhanced effect of mustard gas vapour in hot humid conditions (*see* Chapter III, Section 3), and the use of both this and lethal gases in the favourable meteorological conditions encountered in the jungle. A particular development requirement was for gas weapons to aid in the reduction of the strongly-fortified Japanese bunkers and pillboxes. The following is a list of the weapons which were experimented with, showing how far they had been developed when the war ended.

Frangible Grenades.—These were glass bottles with a fly-off lever fuze, charged prussic acid. Four grenades were carried in a special " haversack " type of carrier. These weapons were not accepted, as accurate throwing was difficult, and, being made of glass, they involved danger to our own troops.

The Grenade AC 600 cc.—This was a light metal cylindrical steel grenade with a content of 600 cc. prussic acid, which was developed from the frangible grenade. It could be fired from a rifle launcher or cup discharger with a maximum range of 170 and 130 yards respectively, or it could be thrown by hand up to a range of 18 yards. It weighed $2\frac{1}{4}$ lb. ($2\frac{1}{2}$ lb. with launcher tail). It was designed for use against covered weapon pits. To have a killing or incapacitating effect on the occupants, two or three grenades had to be burst within a foot or two of the embrasure. Fired from a rifle launcher at 10 degrees, it had a flat trajectory, a range of 60 yards and was quite accurate. The war ended before development was completed.

A grenade of this type but with a more lethal charging, *e.g.*, nerve gas, is a possible requirement for the future.

The Jet Chemical No. 1, Mark I.—This was a development of the jet anti-tank portable, Mark I, made jungle-worthy by the deletion of the hose and other exterior fitments and by a decrease in weight. It weighed 17 lb., and contained approximately one gallon of prussic acid, and had a range of 20 to 25 yards. It was designed primarily for use against bunkers and pillboxes, with the object of killing or incapacitating the occupants. Its occasional use in jungle as an ambush weapon or as an offensive weapon to be used in local attacks, was also visualized. A number were sent to India and Australia for trials to decide whether they were acceptable as a normal infantry gas weapon. The war, however, ended before reports were received. Owing to its very short range, it is improbable that it will be a requirement for the future.

Flamethrowers Charged Blister Gas or Thickened Prussic Acid.—Consideration was given to the use of flamethrowers charged blister gas as an anti-bunker weapon. The Lifebuoy man pack flamethrower proved unsuitable without drastic modification, owing to leakage in the system. The Wasp (carrier-borne) flamethrower gave satisfactory performance, but was discarded owing to the considerable contamination of the crew, except under the most favourable conditions (stationary and a following wind). Further, the Wasp itself was not very suitable for use in jungle country owing to its lack of overhead cover. Ultimately, it was decided that there was no requirement for the use of blister gas from any flamethrower, but a possible requirement remained for the use of thickened prussic acid from a tank flamethrower. Trials with the Wasp had shown that two short bursts directed at a pillbox at 35 yards range produced extremely high and lethal concentrations inside for a few minutes. This use of tank flamethrowers was kept in view, but the war ended before any further progress was made.

Vesicant Spray Containers and Generators.—The enhanced effect of mustard gas vapour under tropical conditions has been referred to. Some experiments were carried out with the object of obtaining a high vapour concentration more rapidly than is possible by evaporation from liquid on the ground, particularly for use in circumstances where weather conditions lend themselves to the slow dispersion of vapour, *e.g.*, in jungle. Two methods were considered—a container which was cordite pressure-operated and discharged its contents rapidly in the form of a very fine spray, and a generator, which was heat operated and drove off the mustard gas in vapour form. The containers or generators were to be loaded into projectiles, *e.g.*, 3-in. or 4·2-in. mortar bombs. Both projects were still in the early research stage when the war ended. A requirement for the future would depend on the degree of efficiency ultimately achieved by such methods of projection.

Chemical Bomb for 2-in. Mortar.—In view of the fact that the 2-in. mortar H.E. bomb is not a very effective weapon when burst in thick undergrowth, a proposal to introduce a chemical bomb for this mortar was considered. Owing to the small size of the bomb, only those gases which were effective in low concentrations, *e.g.*, tear gas or nose gas, made suitable chargings. A bomb was developed and trials were carried out, but although the bomb was technically effective, its introduction was not considered worth while.

Chemical Bomb for 3-in. Mortar.—In view of the fact that the 3-in. mortar was the main mortar armament in the South-East Asia Command, a requirement was stated for the development of a mustard gas bomb for general contamination purposes. Normally the 3-in. mortar had not been considered suitable for this role owing to the small capacity of the bomb. To obtain an increased gas capacity, a loss of range was accepted as inevitable when the development requirement was put forward. A minimum range of 1,000 yards was first stated but this was later amended to 1,500 yards. A number of bombs were developed with ranges varying from 1,500 to 2,400 yards with gas capacities of from 2 to 3 lb. Trials were being carried out in India and Australia when the war with Japan ended and the requirement lapsed.

Training Stores

Tear Gas and Nose Gas Generators

The Generator, Lachrymatory, No. 2, was the standard tear gas generator, and the Generators, No. 21, Mk. I, or No. 23, Mks. I and II, or No. 30, were nose gas generators of varying capacities and times of emission (3 minutes, 3 minutes and 1½ minutes, respectively), used for outdoor training exercises. Instructions for their use will be found in Gas Training, 1942.

Of the nose gas generators, No. 30 was for training only, but the remainder could have been used as a war store. The case for nose gas generators in this role was considered at various stages of the war, but there were no stocks for this purpose. Production for training purposes could have been stepped up had a requirement arisen after gas warfare had started.

Bomb, Ground, Training

This was a white painted cylindrical metal container into which a burster (issued separately) was inserted. There was a filling hole with a screw cap for charging it with training mixture, when required for use. It was used by all Arms to simulate a bomb, ground, 6-lb., for training purposes.

Projector, Spray, Training

The projector consisted of a 9·5-in. steel barrel of 4·2-in. diameter with a spike attached centrally to the closed end. It fired a tin plate cylindrical bomb, 7·5 in. by 4·1 in., charged with training mixture. There was a central tube through the bomb into which was inserted a length of safety fuze ; at the end of this fuze was a bag with a small propellent charge. The bomb was inserted into the projector and the exposed end of the fuze lighted. The propellent projected the bomb about 50 ft. into the air where the burster (in which there was a delay) exploded the bomb, scattering the training mixture as a spray. It was used to simulate air burst shell. It was only issued to the Army Gas School.

The Mine, Chemical, Training No. 1

This was almost identical with the chemical mine and was charged with training mixture, as required. It was painted white. It was issued to certain training establishments and R.E. units in the early part of the war, as a training store, to simulate the chemical mine.

Comparison between British and U.S. Ground Gas Weapon Development

In the United States ground forces, discharge of gas from ground weapons was mainly the responsibility of the chemical mortar battalions of the Chemical Warfare Service, from whom the major ground gas effort would have been forthcoming. Their organization is shown in Chapter II, Section 8.

The United States 4·2-in. mortar was a rifled mortar with a maximum range of 4,400 yards. The bomb weighed 23 lb. and had a gas content of 6½ lb. In gas efficiency it was superior to the British smooth bore 4·2-in. mortar, for which the bomb weighed 20 lb. with a gas content of 3½ lb. Relative ammunition expenditures were in proportion to this difference in gas content.

The United States chemical shell were of the bursting chemical type charged mustard gas for 75, 105 and 155-mm. artillery shell. They would have created a vapour hazard and on like terrain the results would have been similar to those achieved with their 4·2-in. mortar mustard gas bombs, allowance being made in calculating ammunition expenditures for the difference in the gas content of the respective projectiles.

They possessed a Livens projector which was practically the same weapon as our Livens and which they declared obsolete. They had a tear gas hand grenade and also one charged nose and tear gas. Their nose gas generator was a war store weighing 9½ lb. with 3½ minute emission. Their chemical mine was about one-fifth of the size and capacity of its British counterpart and more analogous to our bomb, ground, 6 lb. They also possessed a portable cylinder for discharging choking gas, weighing 54 lb. and containing 31 lb. of gas. We had no parallel Service store because we considered the release of gas from cylinders to have been totally superseded by other methods.

On the experimental side their most notable projects were a 7·2-in. rocket and a 4·2-in. recoilless mortar. The 7·2-in. rocket had a maximum range of 3,700 yards. It was launched from a 24-rail artillery launcher. It was an efficient weapon for putting down a surprise concentration of non-persistent gas, but a company armed with it would have been tactically inconsistent with the probable frequency of its employment. Development of a launcher with double, or more than double, the number of rails was in hand to reduce the manpower and vehicle liability for its discharge.

The 4·2-in. lightweight recoilless mortar had an overall weight of 170 lb. and incorporated new ideas of mortar construction. It had a range of 3,800 yards. The bomb body was the same as that of the 4·2-in. standard mortar bomb. Apart from the weight factor, its main advantages were the highly effective ricochet bursts and the dispensing with the emplacement problems of the standard type of mortar. Its disadvantage was the breach blast, which revealed its position.

Toxic rifle/hand grenades were also in development. As with us, frangible grenades were considered unsatisfactory although these were initially accepted as a service store.

German Army Gas Weapons

These are discussed in Appendix XXIV.

SECTION 2.—AIR GAS WEAPONS

Introductory

The detailed history of the development of air gas weapons during the war will be found in the Air Ministry, Ministry of Supply and Ministry of Aircraft Production histories.

In its relations with the Air Force it is the Army's responsibility to indicate a task but the method of carrying it out and the best weapon to employ is the responsibility of the Air Force. It is, however, essential to record the general development of air gas weapons as a background for the proper understanding both of major policy and the tactical employment of gas. Had gas warfare eventuated, our major gas effort would have come from the air ; it is impossible to divorce the air effort from other aspects when considering the development of gas warfare as a whole.

Air gas weapons fall into two main groups :—

(*a*) S.C.Is.(1) and Spraying Bombs charged mustard gas.
(*b*) Gas Bombs—charged mustard gas and phosgene.(2)

Spray Weapons

These comprised the following, and are shown in the order of their availability during the war :—

250 lb. S.C.I. Type S/G (1939).
500 lb. S.C.I. Type S/G (1939).
1,000 lb. S.C.I. Type S/G (1940).
I.S.C. 400 lb. Type S/L (1942).

S.C.Is. 250, 500 AND 1,000 LB.

One of the main problems in designing spray weapons during the inter-war period was to overcome the shattering of the liquid, caused by the force of its impact with the air stream on discharge from the aircraft, in order to obtain droplets of an optimum size. In the development of high spray this was of vital importance in order to obtain a droplet of sufficient size to reach the ground before it evaporated. To begin with, S.C.Is. were designed to contain a pressure apparatus which forced out the liquid at approximately the same pace as the speed of the aircraft, and therefore theoretically as though it had been discharged from a stationary source. This had many disadvantages, *viz.*, weight of the unit, complexity in design, and inferior gas content to weight ratio. With the introduction of thickened mustard gas the inclusion of a pressure system became unnecessary, since the thickened liquid suffered less from the shattering effect on discharge. The superior gas content to weight ratio of the S.C.I., resulting from the deletion of the pressure device, was an added compensation and advantage.

Note (1).—"Smoke Curtain Installation." The same type of S.C.I. is used for laying smoke curtains and spraying gas.

Note (2).—Mustard gas and phosgene were the only types of gas charged into gas bombs during the war. Certain other chargings which were submitted to exhaustive trials but not accepted are referred to in Chapter III.

All S.C.Is. type S/G available during the war were gravity flow types containing thickened chargings. They are illustrated at Appendix XX. They were operated by the breaking of two bakelite discs by means of electric detonators connected to the aircraft electrical circuit for the release of aircraft bombs ; one disc was on the air inlet assembly on the upper surface of the body of the S.C.I. and one on the outlet-assembly on the underside which was connected to an emission pipe and nozzle. On the breaking of the discs the charging was emitted by gravity flow alone.

The reason for having three types of S.C.I. of different sizes and capacity was the suitability for their use in particular types of aircraft according to the design, lift and stowage capacity of the aircraft. They could be stowed externally or internally according to the type of aircraft. If stowed externally certain fairings were added for streamlining.

The United States " Airplane Smoke Tank M.10 " (which is also illustrated at Appendix XX) operated on a similar principle to British S.C.Is. It was stowed externally and was a requirement for British " Typhoon " aircraft of the Tactical Air Force towards the end of the war. It was at first considered that this streamlined tank was better suited for stowage on these craft than was its existing British counterpart—the S.C.I. 500 lb.—as it was thought it would create less drag at high speeds, but subsequent trials showed that there was little to choose between them when the S.C.I. 500 lb. was fitted with nose and tail fairings.

It is presumed that the characteristics of aircraft spray are known to the reader but for easy reference they are given in Appendix XVIIIA which is an extract from Military Training pamphlet No. 5 (1943) " The Use of Gas in the Field." Technical data regarding the extent of the area sprayed from varying heights with the different types of S.C.I. and aircraft, and the droplet distribution within the area, will be found in the " Chemical Warfare Pocket Book, 1942."

Some residual danger from vapour existed after low spray attack, particularly where conditions of temperature and terrain were favourable to the rapid evaporation of the liquid, and although ground contamination of an area could have been achieved with spray from a number of aircraft flying at very low heights, the role of spray was that of an anti-personnel weapon for direct contamination of troops in the open.

From the R.A.F. point of view, there were a number of disabilities in gas operations with S.C.Is. The major objection was that, during the period of emission of the spray,(1) the aircraft had to fly on a comparatively level course between two points and was therefore very vulnerable to A.A. fire, particularly in attack from low altitudes. Certain technical difficulties were also involved. In some cases, *e.g.* fighter bombers, where S.C.I. had to be stowed externally, aircraft speed and manœuvrability were affected. Special training in spray was needed for aircrews and the time factor in training imposed limitations on the number of squadrons that could be kept fully trained. From the point of

---

Note (1).—The distance which the aircraft flew during emission varied according to the emission time of the S.C.I. and the speed of the aircraft. Spraying speed would have been kept down but was limited mainly by the characteristics of the aircraft. Examples are 580 yards for 250-lb. S.C.I. on a Lysander and 3,900 yards for 1,000-lb. on a Wellington.

view of ground staffs S.C.Is. were awkward weapons to handle on return of aircraft from spraying missions. The wearing of protective clothing would have been necessary since the plane might have become contaminated when spraying, and there was a possibility that the S.C.I. might drip, although the nozzle was designed to obviate this. Special training in recharging S.C.Is. and in decontamination was necessary. Rearming of aircraft after a spray mission would have involved delay. Since gas was not used there was no opportunity of determining the extent to which these difficulties might have been ironed out.

Low spray missions in support of Army operations would have had to be prearranged unless aircraft were kept on call and immobilized from other tasks.

L.S.C. 400 LB. S.L.3

This was given the nickname of "The Flying Cow." It was shaped like a bomb and similarly attached to and released from aircraft. After release it started to emit liquid when about 20 ft. below the aircraft and completed its emission after about 700 ft. of its fall, thereby leaving a "rod" of liquid in the air in the shape of its trajectory. The liquid was ejected through a nozzle in the tail by the pressure generated by a slow burning cordite charge in the nose of bomb. When the pressure build-up reached a certain point it broke a disc in the nozzle and the liquid was ejected.

One object in its development was the requirement for a spray weapon that could be carried and released in the same way as a bomb, and so leave the aircraft free to take avoiding action, which was impracticable when using S.C.I. Other advantages were that the risk of contamination of the aircraft was reduced and the time required to rearm the aircraft was no longer than with any other type of bomb.

It could only be used from heights above 700 ft., by reason of its characteristics as described above, and it was therefore not suitable for low spray attack. Although at the time it was hailed as a considerable advance in spraying technique, it was not a very satisfactory weapon owing to difficulties in aiming common to all high spraying, and it in no way overcame the disadvantages of high spray as a weapon.

Policy in Relation to Aircraft Spray

Policy in regard to aircraft spray, its tactical use and our change in view about its place in our gas armoury are discussed in Chapters I and V.

Knowledge acquired during the war showed high spray to be of little value, and although low spray remained a potentially effective casualty producer, the tactical opportunities for its use would have been few. With low spray attacks the difficulties of aiming spray from S.C.Is. and the dangers to the aircraft must be remembered. It was only infrequently that we should have had the knowledge in sufficient time of the existence of large enough bodies of troops to make such attacks worth while.

Development at the end of the war for direct anti-personnel contamination of troops in the open with mustard gas pointed to the perfection of the airburst bomb (which is discussed below) as the best means to this end, particularly if a fuze can be developed to give a consistent height of burst at the optimum height above the ground. This would ensure a heavy degree of contamination and also overcome in some degree the difficulties of aiming spray. It would provide operational flexibility in that the same weapon could be used both for anti-personnel tasks and for tactical contamination of ground.

On conclusion of the war the 1,000-lb. S.C.I. and 400-lb. S/L spraying bomb were declared obsolete for gas but were retained as an interim measure for spraying smoke and insecticides.

Aircraft Bombs

General

Aircraft bombs comprised the following, in the order of their availability during the war :—

Bomb, Aircraft L.C., 250-lb.(1), charged mustard gas (1939) and phosgene (1941).

Bomb, Aircraft L.C., 30-lb., Mk. I, charged mustard gas (1939).

Bomb, Aircraft L.C., 30-lb., Mk. II, charged mustard gas (1940).

Bomb, Aircraft L.C., 65-lb.(1), charged mustard gas (1941).

Bomb, Aircraft L.C., 500-lb.(2), Mk. I and II, ground-burst, charged phosgene (1943). Mk. I, charged mustard gas (1944).

The 30-lb., Mk. I and the 250-lb. bombs were fuzed bombs of conventional appearance. They functioned on impact on the tail ejection principle in which a small internal burster blew off a lightly welded tail plate, the charging thereby being ejected with minimum loss in the crater.

The 30-lb., Mk. II and 65-lb. bombs were fuzeless bombs which broke open on impact and scattered their contents on the target. They were carried in the small bomb container, which held respectively 8 and 3 bombs, and they were released in salvos of that number. These bombs are described in more detail below.

The 500-lb. bomb was also of conventional appearance but it was not of the tail ejection type. It had two bursters—one in the nose and one in the tail—which fragmented the case. For ground functioning nose and tail pistols were fitted, or the nose pistol could be omitted entirely. For low level attacks an 11 seconds delay action was fitted to allow the aircraft to clear the danger area. When charged mustard gas it was intended that it should be airburst by means of a barometric or clockwork fuze, thereby discharging its contents in the form of spray. It was not, however, finally cleared for airbursting by the time the war ended, and it had not been designed for ground bursting with a mustard gas charging.

All the above bombs are illustrated at Appendix XX, and in the case of mustard gas chargings, their contamination performance, *i.e.*, areas of light and heavy contamination and the anti-personnel areas, is anotated on the photographs. The size of the area contaminated varied according to the size

---

Note (1).—There were two marks of both the 250-lb. and 65-lb. bombs—the Mk. I and the Mk. II. The difference lay in small structural modifications which did not alter the contamination performance of the bomb or affect its method of release.

Note (2).—The two marks of this bomb varied in the method of fixing the burster container in the body. Both Mk. I and Mk. II bombs could be charged with phosgene but, for technical reasons, only the Mk. I was charged with mustard gas.

of the bomb and the strength of the wind and was mainly down wind at the point of impact. The data given in Appendix XX is for a wind speed of 10 miles an hour. In winds of higher velocity the areas became larger and narrower while the lightly contaminated area increased in extent at the cost of reducing the density of the contamination.

Phosgene Bombs—Bombs Aircraft 250-lb. and 500-lb.

Phosgene was not introduced as a charging for aircraft bombs until the winter of 1940/41 when a limited number of 250-lb. bombs were so charged in case it became necessary to retaliate against the German civil population. It was known that the Germans placed considerable reliance on the use of gas proof shelters, and it was the exploitation of this factor that first led to the charging of phosgene into air gas weapons, the intention being to open up the shelters with H.E. before dropping the phosgene bombs. The idea of using phosgene from the air against suitable enemy targets in the field in support of mobile operations followed consequent on this and was a secondary reason for the introduction of the phosgene bomb. Subsequently, large stocks of these bombs were produced (Appendix III).

The 500-lb. bomb was introduced in 1943 as it was more economical in stowage on the types of aircraft available during the latter part of the war.

Mustard Gas Bombs

The 30-lb. Mk. I and the 250-lb. mustard gas bombs were the standard bombs available in 1939. The 30-lb. Mk. II became available early in 1940.

The functioning of the 250-lb. and 30-lb., Mark I bombs has been described above. The 250-lb. bomb gave a larger area of contamination than the 30-lb. Mk. I bomb but the 30-lb. bomb, when showered down on the target area, gave rise to numerous but smaller sources of contamination more evenly distributed over the area. Both these bombs had a high terminal velocity which, particularly in the case of the 250-lb. bomb, was of advantage where roof penetration was required for contaminating the inside of buildings. For this latter purpose the 250-lb. bomb was the weapon of choice.

When discharged from heights above 4,000 ft. against hard targets, both these bombs broke up irrespective of the operation of the fuze. There was no limit to the operational height from which attacks could be delivered against hard targets.

When used against soft targets, grass and arable land, their high terminal velocity led to the burying of the bomb with consequent loss of charging in the crater. This imposed a limit on the operational height from which they could be released.(1) A retarder was developed for use on the 30-lb., Mk. I, and, when this was fitted, there was no limit to the height of release.

Note (1).—The area contaminated to any given density was similar on soft and on hard targets when the height of release was not above 250 ft. but on soft targets it was less from greater altitudes. The maximum height of release to prevent loss of charging owing to burying of the bomb was :

250-lb. Mk. I and II	5,000 ft.
30-lb. Mk. I (without retarder)	5,000 ft.
30-lb. Mk. II	1,000 ft.

The performance of the 30-lb. Mk. I (retarded) was very little affected by the type of terrain on to which it was dropped.

The 30-lb., Mk. II, bomb was a thin-walled cylindrical bomb made of terne plate. It was fuzeless and functioned by break-up on the target, splitting down the longitudinal seam. There was a 9-lb. weight in the nose to improve its ballistics. The tail was a light sheet metal cylinder of the same diameter as the body. It had a better charge/weight ratio than the Mk. I (8½-lb. charge as opposed to 6¼-lb.) and gave rise to a larger area of contamination (*see* Appendix XX). Against soft targets, however, it buried badly from heights above 1,000 feet, and it could not be fitted with a retarder. It was simple and cheap to produce.

Broadly, the object of mustard gas bombs is to make a mess on the target area with the liquid. With the above bombs we could effect this, but, owing to their limitations, they were neither operationally nor tactically flexible. In 1941 the 250-lb. and 30-lb., Mk. I, bombs were withdrawn to reserve and replaced by a bomb which afforded a cheap and simple means of achieving the object as well as greater operational and tactical flexibility.

This bomb originated in a 5-gallon cylindrical oil drum fitted with a suspension lug and lead nose weight. This did not prove satisfactory, but it led to the rectangular " tin can " of light sheet metal fitted externally on the side with a corrugated strengthener—the 65-lb. bomb—which became the standard R.A.F. mustard gas weapon. The bombs burst open on impact and scattered the liquid with some force in splashes and droplets. They were normally discharged from the aircraft in salvos of three or in multiples of three, but could have been released singly if required.

These bombs had many advantages ; they were easy to produce in large quantity ; they had an excellent gas content to weight ratio ; they could be used against hard and soft targets without limitation on height of release ; and their performance varied little on any surface, except that from high altitudes against very soft ground there was some loss of charging owing to cratering. Their disadvantage lay in the fact that they were ballistically poor and were therefore difficult to aim accurately except from very low heights, but when engaging a large target area this was offset by the fact that a high degree of accuracy for any individual bomb was not essential. They were somewhat fragile, which led to difficulties in storage and transport, particularly for despatch overseas.

The 65-lb. bomb was tried out as an anti-tank weapon against a captured German Pz. KW.III and found to be very effective, a direct hit or very near miss causing heavy external and internal contamination. Owing to the lack of " aimability " of the bomb, attack would have been necessary from very low altitudes. The results obtained were therefore likely to have been outweighed by the aircraft casualties incurred in such form of attack.

A development which we never completed owing to lack of priority was of an " aimable " cluster of 65-lb. bombs, which would have opened at a given height above the ground and released the individual bombs. This would have facilitated operational flexibility with United States aircraft since their aircraft could not carry our Small Bomb Container.

The Canadian 50-lb. bomb was similar to our 65-lb. bomb but rather shorter. Its development arose from a can of this size being more readily available. A " cluster " of five was developed which could have been used from United States aircraft, but it did not go into production. The Canadian cluster is illustrated in Appendix XX.

Stocks of a 500-lb. bomb charged mustard gas were available from June, 1944, onwards. These were provisioned for the Tactical Air Force, partly because the 65-lb. bomb was uneconomical in stowage on T.A.F. craft, and partly because it was intended to use this 500-lb. bomb as an airburst bomb with the dual function of spray against troops in the open and ground contamination of tactical targets. When airburst it was a tactical rather than a strategic weapon.

The airbursting of this bomb, however, was not fully cleared by the time the war ended. There was a tendency to inflame on being functioned, although the frequency of this occurrence was small. There were also difficulties over the fuzing, and an optimum and consistent height of burst could not be obtained. Had gas warfare started, it could have been used as an air burst bomb with the M.111 clockwork fuze (which was available) ; it could also have been used ground burst, but in neither case would it have functioned with high efficiency.

Comparison between British and United States Development

The standard United States mustard gas bombs were the 100-lb. bombs M.47A2 and M.70. These were fuzed bombs, with a central burster, which burst on impact. The M.47A2 was thinner walled than the M.70 and contained a larger quantity of mustard gas (71 lb. as compared to 59 lb.)

At the end of the war the United States had certain bombs under development. These included the clustering of a 65-lb. bomb which they liked for its high chemical efficiency and considered an excellent means of developing heavy contamination from low level bombing, and a cluster of thirty-eight 10-lb. tail ejection bombs (the E5 bomb), which they felt to be required for optimum efficiency in dispersion of mustard gas for ground contamination of tactical targets. They were also developing large airburst bombs, but difficulty was encountered in obtaining a consistent height of burst and in establishing an accurate area of dispersion for each bomb.

In the extensive trials carried out with British and United States munitions in 1944–45 (Chapter II, Section 9) it was found that, in general, fuzed bombs of the American type with a moderately large burster shattered the charging into finer droplets than the British bursterless munitions. For production of vapour hazard in a short time the United States bombs were superior, but where the maximum persistence is required bursterless munitions were preferable. When released from high altitudes on very soft ground there was considerable loss in the crater with both British and United States munitions.

The United States non-persistent bombs—the M78 (500-lb.) and M79 (1,000-lb.)—functioned by the use of nose and tail fuzes and a burster, and detonated on contact. In addition to phosgene, cyanogen chloride (CK) and prussic acid (AC) were also charged into these bombs.

United States airplane smoke tanks M.10 and M.33 A1 were of the gravity flow type and were functioned similarly to our S.C.Is. They were charged with unthickened mustard gas, only very low spray attacks being visualized. With unthickened mustard gas the area of contamination is wider than with thickened mustard gas but the density of contamination is less.

German Air Gas Weapons

These are discussed in Appendix XXIV.

Possible Lines of Future Development

The discovery of nerve gases provides a more lethal charging than phosgene, and bombs will be required to be developed for their most efficient dissemination against all types of targets.

These nerve gases may have a marked effect on the development of air gas weapons for use in support of Army operations. Hitherto no fully suitable gas has been available for use in support of an assault. Neither phosgene nor mustard gas was immediate in its effects, and, in the case of the latter, residual contamination was a serious disadvantage. Nerve gases may prove to be a suitable agent for this role, and must be considered for other air weapons besides bombs.

As regards the 65-lb. mustard gas bomb, stowage in relation to aircraft development renders it a less economical weapon than during the war and it is likely to become obsolete for this reason ; the difficulties in aiming as well as those connected with storage and transport must also be regarded as a disadvantage. Moreover, in order to achieve economical distribution, an airburst bomb releasing smaller units over a wider area seems to offer a better solution than a single bomb burst above the ground or on impact with the ground.

CHAPTER V

EMPLOYMENT OF GAS AND TACTICAL DOCTRINE

SECTION 1.—FACTORS AFFECTING DEVELOPMENT OF DOCTRINE

In considering the development of our doctrine for the use of gas it is desirable to summarize the factors underlying the changes. Much of the detail is covered in other chapters of this book but the salient points are summarized below.

There is one factor which overlays all others and regarding which our ideas have never changed—namely that gas is an auxiliary weapon to be exploited in conjunction with other weapons but is not to supersede them.

The Efficiency of Enemy Gas Defence

During the inter-war period we had ourselves put a very great effort into the development of defensive appliances—particularly the respirator and protective clothing. This had been paralleled by a similar effort by almost all other nations. We had no reason to believe, therefore, that other nations had not reached as high a standard of all round defensive efficiency as ourselves, and our main opponent, Germany, was known to be very well prepared. Our gas doctrine, therefore, was based on the assumption that we should use gas against a fully protected and well-organized enemy.

Two deficiencies in enemy protection had a bearing on the planning for gas warfare against the Japanese—their lack of impregnated clothing which facilitated the exploitation of the enhanced aggressiveness of mustard gas vapour in the Tropics (Chapter I, Section 7), and the inefficient protection afforded by their respirator against cyanogen chloride (C.K.) which led to extensive trials being carried out with this gas. (Chapter III, Section 2).

Terrain and Meteorology

During the inter-war period these factors had little bearing on the development of our tactical doctrine. Our gas re-armament which began in the middle 1930s occurred concurrently with the emergence of the Western plan, with Germany as the main opponent, and the conditions of meteorology and terrain were tacitly accepted as being similar to those of the 1914–18 war and were well known. This factor, therefore, introduced nothing new in its bearing on tactical doctrine in relation to the European theatre.

It did, however, have a major bearing on our doctrine for the use of gas in the war against Japan (Chapter I, Section 7, Chapter III, Sections 2 and 3, and Chapter IV).

Battle Experience

Battle experience of gas warfare is confined to the 1914–18 war. The use of gas by the Italians against the Abyssinians has no bearing on the doctrine of modern gas warfare as the Abyssinians were totally unorganized and unprotected against gas.

The static conditions of the 1914–18 war lent themselves to the use of gas from ground weapons, since they permitted the necessary large concentration of gas munitions before any operation. Gas was used as a weapon of attrition and relied for its effect on surprise attacks with a heavy weight of gas, or on prolonged bombardment. As an assault weapon it could not be relied on for day to day use owing to its dependence on suitable weather conditions, but this was in some degree offset in that the target was always there, and in many cases we could wait until the required weather conditions prevailed. They were systematically in our favour.

In applying our 1914–18 experience to modern war it was thought that the same doctrine would still apply during temporary pauses in battle. Our battle experience had also shown us the superiority as a casualty producer and the weight for weight economy of mustard gas over other types of gas, even with the relatively inefficient weapons of the first war. These factors had considerable influence on our pre-war doctrine as improved means of projection became available.

Large Scale Field Trials

Facilities became available during the war for large scale field trials in Canada, Australia and India—as well as in the United States—which had not existed in peace (Chapter II, Section 9). Apart from obtaining much valuable data on the performance of gases when released in large quantities, in particular they afforded us the opportunity of testing our assessment of the efficiency of aircraft spray and of the effects of tropical and jungle conditions on chemical warfare substances and appliances by means of realistic trials on a scale comparable to that which would actually be encountered in war.

Characteristics of the Available War Gases

Details regarding the war gases are given in Chapter III. From these it will be seen that fundamentally, until the advent of nerve gases, the war gases available remained the same as those developed in the 1914–18 war. All round, phosgene remained the most efficient lethal gas. Although the Americans were impressed with the potentialities of prussic acid and cyanogen chloride from aircraft bombs, extensive trials did more to prove the difficulty of using them effectively, owing to risk of inflaming and other troubles, than to establish any superiority over phosgene (Chapter III, Section 2).

Some improvements in mustard gas had occurred resulting in increased persistence ; the addition of thickeners improved the performance from certain weapons. Some more insidious types, *e.g.*, nitrogen mustard gases, had been discovered, but they lacked the overall aggressiveness of the original mustard gas. We never produced them other than on an experimental scale, but enemy development had a bearing on our defensive doctrine (Chapter III, Section 3).

Liquid tear gases were improved in persistence and stability, without change in their physiological characteristics. More effective types of nose gases were developed. Physiologically their effect was the same, and, in general, our own and foreign respirator development matched the risk. Their harassing effects on well trained and determined troops had been over-estimated (Chapter III, Section 1).

(C29058) H

It was not until the advent of nerve gases in 1945 that any major advance in the discovery of war gases occurred which had a marked bearing on doctrine (Chapter III, Section 4).

Although the characteristics of mustard gas had not changed fundamentally, our views on the danger it presented in its liquid and vapour form did change. In the 1914–18 war, the means of projection available were inefficient for direct contamination of personnel with liquid, although this did occur. The weapons of that war contaminated the ground and obtained their effect primarily from the effects of prolonged exposure to the vapour and only incidentally by liquid contact. Avoidance of the vapour in the static conditions prevailing was difficult.

Improved means of projection gave greater scope for direct contamination of personnel with the liquid, involving not only the probability of more certain and severe casualty action but imposing the necessity for the frequent changing of clothing if casualties were to be avoided. There was greater possibility of taking avoiding action against the vapour in the conditions of open as opposed to static warfare.

Our tactical employment of blister gas came to have as its primary object the contamination of personnel with the liquid—either directly by spray from aircraft or shell, or by bringing personnel into contact with liquid on the ground. The vapour action came to be regarded more as a bonus addition except where congested conditions of a static target lent themselves to vapour exploitation. Our perspective changed somewhat during the war and this is discussed later in this chapter.

The conditions of tropical gas warfare against the Japanese led to a swing of the pendulum since these conditions lent themselves to the exploitation of the vapour effects of mustard gas (Chapter III, Section 3).

Means of Projection

It was in the means of projection of gas that the greatest advances occurred after 1918 and it was this factor which had the major influence on changes in doctrine—in particular the advent of the Air Arm for the discharge of gas by spray or bomb. Artillery shell were greatly improved particularly in relation to direct anti-personnel effect. The 5-in. rocket, although by no means a first-class weapon, was an improvement on the Livens projector, owing to its greater range and greater flexibility (Chapter IV, Section 1). The means of projecting nose gases from generators, discovered but not used in 1918, was developed, but not exploited for the reasons given in Chapter III, Section 1.

Special equipment (the bulk contamination vehicle, the chemical mine and the hand contamination bomb) were developed for the contamination of ground to which we had access, but the requirement for these lapsed for reasons given later in this chapter.

Attention was also given during the war to the exploitation of certain gases in an anti-tank role from specially designed weapons (Chapter IV, Section 1).

In the Air Arm a gas weapon presented itself which, to a large degree, did away with many of the disabilities from which projection of gas from ground weapons suffers. The characteristics and development of air gas weapons will be found in Chapter IV, Section 2.

Conditions of Warfare Prevailing

At the time of our gas re-armament in 1936 we visualized a war involving periods of mobility interspersed with considerable periods of static warfare on the 1914–18 model in which the necessary weight of material could be brought up for the use of gas from ground weapons. Further, we had considerably overestimated the power of the defence. It was this general outlook that governed our pre-1939 gas doctrine.

The first six months of the war—the " phoney " war—did little to change this outlook.

The period after Dunkirk turned our thoughts to the exploitation of gas as a defensive and delaying weapon against invasion, a role to which it lent itself and to which our doctrine has always given full weight, but danger to our own civilian population imposed certain limitations as did the probable fluid nature of the course of operations. Anti-invasion plans have been discussed in Chapter I, Sections 3 and 4.

The conditions of the war as they developed in overseas theatres, the change-over to the offensive, and the requirements of amphibious and airborne warfare in connection with invasion of the Continent of Europe were a decisive factor in our attitude to the use of ground gas weapons. Such weapons were visualized as being only of value where weight for weight they were superior to H.E. Reliance for putting down heavy concentrations of gas in the forward area was placed in the R.A.F.

Intelligence on Foreign Offensive Methods

The fact that it was known from Intelligence sources that both Germany and Italy possessed " smoke " troops who could be used to discharge chemical weapons had some influence in justifying our own policy in respect of C.W. troops in the early days of the war. German smoke troops could, however, also fire H.E.—our C.W. troops could not.

The fact that we were the only country without a gas mortar had a bearing on our decision to introduce it into the armament of C.W. troops.

The capture of the German handbook, "Gasschiessen aller Waffen," led us to believe that the Germans had a H.E. chemical shell, "Green Ring 1," very probably charged with nitrogen mustard gas or some other practically odourless toxic material. We experimented with nitrogen mustard gas but did not consider it worth adopting (Chapter IV, Section 3). In 1945, captured stocks showed Green Ring 1 shell to be charged nitrogen mustard gas. Green Ring 3 was not mentioned in the handbook ; this charging was nerve gas and was not identified until stocks were captured towards the end of the war in Europe.

Intelligence on enemy methods, therefore, had little bearing on our offensive doctrine during the war, though it had considerable bearing on our defensive policy and training.

The German discovery of nerve gases, which only became known to us early in 1945, is likely to be a decisive factor in future development.

Summary of Outstanding Factors

The outstanding factors affecting doctrine between 1918 and 1939 were :—

(*a*) The advent of the Air Arm and the potentialities of aircraft spray.

(*b*) Improvement in the means of projection of gas from ground weapons.

(*c*) Our desire to exploit the use of mustard gas in its liquid form.

Between 1939 and 1945, doctrine was modified by a better appreciation of the real dangers of liquid mustard gas—particularly of aircraft spray—and of the value of mustard gas vapour under tropical conditions, and by a more balanced view on the respective roles of ground gas and air gas weapons under the prevailing conditions of the war.

SECTION 2.—DOCTRINE

Pre-War Doctrine

In discussing the employment of gas it is presumed that the reader is conversant with the peculiarities of the gas weapon, *viz.*, the characteristics of the war gases, the effects of weather and ground and the dependence on favourable weather conditions for discharge of gas, the necessity for surprise to achieve effect particularly with non-persistent gases against respirator-protected troops, the harassing effects of gas in relation to morale, the delay period after exposure before phosgene or mustard gas casualties develop, and the unsuitability of mustard gas for the immediate support of an assault.

Desire to exploit the direct anti-personnel use of mustard gas in its liquid form largely underlay our tactical doctrine and weapon development between the wars. The Air Arm in particular lent itself to this and, with a rather highly coloured picture in our minds of the casualty value of small drops of mustard gas, we pressed on with the development of aircraft spray. The " drenching " of columns of troops with low spray was visualized and it was thought that this would result in heavy casualties even among well-protected troops. Attacks from high altitude, particularly by night, in which the spray could reach the ground unseen and without warning stood out as the means of harassment *par excellence* in rear areas where it would also have a considerable casualty effect, particularly if gas discipline was poor. The use of high spray in forward areas was impracticable owing to the large area covered by the spray and the inaccuracy in aiming with its consequent danger to our own troops.

Difficulties in delivering high spray attacks, owing to the effects of wind on the fall of the spray, were appreciated, as were those in " aiming " ; these rendered it an indiscriminate weapon with which attacks could be carried out only against large areas where concentrations of troops were believed to exist. It was also argued latterly that the advantages to be gained from low spray attacks might be outweighed by the loss of aircraft in such attacks. The assumed potency of spray outweighed its limitations, and our pre-war view was that aircraft spray was the best air gas weapon for use against the enemy's forces in the field. (*See* Chapter I, Section 2 for gas plans before the invasion of the Low Countries by the Germans in 1940.) It was unfortunate that owing to the large areas involved in spray attacks it had been impracticable to carry out trials with actual mustard gas in this country. Our pre-war views were based on laboratory experiments with liquid mustard gas on limited numbers of men and on animals, and also on the spraying from aircraft of innocuous liquids with similar physical properties.

Although we were greatly impressed with the value of spray as a direct anti-personnel weapon, we fully realized the value of mustard gas bombs, but more for strategic than for tactical use. Heavy attacks—combined with H.E. and incendiary—against the enemy's vital chains of supply (ports, railway yards, essential war factories, etc.) were visualized. Here the intent was to exploit the effects of vapour as well as the liquid contact danger and to enforce defensive measures. Doctrine in relation to strategic air gas bombing has been discussed in Chapter I. Our views changed little during the period of the war.

The vapour danger arising from contamination of the ground in the forward area was fully appreciated, and the need to select alternative positions when on the defensive, which could be taken up within the time limits before exposure to vapour would be effective, was particularly emphasized in our defensive doctrine.

Our tactical doctrine envisaged the use of gas mainly from the air in the opening phases of a war. This included the use of high spray against rear areas, to harass and impose delay, which in this phase was visualized as likely to have far-reaching effects, particularly when combined with mustard gas bombing on enemy centres of production and lines of supply. Low spray attacks were to be directed against advancing columns.

When contact had been established, low spray was to be used against concentrations of troops in the forward area, whilst high spray and mustard gas bombing of the L. of C. continued.

Artillery gas shell was recognized by its nature as a harassing weapon. The tear gas shell was visualized as an assault weapon and a weapon for harassing (*see* Chapter IV). Both blister gas and tear gas were regarded as potent counter-battery weapons—in particular blister-gas which would have imposed the necessity of frequent moves of battery positions if vapour casualties were to be avoided. Harassment and casualty production would have occurred through liquid contamination of guns and ammunition. When the use of blister gas was undesirable for tactical reasons, tear gas combined with H.E. would have caused considerable harassment.

The use of chemical warfare troops armed with Livens projector charged blister or lethal gas provided the main weight of attack from ground weapons. Lethal gas would have been used to assist in preparation for the assault, but it was recognized that this assistance was in the nature of a bonus owing to its dependence on suitable weather conditions. Contamination of enemy defences either would have been carried out on areas which were not the immediate objective, with the object of partially isolating them, or, if used on the objective itself, would have been put down at least 24 hours or longer before an attack, to allow casualties to develop and the area to weather so that it would not be a danger to our own troops. This latter use would, however, have involved a nicety in timing under varying conditions of weather and terrain, and it was not regarded as a normal use for mustard gas.

In the withdrawal it was considered that mustard gas with its delaying effect came into its own ; it was to exploit this use that bulk contamination vehicles and mines were introduced for the contamination of large areas. Such contamination was visualized with and without demolitions. It was realized that contaminated ground presented no barrier to a vehicle and that contaminated obstacles must be kept under fire in order to be effective.

In all circumstances the use of gas in the forward area—in particular mustard gas owing to its persistence—would have been part of a prepared plan co-ordinated by higher formation H.Q.

Although our training doctrine included the projection of both lethal gases and nose gases by a variety of means including cloud attacks from cylinders and generators, our pre-war policy provided for mustard gas and tear gas chargings only for the ground gas weapons, and for mustard gas as the sole charging for air gas weapons (Chapter IV). The provision of lethal gas (phosgene) was left to be determined by post-mobilization requirements. No action was ever taken to provide cylinders, and nose gas generators were provided after the outbreak of war for training only.

Doctrine—1939–1940

In May, 1939, staff conversations with the French revealed divergence in doctrine. The French, unlike ourselves, had not perfected the means of spraying blister gas from aircraft and, although impressed with our claims, were dubious of the practicability of high altitude spray and whether it would not largely evaporate before it reached the ground. Our spraying technique had been developed on the basis of trials with mustard gas substitutes.

On the other hand, the French considered phosgene as immediately effective as H.E. and stressed that it was able to reach troops protected from H.E. Included in their policy was the provision of gas shell charged phosgene as in the 1914–18 war, whereas we considered such chargings totally unsuited to such small capacity weapons as artillery shell. The French, however, had a better appreciation than we ourselves then had of the value of phosgene from aircraft bombs.

The French were also much impressed with the potentialities of nose gases, and had developed a type of large generator (known as the "Engin Z") for their discharge. They, however, at the time had over-estimated the effectiveness of nose gases to a greater degree than ourselves. The results of our laboratory work and field trials converted the French to our view.

On the whole it appeared that we had advanced further in applying the use of gas to modern conditions than had the French. Our doctrine was based on the impracticability of maintaining the supply of non-persistent gases in the field in mobile warfare, and on the view that, weight for weight, mustard gas was the best gas for all weapons in most conditions.

As a result of these conversations we reviewed our requirements for phosgene and, as the French views tended to be endorsed by the early conditions of the war in the winter of 1939–40, phosgene was introduced as a charging for the Livens projector. It was somewhat later (October, 1940) that it was adopted for air gas bombs (Chapter IV, Section 2). In the case of the Livens projector we adopted a phosgene charging largely in the desire for an assault gas weapon, and it became the primary charging for the Livens, and its successor, the 5-in. rocket. The use of air (phosgene) bombs in this role was not finally decided until 1943.

Modification in Views on the Efficacy of Spray

Trials with actual mustard gas in the technique of spraying it from aircraft, which were held in the Sahara in conjunction with the French at the end of 1939, did not shake our faith in air spray as a weapon. They endorsed the fact, which the French had doubted, that it would reach the ground from high altitudes without evaporation.

More complete physiological experiment carried out during 1940–41, with a larger number of human observers than had hitherto been possible, showed that our earlier assessment of the effect of droplets of blister gas had attached too much weight to the size of droplet alone. It was shown that a minimum density of drops of any given size, involving a minimum weight of mustard gas, must fall on a man for him to be a likely casualty. Broadly about 10 times the weight of mustard gas was required to produce an effective casualty than had previously been assessed. In addition the casualty value of small blisters had been over-estimated. In 1943 when actual spray attacks with real mustard gas (as opposed to substitutes) were carried out against troops at Suffield (in Canada), this reassessment was fully confirmed.

High spray with its wide dispersion of the drops over a large area, its inaccuracy, and the consequent inability to guarantee hitting a specific target, would have been of far less use than had been hoped. It was unlikely to produce a worth-while return for the effort expended.

The heavy contaminations resulting from very low spray were still potentially valuable, but, as the technique of other forms of air attack against troops, vehicles, and tanks, with H.E. and rockets improved and developed during the war, it was questionable whether low spray would be worth the diversion of aircraft from other forms of attack, particularly in view of its disabilities from the R.A.F. standpoint and the infrequency with which worth-while targets for such attacks with S.C.Is. presented themselves (Chapter IV, Section 2). (*See also* Chapter I, Section 5, for views of 21 Army Group.)

Contamination of Ground by Special Contamination Vehicles

With the pace of modern war and the increase in the range of modern weapons such contamination, to be effective, would have had to be laid on a very extensive scale. Apart from the fact that the inclusion in our armament of a purely defensive weapon of this nature, which was of occasional use in the withdrawal only, became unjustified, supply and maintenance factors would have been prohibitive, and it was improbable that the vehicles could have been made readily available in sufficient numbers at the time and place required. Such contamination had no immediate stopping power and would only have been effective against infantry who were compelled to remain in the area. Unless very extensive and gross contamination were laid, involving a prohibitive effort even with special vehicles, the traversing danger would have been small even for infantry, and only of relatively short duration. The timing of the laying of the contamination, in relation to its contact by the enemy in order to obtain maximum effect from fresh contamination, would have been difficult, particularly as the vehicles were un-armoured. More worth-while results would have been obtained in defence or withdrawal by the engagement of selected targets and by the contamination of selected areas from the air, rather than by attempting to impose an extensive mustard gas barrier.

Ground Gas versus Air Gas Weapons

As the war progressed it was increasingly brought home to us that the only practical method of using gas on a heavy scale in the forward area was its use by the R.A.F.

The chemical warfare rocket companies with their existing armament (*see* Chapter IV), had proved totally unsuitable to modern conditions owing to transport and supply problems. A rocket attack with phosgene on an area 1,000 yards square required 500 men and 100 vehicles. Such an attack could put down 8,000 lb. of gas, of which about 5,000 would have actually been on the target area itself. One such shoot only could have been mounted on any one night on a corps front. In terms of heavy bombs this represented 64 250-lb. bombs.

The combined salvo of the two rocket companies of a C.W. group would have been required to engage a battalion area and if a C.W. group were allotted to a corps the total effect on a corps front was not great. The companies took up too much road space to be kept in the divisional area before they were wanted, and the laying on of a shoot required considerable preparation. They would normally only have been used at night owing to the dependence on suitable meteor conditions for the use of phosgene. The incidence of worth-while targets which it would have been practicable to engage in these circumstances in mobile phases was small. The employment of rocket companies against opportunity targets would have been exceptional ; where attempted, it would have been a race against time to bring them into action. Normally a rocket shoot would have been reserved for employment as part of a major fire plan and to provide a softener before an assault, but even so its dependence on suitable weather conditions made it unreliable and of a bonus value only. Overall the potential tactical incidence of the employment of rocket companies was too infrequent to justify their retention, particularly in view of the greater ease with which their tasks could be effected from the air.

The 4·2-in. mortar was, as has been pointed out in Chapter IV, in the main a harassing weapon. It could not have put down a high crash concentration of lethal gas ; with mortars the concentration has to be built up by rapid fire. Surprise effect thus lost before a high concentration develops admits of timely respirator adjustment by alert troops. To state this type of attack was harassing does not mean that no deaths would have resulted. However high the enemy standard of training and respirator drill a number of casualties is always probable. In the main, such attacks would have been harassing owing to continued threat of lethal casualties and the constant need for alertness.

With mustard gas chargings small areas of contamination could be laid around specific targets. Contamination of large areas with mustard gas would have involved the use of an uneconomical number of mortars. In general, therefore, the 4·2-in. mortar could correctly be called a harassing weapon in its gas role.

Artillery had always been recognized as a harassing weapon, since, by its characteristics, it cannot put down a heavy weight of gas attack in mobile warfare. Also, the use of gas would have been at the expense of H.E.

By 1943, therefore, we had reached the stage when we had to accept the fact that the laying of heavy concentrations of gas in the forward area must either be done by the R.A.F. or foregone under modern conditions of war. The role of ground weapons as a whole with the existing war gas and weapon development had to be accepted as harassing.(1)

At the beginning of the war, although the power of the air gas arm had been recognized, close support bombing with gas was not part of our doctrine. The R.A.F. were not fully equipped for this role (they initially had only mustard gas bombs), and, in any case, it is doubtful whether they could have accepted it.

By 1943 it was accepted that, where agreed on by the Army and Air Force commanders concerned in any theatre of war, the following types of gas attack could be delivered by aircraft on a corps front :—

(*a*) Choking gas attacks at night to assist the neutralization of objectives by causing heavy casualties before an attack.

(*b*) Contamination by day or night to such a degree as to enforce either the evacuation of the area or the acceptance of subsequent heavy casualties.

Doctrine at the End of the War

At the end of the war, therefore, the balance had swung away from the use of spray as the main air gas weapon against the enemy in the field, even the use of low spray being regarded as exceptional. If used, it would have been directed at troops in the open in reserve areas, and against columns on roads.

On the other hand, the onus of putting down neutralizing concentrations in the forward area had been transferred from the ground to the air. The main air gas role in support of ground operations would have lain in the use of mustard gas bombs in conjunction with high explosive to make more complete those tasks of interdiction upon which the R.A.F. were engaged. Heavy scale attacks on defended positions, forming-up areas, gun areas, higher H.Q., etc., would also have been delivered either with mustard gas or phosgene bombs, according to the circumstances. The disadvantages of the residual contamination from mustard gas in relation to the movements of our own troops have to be remembered. The criterion for the use of gas would have been that weight for weight it was superior to H.E.

There was little change in the gas role of Artillery, since the purpose of artillery gas shell had always been to harass. Tear gas shell remained available, but no great reliance was placed in them and they were abandoned on the conclusion of the war.

Note (1).—Harassing was defined as the use of gas with the primary object of affecting morale. The term " neutralization " described the use of gas " as part of or connected with the fire plan with the object of reducing the efficiency of enemy fire ". This led to some confusion. " Neutralization ", as defined, included anything from the crash rocket shoot, with its attendant high rate of casualty production and disorganizing effects, down to causing the adjustment of the respirator by the use of tear gas. Neutralization in its dictionary sense means " reduction to inactivity " and in most people's minds this meaning is implicit in the use of the word. With the chargings available during the war, to achieve neutralization in this sense was only possible from ground gas weapons with the Livens projector or 5-in. rocket. To apply the word to tear gas implies a misconception of the effects of this gas on troops possessing respirators. Blister gas with its delayed effects cannot achieve immediate neutralization in the proper sense of the word. Broadly, therefore, the effect of gas from ground weapons, after the abolition of C.W. troops, was " harassing ", and this term could be held to include casualty production and the imposition of delay by means of blister gas.

The use of gas as an anti-tank weapon and/or against small enclosed spaces has been referred to in Chapter IV and would have been exploited as experience of gas warfare had dictated.

The mortar companies of the divisional machine-gun battalions would have had the role of engaging small targets with phosgene, particularly at night, mainly to harass. In addition to the nature of the weapon, its scale of availability, 16 mortars to each division, would have restricted its use, in general, to a harassing role(1). The mortar companies would also have carried out limited ground contamination tasks where these were of a size and nature which did not warrant air action or where pin pointing from the air was difficult.

In defence or withdrawal, the laying of extensive mustard gas barriers was no longer considered a practicable proposition. In other respects our doctrine was little changed except in so far as there was greater flexibility in the selection and engagement of targets owing to the extended use of gas from the air in close support of ground operations. The use of gas was still generally assessed as favouring the defence.

Gas would not have been used indiscriminately ; heavy attacks would have been co-ordinated in a fire plan at higher formation H.Q.—normally not below Army. Harassing attacks with ground weapons would have been co-ordinated by Division within the general plan of operations. In all cases the achievement of surprise in point of time, place or method would have been kept constantly in view.

Doctrine for the Use of Gas in the War against Japan

This has been fully dealt with in Chapter I, Section 7. The employment of gas in amphibious operations is referred to there.

The Effect on Doctrine of the Discovery of Nerve Gases

The effect of the emergence of nerve gases on weapon development has been discussed in Chapter IV. It is not improbable that in the future more emphasis will again come to be laid on the use of ground gas weapons. With nerve gases, effective concentrations should be more easily achieved with practical munition expenditures than was the case with phosgene, and owing to their rapid action and the difficulty of recognizing them, they may prove a suitable assault gas, particularly for use in conjunction with H.E. All the same, the major gas effort is still likely to come from the air, owing to the greater administrative problems which are connected with the discharge of gas from ground weapons. Mustard gas should still retain a high place as a producer of large-scale casualties and general disorganization.

Note (1).—To obtain the requisite concentration with phosgene within the short time limit available if surprise is to be achieved, a minimum of 8 mortars to each artillery square (100 yards by 100 yards) is necessary. With the United States 4·2-in rifled mortar and its larger capacity bomb a concentration of 4 mortars to an artillery square is required. The United States ground forces had, therefore, a capacity for putting down neutralizing concentrations, in favourable meteor conditions, both by nature of the superior gas efficiency of their mortar and the increased numbers of mortars available in the chemical mortar battalions (36). In our case, neither our organization nor the gas performance of our mortar suited it for such tasks. From one standpoint the mortar had an advantage over air weapons : the element of surprise was likely to be greater than with air attacks where warning systems were more likely to keep the enemy on the alert. It is more difficult for tired troops to be on the alert against sudden attacks from ground weapons.

CHAPTER VI

DEFENSIVE EQUIPMENT

INTRODUCTORY

Broadly, we emerged from the first war with a respirator, a gas alarm system and a technique for gas proofing dug-outs and buildings in the forward area.

During the early part of the inter-war period, defensive research and development was mainly directed to improving the respirator. It was not until the early 1930s, when the danger of mustard gas attack from the air—in particular aircraft spray—came to be realized, that the numerous items of anti-gas equipment for warning, detection, and delimitation of attacks, as well as for protection and decontamination, began to see the light of day.

Our defensive doctrine underlying the scales and types of equipment provided was based on the necessity for combating the means of gas attack visualized in our offensive policy as described in Chapter V. Scales of anti-gas equipment and reserves, which did not change greatly during the war, and their carriage in the field, are discussed in Section 4 of this chapter. There was some lack of balance in the early days of the war in the degree of protection provided against gas as compared to that which could be afforded against other battle hazards. Although a better perspective prevailed at the end of the war, it was felt that, pending gas warfare experience, the scales were, overall, a reasonable answer to the risks involved, and that no major modification was worth while. Economies were effected by a reduction in War Office reserves and reserves held in overseas theatres.

The provision of this equipment involved a major production effort. Its magnitude—and its cost—can be gauged from Appendix XXI, in which the totals of equipment for all Services, which passed inspection in the United Kingdom between 1939 and 1945, are shown. In addition much equipment was manufactured in Canada, India and Australia.

It was an insurance policy which, as has been discussed in Chapter I, had a deterrent effect on the enemy initiating gas warfare, but for which the premium in money and effort was high. Initially provisioning was somewhat lavish and based on high rates of wastage—in particular in so far as wastage during " gas peace " was concerned. This was, however, offset during the war by adjustments in wastage rates and the elimination of certain reserves. In 1942, the production of many items of anti-gas equipment ceased and in other cases was reduced (*see* Chapter I, Section 5). By the end of the war, stocks had been reduced to a minimum compatible with safety. Viewed over the whole period of the war, therefore, the premium for the insurance policy was no higher than was necessary.

The majority of our anti-gas equipment, which was common to all Services, was developed before the outbreak of war ; fundamentally, design changed little during the war. Minor improvements were incorporated as a result of user experience, or of new risks engendered by possible enemy war gas development, and in some cases a number of inferior substitutes had to be incorporated during manufacture owing to the shortage of raw materials.

There were, however, a number of outstanding developments during the war, namely :—

(i) Improvement in the range of protection afforded by the container of the G.S. respirator.

(ii) The introduction of the light respirator.

(iii) The introduction of a device to improve speech transmission when the respirator is worn.

(iv) The introduction of impregnated clothing (it had been developed before the war).

(v) The introduction of protective dubbin for boots.

(vi) The introduction of more efficacious ointments, in particular from a preventive, *i.e.* prophylactic, standpoint, and from the point of view of their suitability under tropical conditions.

(vii) The introduction of the pocket vapour detector.

(viii) The protection of tank crews. This was a requirement that was not fully met.

Up to April, 1945, development of equipment, other than the respirator, had been mainly directed to combating the risks engendered by mustard gas. With the advent of nerve gases another major hazard has been introduced, the implications of which have been discussed in Chapter III, Section 4.

No attempt is made in the ensuing sections of this chapter to give either a full description of each item of equipment or the detail of its use in the varying circumstances in which gas may be encountered. This will be found in " Gas Training, 1942 " and in " Advanced Gas Training, 1943 " and certain other relevant pamphlets. Only such detail is given as is necessary for the understanding of war-time development, the necessity or otherwise for retention of an item in post-war equipment, and for purposes of comparison with Allied and enemy development.

The equipment is discussed under the following headings :—

Section 1.—Personal and Vehicle Equipment.

Section 2.—Unit Equipment.

Section 3.—Miscellaneous Equipment not issued on a Unit Basis, and Animal Equipment.

Section 4.—Scales of Equipment and their Carriage in the Field.

Section 5.—Defence of A.F.Vs. against Gas.

SECTION 1.—**PERSONAL AND VEHICLE EQUIPMENT**

Respirators

The General Service Respirator

A description of the different types of facepieces, containers, respirator haversacks, and anti-dimming outfits current on the outbreak of war, will be found in " Protection against Gas and Air Raids—Pamphlet No. 2—Respirators —1939 " (Reprinted with Amendments (No. 1), 1940).

In so far as components other than the container are concerned, little development occurred during the war.

It was during the inter-war period that a major effort had been directed not only to providing a respirator with a container which gave protection against a wide range of war gases, and in particular against the particulate gases, *e.g.* nose gases (*see* Chapter III, Section 1), but also to overcoming those factors in facepiece design which were such a handicap to Service efficiency in the final type of respirator introduced at the end of the1914–18 war, *i.e.* the Small Box respirator. The facepiece of this respirator with its noseclips, mouthpiece, lack of facial ventilation, and the necessity for a tightly drawn head harness, not only made it uncomfortable to wear and particularly harassing if worn for long periods, but imposed a severe handicap on the performance of normal duties in the field. The range of vision was poor, the use of optical instruments almost impossible, and speech was both indistinct and difficult.

In the facepieces current at the beginning of the war(1) these handicaps had by comparison become eliminated. Speech transmission, including speech over field telephones, was good, although still open to improvement. Optical instruments could be used ; range of vision was good ; eyepiece and facial ventilation had been introduced, and the facepiece rested on the face with the minimum tension of the head harness, mouthpieces and noseclips having long been abolished. The sum total of this resulted in greatly increased comfort in wear.

Other than a relaxation in the specification of the rubber mix in order to allow the use of a percentage of reclaimed rubber, which was unavoidable owing to the rubber shortage, there was no development of the G.S. respirator facepiece during the war. A rubber tube without a stockinette covering was introduced for the connecting tube.

Note (1).—These were the Mk. IV (normal issue), the Mk. IV Special T (for persons with abnormal face contours and for spectacle wearers), and the Mk. IV Special T Mic. (which had a mounting on the left cheek for a microphone adaptor for use with wireless sets and power telephones). The Mk. V, which was getting into production when the war started, was introduced partly for production reasons, but also incorporated all improvements of the Mk. IV, including those of the Mk. IV Special T and Special T. Mic. It was made in the normal size only, persons requiring other sizes wearing the applicable type of Mk. IV.

Facepieces fitted with a long connecting tube, as used by the Navy to admit of the method of side carriage which they adopted, were issued to certain Army personnel, *e.g.* anti-tank units R.A., coast defence units R.A., and A.T.S., for whom this method of carriage was more convenient. " Chairborne " staffs also favoured this method.

A later Mark haversack than those described in the respirator pamphlet referred to above—the Mk. VII—incorporated improvements in design of the haversack sling and the whip cord attachment.

A respirator microphone was available for use with the G.S. respirator by certain personnel. Although speech with this respirator was adequate over the field telephone, it presented certain difficulties for switchboard operators and was inadequate over wireless. Officers, switchboard operators, wireless and driver operators of the Royal Artillery, Royal Signals, and Royal Armoured Corps were therefore issued with a respirator microphone which could be attached to the adaptor in the facepiece of the Mk. IV Special T. Mic. or the Mk. V. When not in use, the adaptor was fitted with a plug. There were two types of respirator microphone, the No. 1 for telephone switchboards, and the No. 2, which was suitable for either telephone switchboards or for use with wireless sets. With the introduction of a redesigned outlet valve incorporating an improved speech transmitter, referred to later in this chapter, the respirator microphone became redundant. This redesigned outlet valve was interchangeable with the existing outlet valve of the G.S. respirator facepiece, but it was never fitted, since by the time it was available the Light Respirator was also available and Light Respirators fitted with the redesigned valve and speech transmitter were issued to all personnel whose duties demanded superior speech performance.

The G.S. Respirator Container

It was in the container that a number of improvements were made during the war. A brief résumé of inter-war development is necessary in order to understand the reason why varying types of container were held in the Service on the outbreak of war.

The policy in the early post-war period after 1919 was to have two containers, one for training and the other for mobilization. The training container was the Type A, a modified version of the small box container, containing only charcoal and a very primitive particulate filter. Manufacture and issue of this began in 1922. It gave good protection against all gases except nose gases.

The mobilization container (eventually known as Type D) was a modification of the improved container developed and produced in 1918 ready for use in 1919, but which was never issued. Stocks were retained as the mobilization container and the Type D was produced as replacements were required. There was some modification in the charcoal filling and it contained an improved type of filter. It gave excellent gas protection and a good degree of protection against nose gases.

As the methods of generating nose gases improved, however, it became necessary to improve the particulate protection. In addition, the mobilization container was required to have a storage life of 15 years. These requirements, together with a demand for a low resistance to breathing, led to the introduction in 1929 of the container Type E, Mk. I, as the mobilization container. For some time it was manufactured and held under stringent conditions of security because the filter was thought to be an advance on the filter development of other nations. This assumption subsequently turned out to be wrong and the restrictions lapsed in 1936.

Between 1929 and 1939 improvements were introduced which were directed towards the use of materials more readily available in large quantities and to mass production methods in the event of emergency ; by September, 1939, the current container was the Type E, Mk. V. The Type A was still in use for training and there were some obsolescent type D containers.

The arsine scare of October, 1939, the circumstances which gave rise to it, and the subsequent justification of our original views on the inefficiency of arsine as a war gas, have been referred to in Chapter III, Section 2. Since neither we nor the French had been able to devise efficient means of using arsine for offensive purposes, our protection against such gases had been considered less important than against those which were known to be capable of effective use. The current charcoal filling of the Type E, Mk. V container gave inferior protection against arsine, but it was found that specially treated charcoal gave enhanced protection. Pending the introduction of a container filled with this charcoal, an emergency container filled with lime permanganate granules to improve the protection against this gas—the Type EA container—was produced and issued to all troops in France within a few weeks. It was a shallow container of the same cross-section as the Type E container fitted with a spigot at each end. It was used in addition to the existing container by cutting the connecting tube in half and inserting the EA container between the two halves and re-wiring them at either end. It is illustrated at Appendix XXIII. This was only an emergency measure and production of the Type E Mk. VI container which contained the specially treated charcoal was immediately started. In addition to providing improved protection against arsine the Type E Mk. VI container had an improved filter which had been under development when the war started.

The Type E Mk. V container was in every way a serviceable container. It was painted buff. The Mk. VI was painted red. The publicity given to arsine and the issue of emergency protection led to loss of faith in the Mk. V (buff) container, and troops considered they were inadequately equipped unless they had the red container. This was to be regretted, since an immediate changeover to the red container was impractical for production reasons and the buff container for all practical purposes afforded first class protection except that this protection was inferior against arsine—a gas which, in the end, proved technically too difficult to use in the field.

The Type E Mk. VI remained the standard container for the G.S. respirator until the end of the war. It was found that further special treatment of the charcoal filling for containers gave enhanced protection against prussic acid and cyanogen chloride(1) ; charcoal treated for protection against prussic acid first became available, and later against both these gases. These types of charcoal were only filled into the Light respirator container. The G.S. respirator was declared obsolescent in January, 1943, and all production gradually turned over to the Light respirator. G.S. respirators withdrawn from troops re-equipped with Light respirators were used to maintain those G.S. respirators remaining in service which, by the end of the war, were mainly in the hands of non-field force or L. of C. and base units in all theatres.

Note (1).—The efficacy of these gases as war gases has been discussed in Chapter III, Section 2.

THE LIGHT RESPIRATOR

A description of the Light respirator, its care and methods of carriage, will be found in " Gas Training, 1942," Appendix F, inserted by Amendments No. 1 of July, 1943.

As initially developed, the protection afforded against particulates, *e.g.* nose gases, by the Light respirator container, was somewhat inferior to that of the G.S. type, and the period for which protection was afforded against all gases was less. It was accepted for special troops, *e.g.*, Commandos and assault troops, for whom lightness and ease of carriage were of greater importance than a long period of protection against gas, the degree of nose gas protection being considered adequate.

Owing to the absence of a connecting tube, less rubber was used in its manufacture than for the G.S. respirator, and in 1942, under the spur of the rubber shortage, a requirement was stated for an improvement in the standard of protection to more nearly the equivalent of the G.S. respirator, with the object of adopting the Light respirator as the standard respirator for all arms.

A new type of filter which met the requirement, and which was readily producible in bulk, was developed, and it was decided to introduce the Light respirator as the standard service respirator. Issues were made in priority to airborne troops, special service troops, troops of the Armoured Corps and of other Arms of the Field Forces, 21 Army Group being completed, including reserves, before " D " day.

The period for which the container afforded protection still remained somewhat less than in the case of the G.S. container. This factor was of no great practical significance against concentrations likely to be met with in the field and could, in any case, be offset by increasing the reserves of containers carried. Since the Light container was smaller than the G.S. container, such increase in reserves did not involve any additional carriage problems.

The container in use from 1943 onwards was the L. Mk. II, the earlier L. Mk. I in the hands of troops being withdrawn. In April, 1943, the L. Mk. II (White Dot) container, which gave superior protection against prussic acid, was introduced for tank crews who were the most likely target for this form of attack (*see* Chapter IV, Section 1). Ultimately, in August, 1944, a container filling which gave superior protection against both prussic acid and cyanogen chloride was put into production. No new Mark number was allocated to containers with this filling but they could be distinguished where necessary by the letter embossed on the container.

With regard to the service handicap factors, the Light respirator was the equal of the G.S. respirator as far as speech, range of vision and comfort in wear were concerned, and superior as regards lightness and ease of carriage. There was, however, a tendency for the respirator to slip on the face during periods of intense physical activity, particularly under hot sweaty conditions, to the detriment of gastightness.

For personnel who required superior speech performance, it was fitted with an outlet valve incorporating an improved speech transmitter. (*See* below.)

The Light Respirator Haversack Mk. I was designed especially for assault troops, with a view to minimum weight and size. It had a number of unsatisfactory features and was adversely reported on by users. A Mk. II design was in production by April, 1943, which eradicated the worst faults. In July, 1943, a General Staff requirement was passed to the Ministry of Supply for an entirely new design. Two patterns were offered, each in a range of four different fabrics which were put out to troop trials in all theatres in 1944. In January, 1945, a new design was finalised, No. L.2 Mk. I, but production had not started when the war ended and all production of anti-gas stores closed down. As an interim measure for troops proceeding to the Far East, the canvas and thread of the Mk. II design of haversack had been tropic-proofed and all brass fittings stove enamelled.

Improved Speech Transmitter for Respirators

In 1943 a speech transmitter which gave greatly improved speech performance was developed by the Chemical Defence Experimental Station at Porton for incorporation in a redesigned outlet valve. Another device known as the Tannoy voice valve had been developed previously by Tannoy Products, Ltd. on their own initiative but was less satisfactory.

Speech with the redesigned valve was so much improved that respirator microphones were no longer necessary over telephone systems or wireless. The new valve was incorporated in those light type respirators issued to all officers, to Royal Signals personnel, to Regimental signallers, and to all driver operators. Issues on this basis to 21 Army Group were completed by May, 1944. Issues to other overseas commands then proceeded on a similar scale.

It was left to decide from user reports and from experience of its robustness in the field whether this type of valve should be incorporated in all light type respirators. A decision had not been reached by the end of the war but was so taken shortly after.

Other Types of Respirator

Other types of respirator which existed were the civilian duty respirator, the general civilian respirator, children's respirator, and a respirator for infants. These were issued to the civilian Services and the general public as applicable. Every man, woman and child in Great Britain and Northern Ireland was provided with a respirator. As items of equipment, the Army was not concerned with these except in so far as they were issued to War Department civilian employees and Service families at home and overseas.

A respirator for hospital cases who could not wear the ordinary respirator was also developed.

The Anti-Dimming Outfit

The outfits in use at the beginning of the war (Mks. III, IV and V), consisted of a small cylindrical box containing separately the anti-dimming composition and a rag for its application. In 1943, the Mk. VI outfit was introduced, which consisted of an impregnated cloth in a small screw-top tin.

Comparison between British and United States Respirator Development

Standards of protection of United States respirators were of a similar order to British. Fundamentally, they were similar in design, the General Service respirator having a facepiece, connecting tube and container, and the assault respirator being modelled on our light respirator. The main difference lay in the line of approach to the problems of the requirements of varying users. British policy aimed at providing an article common to all Services and all users. The United States entered the war with three types of General Service respirator for the Army and Army Air Forces—one (the Mask, Gas, Service) for non-specialists, another (the Mask, Gas, Diaphragm) with superior speech performance for a proportion of officers and N.C.Os., and a third (the Mask, Gas, Optical) for individuals using fire control instruments. A training respirator was also issued on a limited basis.

The general use of the diaphragm mask was eventually discontinued and lightweight models of the Service and optical masks were introduced. The assault mask was designed primarily for paratroops and troops participating in amphibious and jungle operations.

The U.S. Navy had their own separate types of masks fitted with head-type containers.

Certain British and American components were interchangeable, *viz.*, the containers and facepieces for the British Light and the U.S. Assault respirators. British G.S. containers could also be fitted to United States connecting tubes or " hoses ". British respirators could not be carried in United States haversacks nor would U.S. respirators fit into British haversacks.

German Respirators

Standards of protection were of a similar order to those given by British and American respirators. The German Army respirator did not have a connecting tube, a drum container being fitted to the front of the facepiece. A proportion of containers were of a non-magnetic type for use with directors. Facepieces were initially made of a composite fabric material, but from 1940 onwards a rubber facepiece was introduced which was carried in a metal respirator carrier, except for parachutists, etc., for whom a canvas haversack was provided. The Germans also had an optical mask, with a box container and connecting tube, and an anti-gas hood for head wounded. Civilian respirators were available largely on a repayment basis, but in some especially vulnerable areas they were issued free. By no means were the entire population in possession of respirators ; there were no issues in rural areas and only partial issues in urban areas.

A development of interest was the plastic emergency mouthpiece for attachment to the spare container of the standard service respirator. It was for protection against rapid acting gases (such as AC) in sudden concentration. It was intended for use in conjunction with a nose clip, and, in conditions of C.W., as soon as aircraft gas attack was suspected the soldier would have breathed through the spare container. There was only small production and it never became a general issue.

Future Respirator Development

War-time respirators provided protection against nerve gases, but owing to the low concentrations in which these gases are effective it is important that the gas tightness of the facepiece should be perfect. Owing to the difficulty in detection and the insidious character of these gases, the wearing of the respirator for very long periods may be necessary and a further reduction in the service handicap factors must, therefore, be aimed at, *i.e.*, decrease in weight, increase in comfort, minimum disability when worn, even better speech characteristics, and the minimum loss of identity to the wearer. Respirators must also protect against the possible use of biological agents.

Broadly the types of respirator existing at the end of the war—the General Service and Light Types—have reached their final stage of development in their present form, and a return to fundamental research will be necessary to provide the most efficient form of respirator for protection against both chemical and biological agents.

Eyeshields

Eyeshields protected the eyes against liquid droplets but not against vapours or gases. When introduced in 1937, they were intended primarily to protect the eyes against the hazards of aircraft spray, in particular against high altitude attacks in which the spray would reach the ground without warning. The continuous wearing of the respirator to counter this risk would have been impracticable. Teaching was that, when gas warfare started or was imminent, eyeshields would be worn at all times when not under cover. They also protected the eyes from droplets arising from any type of liquid gas shell or bomb on burst. During the war, as the likelihood of high altitude spray attacks receded, another hazard to the eye arose, namely the German development of the H.E./chemical type of shell.

Eyeshields were a cheap expendable store ; fundamentally the design remained unchanged throughout the war. A number of modifications were introduced to increase comfort in wear, and latterly the elastic band for attaching the eyeshield to the head was replaced by a tape and elastic band. An improved carton was also introduced.

The use of eyeshields for anti-dust purposes, if so desired, was permitted. Tinted eyeshields were introduced for tropical and sub-tropical theatres to increase comfort by combining anti-glare with anti-gas and anti-dust protection. They did not take the place of anti-glare goggles for specialists, *e.g.*, A.A. gunners. They were issued in the proportion of two tinted to four plain.

The U.S. Forces, when they entered the war, were not provided with this item of equipment. To cover the risk of liquid in the eye, they relied on the rapid adjustment of the respirator on warning of attack. They were impressed with the British eyeshield and rapidly saw the need for it. They adopted a pattern identical to the British type.

The Germans latterly introduced a set of anti-gas eyeshields, two plain, two coloured—more in the style of motor cyclists' folding goggles. Their secondary role was as dust, sun or snow goggles, but from their elaborate nature the so-called secondary role appears to have been the main consideration.

A drop of liquid nerve gas in the eye can be lethal ; this emphasizes the necessity for the continued inclusion of eyeshields in our anti-gas equipment.

The Cape, Anti-gas

In the early 1930s, as the potentialities of aircraft spray came to be appreciated, the need arose for some form of rapidly adjustable protection for the body. Impregnated clothing (*q.v.*) showed no probability of affording adequate protection against liquid droplets. Protection by means of a cape was developed, early designs being of a sleeveless variety which permitted of purely passive defence by rapidly covering the body with it. Troop trials during collective training in the middle 1930s brought out that the adoption of solely passive measures of defence was unacceptable. The threat of spray attack could not be allowed to hamper movement and it became obvious that the best means of protection against low spray attacks was efficient anti-aircraft action. A sleeved cape for all ranks was therefore introduced so that anti-aircraft action could be taken while gas protected. This design had also other obvious advantages.

Some consideration was given in 1935 to a combined anti-gas cape and water-proof sheet, but a design was not forthcoming which did not involve some sacrifice in efficiency in both roles, and the spray danger was then considered so grave that no loss of anti-gas protection was acceptable.

The sleeved cape was made of light oilskin. Originally, it was of a light yellow colour, but this was changed to drab, and later, by the beginning of the war, an olive green and brown camouflage pattern was available.

Once a sleeve pattern had been decided on, there was little fundamental change in design. A pocket was added on the outside to hold a tin of anti-gas ointment and press studs were fitted in such a way that the forward ends of the cape could be drawn back to give ease in marching. In 1943, a special cape was introduced for stretcher bearers which had press studs so arranged that the front and rear edges could be drawn together between the legs to facilitate kneeling and the handling of gas casualties.

The cape was the main item of personal equipment for the protection of the body against liquid blister gas. Although originally introduced to cater for the spray risk, it could be used against all liquid contact hazards, whether from bursting projectiles, or from handling contaminated equipment, or from contact with liquid on the ground. Methods of dealing with the cape when contaminated were included in the drill for personal decontamination, and these, as well as the methods of using the cape, will be found in " Gas Training, 1942."

It is still improbable that impregnated uniforms can be developed to protect against all liquid gas hazards ; the retention of a form of cape in post-war equipment will be necessary, but it will require modification in design.

Although the war design can be used as a water-proof—and it was popular in this role—there were limitations in its use for anti-weather purposes, since it had been designed solely as an anti-gas article, and other water-proof equipment also had to be provided. It was insufficiently strong and unsuitable in shape for use as a ground sheet, or as part of a bivouac shelter.

In the future, the standard of gas protection must be balanced against the administrative difficulties of maintaining a variety of anti-weather equipment in the field.

A garment which will serve as a water-proof, a ground sheet, a bivouac shelter and a gas cape is required. The materials should be such that it is durable in wear for general use and, as far as its gas resistant properties are concerned, ease of decontamination rather than long resistance to penetration of liquid gas should be aimed at.

In 1945, a garment of this nature—the poncho—was a requirement for troops earmarked for the war against Japan, but VJ Day came before the project was implemented.

The Cape, Cotton Porous

This was introduced and manufactured only in India. The Indian type was based on a design developed experimentally in the United Kingdom which was made wholly of pervious material, *i.e.*, twill, having a "Tippet" to reinforce protection in the shoulder area. The principle of protection was to have an air space between the cape and the clothing underneath. This automatically occurred where the cape was not in close contact with the clothing and was created artificially in the shoulder area by means of a piece of rope sewn criss-cross over the shoulders of the body of the cape to separate the "tippet" from it. The material was such that light contaminations of mustard gas, other than the HT type, weathered out of it rapidly, particularly in hot weather.

Trials in India showed that when this cape was worn over unimpregnated tropical clothing, penetration by even small drops in the area of the upper arm was almost immediate.

In the Indian pattern, therefore, oilskin patches were added on the upper arms to reinforce protection, the tippet also being made of oilskin. These items were detachable for the reason that, in the early stages of production, insufficient oilskin was available, and a wholly porous cape was produced with which the oilskin components were matched up as they became available. The Indian Cape also had a hood attached to it which was the counterpart of the curtain anti-gas (*q.v.*).

It was introduced in India in the spring of 1941, largely owing to the scarcity of oilskin. Manufacture of light oilskin was only just beginning as an entirely new project, and production forecasts were so unpromising that there was no alternative to adopting the porous cape if the Army in India was to receive protection of any sort within a reasonable time. Also, the oilskin cape was then considered too hot to wear in a tropical climate. At this juncture, the entry of Japan into the war was not foreseen and eyes were turned westwards from India as the battle area.

For use in the rainy conditions of Burma and the Arakan, this cape became a liability, since it afforded no weatherproof protection and would have had to be carried in addition to water-proofs. In conditions of fighting as they developed, where every additional ounce of weight on the man was a consideration, it was an encumbrance. Early in 1943 the porous cape was made obsolete as a war store and relegated to a training substitute for an oilskin cape of dark olive colour, which was introduced in lieu.

A disadvantage of light oilskin was its fragility. Under jungle conditions it is easily torn, with a consequent high rate of wastage. The ultimate policy was to withdraw the oilskin cape to reserve for use if gas warfare started, but the overall shortage of water-proof raiment precluded this in practice.

At the end of the war, policy in India, as in the United Kingdom, veered towards the provision of a durable multi-use type of garment, such as the "Poncho" (*see* Part V, Chapter XXVII, Section 1 in Volume II).

United States Forces' Cover, Protective, Individual

The U.S. Forces introduced a "Cover, protective, individual." This was a specially designed accordion folded bag, packed in a sealed packing. When unfurled, it fitted over the individual's head and body down to the ankles. The upper half was transparent cellophane, the lower half being coloured olive drab. Owing to the passive nature of defence against spray engendered by its design and method of use, it did not appeal to us. The material used became brittle and tore in cold weather. A winter type was introduced, which was reinforced with rayon scrim to prevent tearing. It was issued on a scale of 2 to each individual and carried in the respirator haversack. It is illustrated at Appendix XXIII.

German "Gasplan"

The Germans had a "Gasplan," or anti-gas sheet, about 6 ft. by 4 ft. initially made of specially treated paper. Later types were made of material coated with synthetic rubber or of plastics of the nylon variety. It is illustrated at Appendix XXIII. Protection was obtained by crouching or kneeling under the sheet. Here again the approach to the problem of individual protection against spray attack was of a passive nature.

The Curtain, Anti-Gas

The curtain anti-gas fitted over the steel helmet and hung round the neck, being buttoned under the chin. It was provided to afford protection to the neck against aircraft spray—particularly low spray. It was first introduced in 1940 for A.A.L.M.G. teams of all units, and later extended to all personnel to whom the cape, anti-gas was issued. In 1941 it was withdrawn to Ordnance charge, except in the case of A.T.S. employed as instrument numbers on A.A. sites. It was in large degree an over-insurance, and in any case the advent of ointments suitable for prophylactic use makes it redundant.

Anti-Gas Ointments

The requirement for protective clothing to counter liquid mustard gas hazards was paralleled by that for a handy and portable means of decontaminating the exposed surface of the body not covered by clothing, and also for decontaminating personal weapons.

For this purpose anti-gas ointment was introduced. For all ointments the time limit after contamination, within which the ointment must be applied if it is to be effective, is governed by the rapidity with which liquid blister-gas penetrates the skin rather than by the properties of the ointment. For practical purposes immediate application after contamination is the answer, although in temperate climates there is a safety margin of up to five minutes against mustard gas, which is reduced to even less than two minutes in the tropics. The safety margin against lewisite is nil.

The first ointment introduced was Ointment, Anti-gas, No. 1. It was effective against liquid mustard gas only; it was irritant after repeated applications, and corrosive to the metal portions of weapons if left on for too long. Being irritant it could not be used prophylactically to protect the skin against mustard gas vapour. It was issued in a two-ounce lever-lid tin. In 1939 it became obsolescent and was superseded by Ointment, Anti-gas, No. 2. This was an ointment in a vanishing cream base which was effective against both mustard gas and lewisite. It was far less irritant than No. 1, but not by any means non-irritant. It was possible to use it prophylactically against vapour, but it was irritant to those parts of the body where the skin is more delicate. This was emphasized under tropical conditions where it was particularly irritant if applied to the scrotal area. It could be used instead of gloves for handling contaminated articles, if a visible layer was left on the hands and subsequently removed. It was slightly corrosive to metals, and its stability in the tropics was inferior. It was issued initially in a two-ounce porcelain jar with a screw-on lid, but later packing was eight ¼-ounce "toothpaste" tubes in a tin box. This latter method of packing was adopted for personal issues of all subsequent ointments.

Scientific development of ointments had for long aimed at producing an ointment which was suitable in all climates, entirely non-irritant, and of high prophylactic value for long periods after application, and which could therefore be used repeatedly and extensively for protection of the skin against blister gas vapours. The requirement for this was enhanced by the greater aggressiveness of mustard gas vapour in hot humid conditions as evidenced by trials in tropics, and also by the need for protection against the "initial cloud" effect of the German H.E. chemical shell.

A series of ointments—Nos. 3, 3A, 5 and 6—were introduced successively. These ointments contained Antiverm, the chemical used for anti-gas impregnation of clothing (*see* below "Impregnated Clothing"). Nos. 3 and 3A (the tropical form of No. 3) were incorporated in a fatty base, and Nos. 5 and 6 were of the vanishing cream type. They were all superior to No. 2 in that they were non-irritant and less corrosive.

No. 6 was the best ointment developed by the British for both decontaminating and prophylactic purposes under all conditions including hot humid conditions. Used prophylactically, where the skin is hot and sweaty, the ointment tends to get washed off and loses its value in less than half an hour. It is also rubbed off skin by clothing, particularly round the neck. Frequent re-application is therefore essential. The American M.5 ointment was slightly superior to the British No. 6 in prophylactic value but the mechanical factors still applied.

Many valuable trials with a wide range of experimental prescriptions were carried out in Australia in an endeavour to find an ointment which would give protection for long periods after application in hot sweaty conditions, but this was not attained.

For the protection of exposed surfaces under such conditions other means were considered (*see* Impregnated Clothing in Tropical Theatres—impregnated hoods). In cool conditions, or where the hazard from blister gas vapour is known to be only slight, ointment afforded reasonable protection.

The American ointment M.5, although not so pleasant in use as the British ointment, was in some quarters thought to be the best ointment produced by the Allies, particularly for prophylactic use under hot humid conditions, but opinion was divided. It was decided to obtain M.5 ointment for all British troops operating in the Far East. The Americans, however, did not consider M.5 sufficiently superior to the British No. 6 to warrant the increased production necessary to meet all our demands. At the time the war with Japan ended the position was that formations sent to Burma and the Far East from the United Kingdom would have been equipped with the American M.5 ointment, but troops based on India would have had British No. 6.

The Germans and Japanese had bleach tablets or powders for mixing with water. The German " Losantin " tablet was an efficient decontaminant but could not be used prophylactically. Not being ready mixed it was less quick to apply than ointment. It could not be used on weapons and a separate weapon decontaminant was issued. The German weapon decontaminant was put in " squeeze side " pocket flasks for personal issues, an outfit for larger weapons being issued in a tin case which also included cleaning oil, rags and brushes. Allied development of personal decontamination equipment was unquestionably superior to the enemies', although enemy equipment was simple to produce.

The advent of the nerve gases raises a requirement for an ointment that will be effective against both blister gases and nerve gases. A return to fundamental research is necessary in order to obtain an ointment that is stable in all climates, non-irritant, effective against liquid and vapour, will not be mechanically removed by sweating or the rubbing of clothing, and will be effective on weapons, clothing and equipment as well as on the skin.

BAL Ointment

Liquid lewisite in the eye is more likely to lead to total loss of sight than liquid mustard gas. Irrigation of the eye with water, which in the case of liquid mustard gas is a suitable treatment if applied early, is useless in the case of liquid lewisite. Research at the Chemical Defence Experimental Station at Porton, and by Professor Peters and his team at Oxford University, led to the development in September, 1943, of a substance designated British Anti-Lewisite (BAL), the adoption of which was recommended by the Panel of Ophthalmic Specialists for acceptance by the Services as a personal issue. Originally it was used in droplet form, but an ointment for easy application was developed.

Theoretically, the use of the eye shield should totally obviate any liquid risk to the eye, but gas discipline will not invariably be so good that the shield will always be worn when it should, particularly in hot climates where the continued wearing of it is more irksome.

There was no requirement for BAL ointment as a personal issue in the European theatre, because the use of lewisite by the Germans was unlikely, and the issue of yet another item of personal anti-gas equipment which had to be maintained in the field was deprecated. It was agreed to as a medical store in 4-oz. jars.

The Japanese, however, were known to possess a mixture of mustard gas and lewisite, and there was therefore a potential requirement for BAL in the Far East as a personal issue. By July, 1944, trial production orders had been completed and the method of packing decided. A small tin, holding three 3-gramme tubes, was decided on, the 3-gramme tube to be of a size to fit the U.S. M.5 ointment outfit. At this time it was proposed to obtain the American M.5 outfit (which included a small tube of BAL) for all troops in the Far East. When the United States were unable to meet our full demands for these outfits large-scale production of BAL ointment was reconsidered, but before production could be started the war ended.

Although BAL was a product of British research, it was the Americans who produced it on a large scale as a personal issue.

Impregnated Clothing

A/V Battledress

A number of impregnites and processes were experimented with and shortly before the war an efficient impregnite was found to which the name "Anti-Verm" was given. It had no anti-vermin properties, the name being a security cloak.

Since the amount of impregnite which can be held in a given area of cloth will only neutralize a given amount of liquid mustard gas and the impregnite is destroyed in reaction with it, A/V clothing, although giving a high degree of protection against mustard gas vapour, afforded little additional protection against liquid droplets as compared to un-impregnated serge. The severity of the burns, however, from the larger droplets would have been reduced.

Serge battle dress impregnated with anti-verm had good stability in storage in temperate conditions. Its protective life under battle conditions was about three months. On first issue it was somewhat stiff and tacky and had a slight smell of chlorine—all of which wore off after a few days wear.

Cleaning when dirty presented administrative difficulties. Laundering and cleaning with solvents destroyed the impregnite. A special process was necessary and in the United Kingdom the clothing was withdrawn to Ordnance who passed it in bulk to specially selected dry-cleaning firms. If disinfected in a steam disinfector, the reaction of the impregnite destroyed the stitching so that the garment fell to pieces and had to be scrapped for salvage. Disinfestation in hot air caused no such destruction but hot air disinfestors were not readily available and disinfection by steam methods, with the resultant scrapping, had to be resorted to.

The position in August, 1939, was that a reserve of impregnated socks was held for the first contingent of the Field Force ; the decision to extend this to the issue of impregnated uniforms for all forces in the field was in process of being taken when the war broke out. The provision of plant, to provide the increased amount of A/V required, involved a time lag of at least six months. This initial policy did not envisage the issue of impregnated uniforms to troops in the United Kingdom, but only to the Field Force in France. Factory production started in June, 1940.

After Dunkirk it was decided to issue every man in the United Kingdom, and also the garrisons of Malta and Gibraltar, with two suits of A/V battledress and all issues of socks were also to be impregnated for foot protection. Capacity had to be increased accordingly.

The ultimate production effort planned was a major one involving the production of 45 tons of anti-verm a week, impregnation of suits being carried out by 14 firms in different parts of the country. This capacity was sufficient for 300,000 battle suits, or 5,000,000 socks a week. It included R.A.F. requirements.

Life of impregnated socks was limited owing to deterioration of the impregnite through sweating and to the need for hygienic washing. Maintenance involved the weekly withdrawal, replacement, collection, etc., of the socks for re-impregnation. The administrative difficulties in this proved so great that the project for impregnated socks was dropped entirely at the end of 1940 after some 1½ million pairs had been impregnated. Protective dubbin for boots was adopted in lieu as the standard method of foot protection.

The impregnation of battledress lagged, partly through initial " teething troubles " and partly because, being a purely defensive item of equipment, production was severely handicapped by plant having to be obtained on the lowest priority only. By July, 1942, the issue of one suit for each man was not complete, but was still proceeding. In September, 1942, it was decided that, in view of the considerable administrative difficulties involved in the provision and issue of anti-verm battledress on the scale of two suits for each man, the issue should be suspended forthwith, and that a reserve of 500,000 suits should be maintained to enable selective issues to be made to a proportion of the forces proceeding overseas for operations in the European theatre. It was so issued to all troops in operation " Overlord." A reserve for the garrisons of Malta and Gibraltar continued to be held in those stations on the basis of one suit for each man.

Sufficient impregnation plant only was retained to maintain and replenish these stocks as necessary, and for the policy of 100 per cent. issue to be resumed should gas warfare begin. This policy held good until the end of the war.

Development was put in hand of an impregnated " pantee " to give protection to a part of the body which is readily affected by mustard gas vapour.(1) The supply and maintenance factors connected with a small garment of this nature would have been less than with complete battledress. Subsequent developments in relation to anti-verm arising from trials in Australia, which are referred to below, led to the proposal being shelved.

Note (1).—In temperate climates the genital area in particular is normally more sweaty than the remainder of the body surface and therefore more vulnerable to mustard gas vapour. Exposure of respirator-protected individuals to a dosage which would be insufficient to affect the dryer body surfaces would cause blistering in this area. An impregnated " pantee " would have led to a reduced casualty rate in many circumstances. As the temperature rises the disparity in the vulnerability of the genital area to the remainder of the body surface decreases and is least marked where the whole body is wet with sweat in hot humid weather. In tropical climates therefore the issue of a " pantee " only is less likely to achieve worth-while casualty reduction, although, failing other methods of protection, it might have been worth considering for reasons of morale.

Cottons were more difficult to impregnate than serge because cottons did not hold the anti-verm so well. Life of A/V cotton garments in wear was therefore basically shorter but in any case was limited by the hygienic needs of washing tropical uniforms. Until 1942, tropical battledress consisted of shorts, and open-necked and short-sleeved garments, and therefore would not have afforded protection against vapour even had it been impregnated. As the war against the Japanese progressed, this type of tropical clothing was found unsuitable under malarial and jungle conditions, and uniforms which covered the majority of the body surface were introduced. About the same time the offensive potential of mustard gas vapour in hot humid climates became better appreciated (*see* Chapter III). Impregnated clothing for the Japanese war therefore became an urgent requirement.

A serious set-back to development occurred in 1943 when, as a result of trials under tropical conditions in Australia, it transpired that under hot humid conditions clothing impregnated with anti-verm was toxic in wear. Anti-verm clothing had been tried out under hot dry conditions in India before the decision to accept it as the standard British impregnite had been taken and this defect had not manifested itself. Under temperate conditions A/V clothing was not toxic. Its failure under tropical conditions can in part be ascribed to the lack of adequate tropical testing facilities before and in the early days of the war. No other British impregnite developed before the war was suitable and there was insufficient time to return to fundamental research. The only course open was to obtain the U.S. type of impregnite and impregnating plant from America.

In the U.S.A., neither plant nor impregnite were available to meet the unexpected demand for our full requirements for the war against Japan in addition to the U.S. Forces' own requirements, nor could the U.S. processing companies in the Far East or South-East Asia theatres cope with our requirements in addition to their own. The question of production and availabilities had not been settled by the time the Japanese war ended.

The United States had two methods of impregnation—the M.1 process using solvents and the M.2 process using water suspension methods. There were different plants of varying capacity—the Z. of I. (Zone of Interior) M.1, the T. of O. (Theatre of Operations) M.1, and the T. of O. M.2.—which were operated by Processing and Base Processing Companies. (*See* Chapter II, Section 8). British mobile laundry plant could not be used for the M.2 process without a conversion unit. Such a unit was developed in 1945 in case of a breakdown in supply of U.S. plant.

The M.1 process gave an article with better storage life. Clothing impregnated by the M.2 process was slightly irritant but not sufficiently so to warrant rejection of the process.

The Americans also had an emergency field impregnation set consisting of a collapsible canvas bucket with hoops, a paddle and the necessary supplies of impregnite for use by the water suspension method.

United States tropical garments were specially designed to prevent leakage, *e.g.*, at cuffs, flies, etc. The life in wear of these garments was in general two weeks with two cold washes without losing an appreciable degree of protection—

after that they required re-impregnation. This meant plant re-impregnation at the base since field methods were not favoured except in emergency. Re-impregnation problems in the field were therefore formidable—and this too in theatres where problems of normal supply and maintenance were already acute. In the British view they were such that, if an entire force were to be equipped with impregnated clothing, maintenance problems would have been insuperable.

Apart from this, technical problems arose in developing suitable clothing. Materials most suitable from an anti-malarial point of view were not in all cases the best suited to impregnation against mustard gas. In addition to impregnation factors, designs of clothing had to be modified to prevent leakage. Exposed surfaces of the body (face, where not covered by the respirator, and neck), which under European conditions could be protected with ointment, required special measures of protection in the tropics. Problems connected with disinfestation also arose.

In August, 1944, it was decided that no attempt would be made to impregnate the standard light mosquito-proof jungle garments. Until gas warfare started the mosquito was a greater danger than gas. Policy formulated in the United Kingdom, which was dependent on availabilities from the U.S.A., was to hold a reserve of impregnated clothing at the base in each theatre as an insurance against gas warfare starting or for issue if required before any particular operation.

The extravagance of providing every man in the force permanently with this clothing was considered unwarranted, not only on account of the problem of maintenance, but also because large areas at the base would be out of range of gas attack. A percentage of base and L. of C. troops within range of attack could in any case temporarily evacuate contaminated areas, and the situation could be met by equipping large and efficient decontamination squads. A 100 per cent. scale of protection, therefore, was only visualized for troops in contact.

The items of clothing which it was decided should be held in this reserve were shirts, trousers, and drawers (especially designed to obviate leakage), socks, vests and a hood. Material for the shirts and trousers was to be herring-bone twill if of U.S.A. manufacture or of Drill No. 2 if of British. The necessity for impregnating the gaiter or short puttee was left to be determined by experience.

The hood was to be made of thin cotton stockinette material, to be loose fitting, and drawn round the edge of the respirator facepiece by a draw string. Trials had shown that, contrary to expectations, the wearing of an impregnated hood as well as the respirator did not lead to any great additional distress. It was the wearing of the respirator itself which was decisive in this respect.

To cover the re-impregnation cycle, five sets of clothing were to be provided for each man provisioned for, *i.e.*, one suit on the man, one suit on the way back to the impregnation plant, one suit in the impregnation plant, one suit on the way forward from the impregnation plant, and one suit in reserve. Provision for 100 per cent. replacement of all impregnated items within twelve months was to be made in the first place. Since the U.S.A. was unable to meet our plant requirements it was proposed initially to obtain a supply of garments of British pattern ready impregnated from the United States.

Policy in India and the South-East Asia Command conformed with the above. When the war ended impregnating companies were about to be raised in India for which it was hoped plant from America would shortly become available.

The Germans and the Japanese

Neither the Germans nor the Japanese had impregnated clothing. The Germans came to know of British and American types through the capture of British anti-verm socks in France and of British documents from the French, and also of some American impregnated gloves in North Africa. They carried out some experimental work with impregnites similar to the British type but a general issue was not contemplated owing to shortage of raw materials for making the impregnites. In any case the work they did showed a lack of originality and was of no great value.

Future Requirements for Impregnated Clothing

With the advent of nerve gases there is no requirement for battle dress impregnated with existing British or American impregnites. Impregnated clothing is, however, a definite requirement. It must protect against both nerve and blister gases, and, in particular, against the " initial cloud " effects arising from the burst of H.E. chemical projectiles. The impregnated clothing of the future must overcome the administrative difficulties connected with re-impregnation and supply in the field which handicapped both British and American types as available during the war.

Protective Dubbin

After impregnated socks had been abandoned in 1940, a protective dubbin for boots was subsequently introduced for foot protection. Protective dubbin did not contain any impregnite but was a mixture of oils or greases that, while fulfilling its function as a boot preservative, delayed the penetration of liquid mustard gas. The liquid tended to remain on the surface as a discrete drop for a sufficient period for it to be removed during the process of personal decontamination before it had soaked in. Ordinary dubbin was less effective in this respect.

Protective dubbin was to replace ordinary dubbin as stocks of the latter became exhausted, but a complete change over was never effected owing to difficulties over availability of materials. U.S.A. " shoe impregnite " was a soft yellow grease used in a similar manner.

Wallet, Anti-Gas

This was introduced in 1939. Its purpose was to protect the man's paybook and personal valuables from contamination at all times. Many such items could not be readily decontaminated, and the wallet also facilitated their safeguarding during the process of the casting and replacement of clothing during decontamination. Issue was discontinued in 1942, much to the regret of many in the Services as it made an excellent tobacco pouch. It was a " frill " which belonged to the era of over-insurance against gas. The United States, who in some respects provided a wider range of protection than we did ourselves, had no such article.

Sleeve Detectors

From 1938 onwards, the basis of the whole of our organization against gas has been that the individual is responsible for his own personal protection. In view of the characteristics of high altitude spray (*see* Chapter IV, Section 2) and of the fact that small drops of spray were not easily visible on khaki uniforms, it was necessary to provide him with some form of liquid spray detector.

A number of devices were considered for this purpose, such as the painting of a patch of detector paint on a steel helmet. Trials, however, showed that with high spray a small patch of paint was unreliable as a detector since droplets from high spray would fall fairly far apart. Further, the detector was required in a place where it was easily visible to the man himself. Eventually, in 1939, a pair of " Detector, gas, sleeves " were introduced. They were made of paper and painted with detector paint. They had a loop at the top end which fitted over the shoulder strap. One was worn over the upper portion of each arm. To serve the same purpose when the cape was worn, the upper arm of the cape was painted in the unit as required with detector paint (*q.v.*).

In an effort to produce an article with a longer life than a paper sleeve and also to obtain the largest practicable area for detection purposes and to combine in some degree the functions of protection and detection, a " Tippet " to be worn over the shoulders and which could be painted with detector paint was introduced in 1940. It was abandoned shortly after but not before a number had been produced.

As the usefulness of high spray as a method of attack receded the necessity for these detectors also diminished. Against low spray no detectors were required as contamination would be obvious. Against an H.E./chemical shell the detector did not react sufficiently quickly or obviously to the very minute droplets of the " initial cloud " to admit of adjustment of the respirator before a damaging concentration was inhaled.

As it seems most improbable that an individual detector of this nature will ever be developed to assist in the timely detection of nerve gases (*see* Chapter III, Section 4), the requirement for it has lapsed.

Neither the United States nor our enemies developed a corresponding item of equipment.

Detectors, Gas, Ground

Detectors, gas, ground, were small booklets of 25 pieces of paper painted with detector paint (*q.v.*) and slotted to fit on to a bayonet or stick. They were issued to every officer, warrant officer, and N.C.O. Their purpose was to detect liquid blister gas on the ground or any other surface. They did not react to vapour, nor did they react to liquid which had soaked into the ground. They confirmed the sense of sight and smell when detecting blister gas liquid, and were of assistance in delimiting areas of contamination. They were not specific to blister gas liquid, *i.e.*, they reacted to, for example, liquid tear gas. This was of little consequence since the presence of tear gas would have been obvious by its effects. They remained a useful adjunct to other means of detection of blister gas (*see* Pocket Vapour Detector) when developed. They will react to liquid nerve gas and confirm its presence but will not serve for timely warning of attack (*see* Chapter III, Section 4).

The U.S. Forces had a very similar article—Paper, liquid vesicant detector, M.6—but the scale of issue was less wide (two booklets for each company, battery, etc.).

The Germáns developed detector powders for testing for contamination on any surface. The earlier type was a pink powder, the later type was an ochre powder. Both these powders showed different colour reactions for mustard gas, lewisite, and nitrogen mustard gas. They were put up in sprinkler type containers. A detection pump was also available for blowing the powder against surfaces which could not be covered by the sprinkler container, *e.g.*, on vehicles. Comparative tests of the ochre powder showed it to be of about equal value to the British ground gas detector.

Considerable developmental work was carried out by the Experimental Station at Porton on detector powders and they were put out to troop trials in 1943. They were not adopted since the pocket vapour detector (*q.v.*) which was then becoming available was capable of differentiating types of blister gases more certainly than were powders, and for liquid detection the British detector, gas, ground, was a simpler and less cumbersome piece of equipment. The scarcity of dyestuffs was also a determining factor.

The United States Forces developed a " crayon vesicant detector M.7 " which turned blue in the presence of blister gases.

Vehicle Equipment

Certain equipment, for the detection of contamination and for decontamination, was issued on a vehicle basis for every type of vehicle on the establishment, and for launches and power barges.

It consisted of detector paint ; trays for detector paint for forward control vehicles ; sacks, paper, gasproof, for back loading of contaminated clothing and equipment ; sandbags for decontamination of the vehicle ; and, for armoured vehicles, a 2-lb. tin of bleaching powder. Details of scales are given in Appendix XXII, Part II.

SECTION 2.—UNIT ANTI-GAS EQUIPMENT

Unit anti-gas equipment consisted of certain items of oilskin clothing, equipment for detection and warning of gas attack, and certain miscellaneous items for decontamination or protective purposes.

Oilskin Clothing

Various items of oilskin clothing were issued on a unit basis for decontamination tasks and also for use by certain individuals to combat liquid contact hazards peculiar to their duties.

OILSKIN CLOTHING FOR DECONTAMINATION TASKS

The position on the outbreak of war was that a number of sets of light anti-gas clothing(1), varying according to the size and role of the unit, were issued to all units and headquarters for normal decontamination tasks. In addition certain units, such as field park companies and dock companies, received a heavy anti-gas suit(2) for heavy decontamination work, *e.g.*, handling contaminated bridging materials, for which the light suit was thought to be insufficiently strong and where a maximum protection against penetration by the liquid was deemed necessary. The heavy suits were abandoned in 1941. Not only were they more exhausting to wear than the light suit but also the standard of liquid protection they afforded was unnecessarily high, and the weight and bulk (particularly of the rubber gumboots) increased the problems of carriage in the field. Henceforward all units were authorized sets of light clothing, only a few units continuing to receive a small proportion of rubber boots for special purposes.

When first introduced oilskin clothing was required for both liquid and vapour protection while working in a contaminated area. The vapour protection afforded was only partial since a certain amount of vapour was sucked in, particularly during heavy work, and rest periods were necessary to allow for the suit being opened up for dispersion of any vapour—as well as for reasons of exhaustion. Oilskin clothing, being impervious to air, did not admit of the body ventilating itself normally, and rests of 10 to 15 minutes were therefore necessary after periods of work varying from 20 minutes to one hour, according to the temperature and type of work—the lower limit being applicable to tropical climates.

Note (1).—A set of light anti-gas clothing consisted of light-weight jacket, trousers and hood, with gloves (3 pairs) and overboots of heavy oilskin. Rubber gloves (2 pairs to each set in lieu of 3 oilskin) were introduced in 1941. They were less clumsy in use than the oilskin type, but owing to the rubber shortage only a small percentage of the total unit entitlement was authorized to be rubber type. Over-mittens were introduced to protect both types of glove during heavy work. A few overboots with rubberized canvas uppers also became available.

Note (2).—The suit, anti-gas, heavy, consisted of a heavy oilskin jacket and trousers as in use in the Navy for weather protection which were worn with a hood, gloves and rubber gumboots.

It was when the oilskin hood was worn that the exhaustion factor was at its greatest, and the hood was therefore only used in strong vapour concentrations where maximum protection was required. In justification of this hood it must be remembered that in the early days anti-gas ointments could not be used prophylactically. It was withdrawn in 1942, by which time prophylactic ointments were available. Where any task engendered an abnormal hazard of liquid contact to the neck, in addition to ointment the curtain, anti-gas could have been used in lieu of the hood.

In temperate climates, after impregnated clothing became available, oilskin jackets and trousers were only required where there was a real danger of contact with the liquid. Many decontamination tasks could have been tackled with safety with the use of ointment, respirators, A/V battledress and dubbined boots, and the wearing of the cape, overboots and gloves in addition would have sufficed for most liquid contact risks where they existed. The use of oilskin suits would not normally have been resorted to except in special cases where the liquid contact danger was unduly high.

For tropical climates the ultimate intention was to provide impregnated clothing (*q.v.*) but as the U.S.A. could not give our requirements a high priority this did not become available by the time the war ended.

Without impregnated clothing, decontamination tasks in tropical conditions presented a somewhat more formidable problem than in temperate conditions, particularly before the availability of non-irritant ointments that could be used prophylactically on any part of the body. The exhaustion factor in wearing the complete oilskin suit was at its maximum in tropical climates, and tasks necessitating the use of this clothing would have involved a far greater number of man hours than in temperate climates. In hot dry climates the wearing of a wringing wet drill garment or sacking over the light suit gave a considerable measure of relief, but where humidities were high the wearing of such over-garments made little or no difference.

A factor to be remembered, however, is that although the vapour danger is enhanced in the tropics the most dangerous period in many circumstances is during the half to one hour immediately following contamination. After this period had elapsed many decontamination tasks could have been undertaken with reasonable safety by men protected with ointment and wearing respirators and such items of oilskin clothing, *e.g.*, capes, gloves, overboots, as were dictated by the degree of liquid danger. If the men used commonsense and had a full knowledge of the risks involved, minor casualties should have been few.

Under the worst conditions that might have been encountered, *e.g.*, the continued contamination of a beach area through which the movement of stores had to be maintained, it is certain that the lack of impregnated clothing would have involved us in serious difficulties.

Special Issues of Items of Oilskin Clothing

Certain special issues of light suits and gloves, in some cases with, and in some cases without overboots, were issued on a unit basis for use by motor cyclists, transport drivers, signals personnel, etc. A complete list is given in the notes to Part III Table 1(*a*) at Appendix XXII. They were issued in addition to the anti-gas cape, as it was considered that the special duties of these men either could not be performed efficiently when wearing the cape or demanded the additional trouser protection.

Apart from the question of the load on production resources, the problems of carriage and maintenance in the field make it necessary that such personal issues should in future be cut to the minimum if a reasonable balance is to be maintained between protection and efficiency. In most cases impregnated clothing plus ointment and the cape (and, of course, the respirator) will suffice for normal hazards both of mustard gas and nerve gases, and in any case a higher standard of protection against gas than against other battle dangers is not justified. Some risks must be taken. Certain special cases exposed to special hazards *e.g.*, motor cyclists or ordnance personnel handling chemical munitions, may require consideration.

In 1941 an anti-gas apron was introduced, and issued together with gloves and overboots to regimental medical officers for the handling of contaminated casualties. Medical units also received special issues of these items (*see* Appendix XXII, Part III and Part VI).

United States impermeable protective clothing was made of oilskin and used for similar purposes to British items. Initially they had a two-piece suit which was latterly superseded by a one-piece suit.

The Germans had a light suit of somewhat novel design (sleeve-apron-shorts, thigh-wader overboots, gauntlets and hood) made of textile coated internally with synthetic rubber. It was issued primarily for reconnaissance work to gas scouts, but could also be used for light decontamination work. They also had a heavy rubberized suit (initially one-piece and latterly two-piece with hood attached and worn with gloves and rubber knee-boots) for special decontamination tasks.

Unit Equipment for Detection and Warning of Gas Attacks

Gas Alarms

Gas alarms were of two types, the local alarm (a trench rattle) and the general alarm (a siren).

The rattle became the standard local alarm during the 1914–18 war and has continued to be used for this purpose to date. In 1941 an improvement was made by fitting it with a sound box so that it could be heard at a greater distance. It was issued down to platoons and equivalent sub-units.

A siren for purposes of general alarm was in use in the 1914–18 war to give warning of cloud gas attacks which were effective for considerable distances from the source of discharge. The 1914–18 general alarm was a Strombos Horn, which was operated by compressed air and was therefore heavy and not mobile. In our 1936 gas rearmament, a general alarm was still considered necessary but a more portable type was required. A hand operated siren was therefore introduced and authorized for all units.

Initially, there was a different series of siren signals to denote spray attack and other forms of gas attack. In 1940, the spray warning was discontinued in the United Kingdom owing to the risk of confusion with civilian air raid alarms and with the general gas alarm. At the end of the year, the general alarm siren was withdrawn entirely owing to the confusion with the civilian warning system. Henceforward, spray warning was by word of mouth and warning for all other types of gas attack was by means of the rattle only. This became standard practice in all theatres of war. The siren for general gas alarm was declared obsolete in 1944. In 1941, Strombos Horns were issued to South Eastern Command to cater for the somewhat remote possibility of cloud gas attacks from the sea.

The type of warning equipment required in the future may be influenced by equipment developed for the detection of or confirmation of the presence of nerve gases. Unless such equipment is readily portable and issuable to sub-units, alarms which can be heard at greater distances than the rattle may be necessary.

The use of screaming rocket gas alarms was considered in December, 1942, initially as a general civilian gas alarm to avoid possible confusion with air raid sirens. Some experimental rockets mounted on a stick were tried out and appeared to fulfil the requirement, *viz.* audible in still air for 1,000 yards and sound persisting for half to one minute. They were tried out by a division but were found to have many weaknesses as well as being unreliable and complicated in operation. As no value could be seen for them as a military store the matter was dropped.

The Germans developed a whistling cartridge fired from a signal pistol as a gas alarm. They also utilized hand signals, *e.g.*, taking out the respirator and waving it up and down, and sound signals—bells, empty shell cases, etc. but not sirens or klaxons.

The American gas alarm was a U-shaped cast-iron pipe with a striker in the open leg held in place by a pin and chain.

Detector Paint

In addition to its use on vehicles, detector paint was used on spray detectors, ground detectors, and on the sleeves of the cape and of the light suit. It was also used at factories on possible leakage points of weapons charged blister gas so that any leakage during storage could be detected.

Detector Paint No. 1 was yellow-green in colour. It was superseded by Detector Paint No. 2 which was a khaki-brown colour. Both these paints turned red in contact with liquid gas. They did not react to vapour.

The Americans had a similar detector paint : "Paint, liquid vesicant detector, M.5."

Spray Detectors

Spray detectors initially consisted of metal squares slotted at the corner, into which a paper painted with detector paint was inserted. The metal squares were 6 in. square and were used in groups of nine to give a large square of 18-in. size. Later the spray papers were abandoned and the metal squares painted with detector paint, and in due course one large metal square with 18-in. sides was substituted for the small squares. They were for use in camps or bivouacs to detect high spray attack, and were put out in the open in irregular lines about 200 yards apart. As the value of high spray receded and the possibilities of this type of attack became less, the need for these detectors decreased. Their use in the future would be more to aid the speedy delimitation of attack, particularly in relation to dumps of material, etc., rather than to give warning.

The Americans had a " Paper, liquid vesicant detector, M.6 " in books of 25 sheets which could be used as a spray detector or as a Detector, Gas, Ground.

The Germans had packets of spray detection cards painted with a pink detector paint which gave varying colour reactions from violet to orange to the liquid of varying groups of blister·gases (H, L, and HN).

Detector, Paper Type A

Shortly after the outbreak of war, when the use of arsine by the Germans was considered possible (*see* Chapter III, Section 2), a special detector was required to detect this gas, owing to its lack of immediate effect on the body and of distinctive smell. Detectors, consisting of pieces of white paper size 3 in. by 2 in. specially treated so that they turned yellow in the presence of arsine, were issued on a wide scale and were hung up in such a way that they could be seen by a sentry and were protected from the weather under petrol tins cut in half. The teaching was that they should be a normal part of every sentry's equipment. When it became obvious that the use of arsine was improbable, these detectors were withdrawn.

Gas Contamination Labels (Army Form A.2028)

This was a gummed oval-shaped yellow label with a purple diagonal cross and the word " Contaminated " printed in black. It was introduced in 1940 for marking stores in transit and other articles or vehicles which might be moved away from contaminated areas. It was standardized between all Services, except that the railways' established system for indicating contaminated rolling stock was a label reading " Gas."

Signs, Warning, Gas

The purpose of these signs was to delimit areas of contamination and to mark dumps of contaminated stores and material : they were triangular metal signs bearing the word " Gas." They suffered many changes in colour and wording, before it was agreed by all Services that they should be painted yellow on one side and white on the other with the word " Gas " in black letters on both sides. They are a requirement for the future.

The Americans had no standard gas warning sign. They relied on improvisation in the field.

The Germans had warning flags—little triangular yellow flags with a skull and crossbones marked in black—mounted on a small steel signpost type rod. They were carried in an outfit which also contained 50 yards of yellow marking tape for laying between flags where the flags themselves might not be easily visible.

Pocket Vapour Detector (PVD)

Until 1943 there was no apparatus in general issue in the Army which was capable of detecting blister gas vapour, and reliance was placed on the sense of smell. The disadvantage of this latter method of detection is that it is impossible to assess the amount of blister gas vapour in the air thus making it necessary to take full precautions whenever mustard gas is smelt unless the risk of casualties is accepted. Furthermore, all blister gases cannot necessarily be detected by smell.

The reason for the lack of such apparatus was that the development of simple and foolproof chemical detectors suitable for use by units in the field for the rapid identification of types of gas as well as of the concentration in the atmosphere was, and still is, one of the major problems presented to us by gas warfare, and has not been satisfactorily solved to date by any nation. Lack of space precludes its discussion in detail here. Details regarding the chemical identification of war gases (other than nerve gases) will be found in the Ministry of Home Security pamphlet "The Detection and Identification of War Gases." A perusal of this pamphlet will make obvious the difficulties in development of field sets for use by non-scientific personnel.

The development of the Pocket Vapour Detector (PVD) was proceeded with at high priority in 1942 owing to the possible use by the Germans of an odourless blister gas of the HN type (*see* Chapter III, Section 3). The PVD was introduced in 1943.

A description of this detector and its method of use will be found in "Advanced Gas Training, 1943." Briefly, it consisted of a suction pump on which a test paper was mounted. There were two types of test papers and two glass tubes of agents for use in differentiating between different types of blister gas. In testing mustard gas vapour only, the number of strokes of the pump required to cause a reaction on the paper indicated the approximate vapour concentration present. This was related on a chart in the lid of the box to the safe time of exposure with and without respirators. It was issued on a company or equivalent basis and used by gas officers and N.C.Os.

As a qualitative detector for differentiating between types of blister gas it was an efficient piece of equipment. As a quantitative detector for assessing the vapour concentration of mustard gas it had certain defects. Under temperate conditions it functioned comparatively well but under tropical conditions it was more erratic since one of the papers was effected by strong sunlight. In any case it was found that misleading results might be obtained owing to minor changes in wind direction affecting the concentration in the air while the test was being made. Also it did not react to extremely low concentrations. This was of little significance in the open, but in an enclosed space, where there is no air circulation, such concentrations for long periods can be effective. During the unloading of the hold of a ship containing mustard gas ammunition an accident did occur because the PVD showed no reaction to a very low concentration. Nevertheless it was a useful piece of equipment and was at least a guide in confirming the presence or the absence of danger in most circumstances.

Nerve gases cannot be detected with the PVD. At the time of writing, a detector which will react sufficiently quickly to give timely warning in case of bombardment with H.E./chemical shell seems unlikely, but one which will serve when the time factor is less critical seems a probability.

Future requirements of gas detectors are that they should be usable by any soldier with a minimum of training, but they need not necessarily be an individual issue; in principle, they should be able to detect all known war gases. The most urgent requirement is for a suitable nerve gas detector.

The Americans had initially a "Kit, 'H' vapour detector, M 4" for detecting mustard gas vapour. It was replaced later by a "Kit, chemical agent detector, M 9" to detect and identify dangerous concentrations of seven chemical warfare agents. This set involved five sets of detector tubes and three bottles of liquid agents. It was mainly an outfit for the identification of the type of gas present and although the concentration and the degree of danger could to some extent be determined it was less satisfactory in this respect than the British PVD. They were issued on a company or equivalent basis.

The Germans developed a gas identification set—the " Gasanzeiger." It was issued to certain specially trained gas personnel in units. It was complicated to use ; it was an identification set only and did not indicate the vapour concentrations present. It identified a wide range of possible war gases in addition to blister gases. There was a special " Gasanzeiger-Fest " for use in fortifications. They also introduced a set of gas recognition equipment (" Satz Gaserkennungsgerat ") which included a Gasanzeiger, detector powder and sampling equipment, so that the operator had all means available to confirm or sample gases either in the atmosphere or on the ground, including contaminated samples of earth.

Unit Equipment for Miscellaneous Purposes

Bleaching Powder

Bleaching powder, or chloride of lime, was issued to all units for decontamination from mustard gas of ground, buildings and such equipment as would not be harmed by it. Details of its methods of use will be found in " Gas Training, 1942 " and " Advanced Gas Training, 1943." It was issued in hermetically sealed tins which were only opened when required for use. Normal packing of unit issues was in 2-lb. and 7-lb. tins ; packing in larger bulk was also available. Bleaching powder held in pre-gas warfare reserves was of a stabilized grade known as super-tropical bleach which had a storage life of at least five years under all conditions. Had gas warfare started, lower grades of bleaching powder with less storage life could also have been used if necessary.

Initially the amount of bleach required for decontamination of ground was much over-estimated ; it was also mixed with earth or sand owing to the danger of its inflaming if brought into contact with liquid mustard gas. Decontamination of ground by this means was so laborious as to be almost impracticable, particularly when, as in the early days of the war, the wearing of the full oilskin suit was visualized during such tasks. Teaching then was that 10 tons of a mixture of earth and bleach was required to decontaminate 180 square yards. This also sealed the area and prevented the escape of vapour.

In 1942, as a result of more extensive trials, it was found that application of 1 lb. of dry bleach to each square yard of heavy contamination was sufficient. Mixing with earth was in general unnecessary ; the danger of inflaming could be accepted in most cases. This revised method did not mean that decontamination of ground would be more widely undertaken but that, in places where it was essential, *e.g.* dockyards and depots, it could be undertaken with less effort and less labour. In the forward area in the field large scale decontamination of ground was never regarded as practicable.

Each Service was responsible for the decontamination of its own areas and before 1943 each Service had provisioned separately for its own commitments, but these stocks were available locally for mutual aid in emergency.

In 1943 reserves of bleach were reviewed. The scale of possible German air attack with mustard gas on the United Kingdom and European theatre was recalculated as was the quantity of bleach required to neutralize it.

Rules of thumb used for purposes of appreciations of provisioning requirements were 2 lb. of bleach for every 1 lb. of blister gas or 1 lb. of bleach for 1 lb. weight of gas bomb. In the early days of the war calculations for provisioning of bleach powder had been based on a much higher figure. To prevent over-provisioning, co-ordination was effected between the Ministry of Home Security, the Air Ministry, the War Office, the Admiralty and H.Q., E.T.O.U.S.A., on the basis of each Service holding its own unit scales backed by reserves, plus a common reserve which was to be allocated by the London Munitions Assignment Board. Overall expenditure assumed after chemical warfare started was expressed in terms of monthly replenishment of the common reserve.

Decontamination Methods—British, American and German

The British relied on the use of bleaching powder, either in the form of bleach paste or dry, for most decontamination tasks. Where damage might occur if bleach were used, *e.g.*, delicate instruments, parts of guns and vehicles, swabbing with petrol was the standard method. Application was by simple hand methods.

The U.S. Forces had, in addition to bleaching powder, two types of " Agent, Decontaminating " which were non-corrosive and were for use on items which might be damaged by bleach. They also had a number of decontaminating apparatuses, some hand-operated and some power-driven, for use with non-corrosive agents and bleach paste. Power-driven apparatus could be applied to other uses such as crash fire apparatus, field shower baths, and sanitary control. They also introduced a spreader mounted on a trailer for decontaminating large flat areas with dry bleach.

Comparative trials showed that American methods were superior to British in one respect in that vehicle decontamination was quicker and more effective. In other respects there was little to choose, and British methods had the advantage of simplicity, a standard decontaminant, and absence of mechanical apparatus. Differences in decontamination doctrine are evidenced by the respective equipments provided.

American bleaching powder was not of the stabilized super-tropical grade, and they obtained the majority of their reserve stocks, both for the European and Far Eastern theatres, from the United Kingdom on reverse lend-lease.

The Germans relied mainly on bleaching powder. For decontamination of ground they had a sieve box holding 22 lb. of " Losantin " powder (or bleach). They also had small and medium semi-tracked vehicles equipped at the rear with hoppers for distributing the powder.

Pathway Anti-Gas

This consisted of 50-ft. rolls of bituminized paper 3 ft. wide. It was for laying pathways across contamination. It was introduced in 1938 as a unit store for Infantry, Royal Engineer, Ordnance and R.A.S.C. field units. It was withdrawn as a field store in 1941 and stocks were held in Ordnance charge for selective issues if required. It was a cumbersome over-insurance and is now superfluous.

The United States Forces did not possess a corresponding item. The Germans developed a similar item which they called a " gas runner."

Covers, Stretcher, Anti-Gas

This was a bituminized paper sheet which could have been used either to prevent contamination of the stretcher by contaminated casualties, or to prevent the stretcher coming in contact with contaminated ground. Early models were an envelope type of oilskin cover. Although means of stretcher protection could be improvised with, for example, capes, anti-gas, it was preferable to have an expendable article for this purpose rather than a more expensive article which required recovery and decontamination. The United States Forces did not develop a corresponding item of equipment.

Sacks, Paper, Gasproof

These were for back loading contaminated clothing to decontamination centres. They were originally known as " bags, contaminated clothing " or " bags, anti-gas." Paper sacks were made of multi-layer bituminized paper and were expendable. They are a requirement for the future. The United States " sack, gas resistant M.1 " was a similar article but rather heavier than the British pattern.

Cloth, Union, Anti-Gas

Material for gasproofing (cloth, union, anti-gas) was held by the R.A.O.C. for issue as required for gasproofing in special cases under static conditions. Gasproofing of dugouts and buildings had been carried out on a wide scale in the field under the static conditions of the 1914–18 war. In 1939 it was still visualized as a requirement during static periods but it was not included in unit equipment scales. Under mobile conditions there is little scope for gasproofing. Policy latterly was to hold cloth, union, anti-gas available for issue as required for the gasproofing of portions of headquarters, hospitals and signal centres, reliance being placed on improvisation for any other forward requirements that might occasionally arise. It must be remembered that gasproofing without blast proofing is largely ineffective. Details regarding methods of gasproofing will be found in " Advanced Gas Training, 1943."

Effective gasproofing implies a total absence of ventilation ; gasproof shelters of a given size can only be occupied for limited periods by a given number of people.

The U.S. Forces had a " Curtain, Gasproof M 1," which was made of high grade light-weight blanket material covered on one side with impermeable cloth. It was issued on a scale of two to every twenty men.

SECTION 3.—MISCELLANEOUS EQUIPMENT AND ANIMAL EQUIPMENT

Miscellaneous Equipment

Cases, War Gas, Testing

These were issued to technical officers C.W. to carry out tests of samples of enemy war gases in order to substantiate the type of gas which was being, or had been, used. They were subject to certain limitations ; they could only be operated by personnel who had had scientific training ; they were dependent on the obtaining of samples, and they could only identify known substances which either intelligence or our own research indicated might be used as war gases and for the identification of which the necessary chemical reagents were provided. Most other common chemicals could also be identified. As far as the known gases were concerned the type used could have been determined with reasonable certainty with the restricted apparatus and chemical reagents available ; final substantiation by an anti-gas laboratory would, however, have been necessary.

Had nerve gases been used the case, war gas, testing could not initially have definitely established the identity of the chemical as it was unknown to us and the necessary reagents were not included in the case. Anyhow, the obtaining of samples after bombardment with H.E./chemical shell would have been difficult.

There were three marks of this set—the Mks. I, II and III. Each mark of set embodied improvements in bulk and weight and included additional reagents to enable tests to be made for a wider range of possible war gases. The illustration of the Case, War Gas, Testing in Appendix XXIII is, in itself, evidence of the complexity of the problem of chemical identification in the field.

A much less comprehensive equipment was the G.I.O.'s set issued to civil gas identification officers. With this, only the common war gases could be identified.

Initially in 1940 the G.I.O.'s set (which is also illustrated in Appendix XXIII) was issued to technical officers (C.W.). The Case, War Gas, Testing was developed in 1941 and initial issues to technical officers were completed by early 1942. Chemicals and apparatus for its maintenance were held by anti-gas laboratories.

Gas Filtration Units

These were available for supplying gas-cleansed air in permanent blast proof buildings or in ships. They were only provided where continuance of work without loss of efficiency was vital. They were standard equipment in the Navy in all capital ships and cruisers, and also in certain R.N. hospitals and first aid stations. Army use was limited almost entirely to gasproof operational rooms in coast defence fortress systems, mainly in defended ports abroad. They were also fitted in R.A.F. establishment war rooms.

Filtration units consisted of casings, varying in size according to the flow rate required for adequate ventilation of the room concerned, into which special containers, the container Type G, were normally inserted. Air was drawn through by suction fans. Where forced air ventilation existed their incorporation was simple. The Type G container was essential for the Naval design of filtration unit, but the Army and R.A.F. could have used Type E containers with a suitable modification to the unit.

The Americans had a wide range of " Protectors, Collective," both portable and non-portable. They visualized gasproofing on a rather wider scale than we did, including the use of the applicable " collective protector " for semi-permanent and temporary shelters. The fundamental differences in British and American doctrine in regard to non-persistent gases referred to in Chapter II, Section 9, is evidenced in the scale of provision of collective measures of protection against them.

The Germans developed elaborate gas protector installations utilizing expensive equipment. They also had installations for testing the serviceability of gas protectors by measuring resistance to air flow, and also for their regeneration if heavily contaminated or if affected by the absorption of water. Numerous installations were found in N.W. Europe which are evidence of the degree of anxiety felt by the Germans about our possible use of gas.

Animal Equipment

Horses and mules are susceptible to war gases in a similar manner to human beings except that they are generally unaffected by tear gases, unless liquid gets into the eyes, and are little affected by nose gases. Details of the effects of war gases on animals will be found in "Advanced Gas Training, 1943 ". Protective equipment is therefore required for animals similarly as for human beings.

The timely adjustment of animal equipment, *e.g.*, respirator, or the application of it, *e.g.*, ointment, has always been a somewhat contentious problem. A man first has to obtain protection himself and then attend to his animal—which is likely to be restive—or in some cases more than one animal. First aid or decontamination of contaminated animals is not an easy matter owing to the size of the animal and the area of the body surface to be covered within the time margin for penetration of the skin by liquid blister gas, which, as in humans, is short. The continuous wearing of protective covers is as exhausting to animals as it is to human beings. Compromise in matters both of protection and of decontamination is, therefore, necessary.

The heel of the horse or mule just above the frog is most susceptible to liquid mustard gas. The development of " leggings " or " boots " has always presented difficulty in design. Ointment gives effective protection but is liable to be rubbed off. In general, the most practical means of protecting the heel and lower leg when crossing known contamination is the use of either ointment alone, or a combination of ointment plus improvised protection, *e.g.*, sandbags, where circumstances warrant it.

In practice, the animal when loaded is in some degree protected from liquid contamination by the saddle and load. The blistering of certain portions, *e.g.*, rump, is unlikely to put the animal out of action and can be treated in the unit. If contaminated, early application of ointment to the most vulnerable parts not covered by the saddle and load should, therefore, obviate a severe casualty, unless contamination is massive. Strict discipline in covering bare back animals with rugs, etc., when in the open, and their early removal if contaminated, will go a long way to reduce casualties. According to the type of country in which operating, concealment, so as not to present a target, or dispersion are among the best methods of casualty reduction amongst animals.

Animal gas casualties in the 1914–18 war were not as high as would theoretically have been expected. Animal respirators were not issued and improvised means of protection against non-persistent gas was resorted to, *e.g.*, damp hay in a feed bag. Only battle experience of modern methods of projecting gas would have shown the precise vulnerability of animals in the field, the casualty rate to be expected and the actual degree of protection necessary. Lacking this, the best that could be done was to devise the most efficient and practical means of protection until experience proved or disproved the need for any item or indicated a requirement for some particular development.

The items developed for animal protection are discussed below. All development up to the end of the war was based on the protection of the breathing tract from the same range of gases as for human beings, and on protection against and decontamination from liquid blister gases. Up to the time of writing, the effect of the advent of nerve gases on requirements for animal protection has not been fully examined.

Horse and Mule Respirators

These animals' eyes are unaffected by tear gases ; they are affected by blister gas vapour, but less readily so than the eyes of human beings. If vapour protection to the eyes is provided it necessitates covering the whole head and interfering with the adjustment of the bit and bridle. For protection of the respiratory tract horse and mule respirators are made to cover the nostrils and upper jaw only, since these animals do not breath through the mouth.

Two types of respirator were developed during the war. The earlier type was an impregnated fabric bag which gave protection against a limited number of war gases and had a number of unsatisfactory features. A later development, which never got into issue, was of a greatly improved design and was made of rubber and rubberized fabric. It had a container at the forward end and inlet and outlet valves. Both types are illustrated at Appendix XXIII.

Earlier United States types were similar to our impregnated bag type. Later types were of somewhat different design and had a rubber face-piece with two canisters carried in front of the shoulders and connected to the face-piece by two long corrugated connecting tubes.

Eye Shields, Horse and Mule

The smallest drop of blister gas liquid in the eye will cause serious injury and temporary blindness in a few hours. Heavy liquid contamination leads to permanent injury and blindness. Eye shields are, therefore, as necessary for these animals as for human beings. They are illustrated at Appendix XXIII. As with the human eyeshield, the United States had not developed a similar store before they entered the war. They would have used the British item had they required it.

Covers, Horse and Mule

This consisted of a drill cover with detachable impregnated lining which covered the animal from head to tail. It was authorized for the Middle East only, but in that theatre it was generally preferred to rely on dispersion rather than to carry this additional equipment. It is illustrated at Appendix XXIII.

In India the impregnated lining was dispensed with, partly owing to the lack of impregnite and to maintenance problems, and partly owing to the excessive distress caused in wearing a double layer cover. A twill cover was developed of which the fore and aft portions were detachable from the centre portion. The centre portion fitted under the saddle and the fore and aft portions were rolled back when not in use. If contaminated, the fore and aft portions could be readily removed, the centre portion either being protected by the load or not touching the animal where not so protected. A number were produced for the first line transport animals and held in reserve for issue if required.

The United States developed a double layer impregnated type, but no issues were made.

Ointment, Bleaching Powder, and Miscellaneous Equipment

Items and scales of provision are given in Appendix XXII, Part V.

Leggings, Horse

An impervious legging was developed for the protection of the lower part of the leg and hoof with a special anchorage at the heel. It is illustrated at Appendix XXIII. There was no production. The United States developed a similar item.

Respirators for Army War Dogs

This problem was initiated in 1943, but on a low priority, as the probable requirement was, at that time, only for guard dogs.

In the middle of 1944, the position was reviewed on the introduction of mine detecting dogs, and a suitable design was developed, but it did not go into production. There was no general requirement and, with the end of the war, the matter lapsed. The design exists should it be required in the event of the use of dogs for mine detection being extended. It is illustrated in Appendix XXIII.

Pigeon Basket Cover

A " Bag, pigeon protection " was developed by the United States Forces. A British pigeon basket cover was designed, but never put into production.

German Development

German development of animal protective equipment was, as would be expected, both logical and thorough. For horses it comprised respirators, goggles and leggings. Their latest type of horse respirator was novel and ingenious. It consisted of a pair of hollow cones to which containers were attached. The cones were inserted into the nostrils and held in position by a head-harness. Examination showed it to afford efficient protection, but wearing was liable to cause damage to the animal and harness tension was critical in this respect. The cones were somewhat fragile and easily broken in wear.

For dogs, in addition to a respirator, an anti-gas suit of impregnated textile was developed which completely covered the animal and included sewn-on foot-covers of rubber. The Germans also visualized the use of dogs for gas detection purposes, and special equipment existed for their training.

SECTION 4.—SCALES OF ANTI-GAS EQUIPMENT AND RESERVES, AND THEIR CARRIAGE IN THE FIELD

Scales of Equipment and Reserves

Full details of War Office scales authorized during the war, which applied to all theatres except India and South-East Asia, will be found in the document—Security B.498—" Scales of Anti-gas Clothing and Equipment for all Units, 1941," as amended from time to time, extracts from which are given in Appendix XXII. It was a most comprehensive but complex document.

Personal Equipment and Vehicle Equipment

Scales of this equipment will be found in Parts I and II of Appendix XXII. The reserves allotted to unit transport should be noted.

Unit Anti-Gas Equipment

Scales of this equipment, including the reserves to be carried in first line transport, are given in Parts III, IV, V and VI of Appendix XXII.

A glance at these tables will show the complexity of the method of allocation of this equipment. It was never fully understood by the Army as a whole and gave rise to much unnecessary work. The probable requirements of each unit were assessed individually according to its size, organization and role, on the basis of a theoretical appreciation of the hazards involved. Pending battle experience under gas warfare conditions, allotment on a strength percentage basis would have adequately covered the risk and would have been straightforward, simple to apply and no less economical. Any minor maladjustments arising from such a method of allotment would have been cushioned by the reserves referred to below, or if necessary, in particular cases, by the allotment of special bulk reserves.

Reserves

In addition to the 1st line reserves of personal and unit anti-gas equipment, certain reserves were allocated to 2nd and 3rd line and more rearward echelons. These reserves were of two types, firstly, a reserve of items of anti-gas equipment, commonly known as Reconditioning Reserves, to replace expended, unserviceable, or contaminated items ; and secondly, a reserve mainly consisting of ordinary clothing and web equipment to replace contaminated articles, and styled Decontamination Reserves. Document B.498 contained tables which could be applied to any size of force, the actual quantities of the various stores being given for an infantry division, an armoured division, an independent brigade group and an Army tank brigade, based respectively on strengths of 15,000, 12,000, 6,279 and 2,680 all ranks. Those for an infantry division are given in Parts VII and VIII of Appendix XXII, other formations being *pro rata* to strength.

On the outbreak of war the reserves to be held in each echelon were considerably greater than those shown in Appendix XXII, in particular in regard to the decontamination reserves. The whole problem had been thrashed out in 1939 and allocations were then based on the assumption that the average daily rate of contamination from all means including spray, was likely to be 15 per cent. of the whole force, *i.e.*, the equivalent of every man once a week. This connoted the same daily rate of decontamination. As it was estimated that it would take 14 days to decontaminate clothing and re-introduce it to the reclothing chain, reserves of nearly all items of service clothing were fixed at 200 per cent. of the establishment of a force, 25 per cent. being carried in unit 1st line transport.

The appreciation underlying this allocation dealt mainly with the risks likely to be encountered by the first contingent of a British Expeditionary Force in a campaign against Germany, when it was considered quite likely that they might be subjected to intensive gas attack in the early days of a war. This possibility, combined with the small size of the force, necessitated the high percentage of the reserves.

It was realized after the outbreak of war that, if the figure of 200 per cent. continued to be applied at a flat rate as the B.E.F. grew, it would lead to astronomical figures. A sliding scale was therefore introduced. The full percentage was applied for a strength up to 200,000, reductions in percentage being made for each excess 100,000 up to 400,000

In March, 1940, the position was re-appreciated, and the probable scale of attack re-estimated. As a result, the average daily contamination rate for the whole force was re-assessed at 2 per cent. Reduced scales of decontamination reserves, as in Appendix XXII, were accordingly authorized in June, 1940, some modifications also being made in the reconditioning reserves which arose partly from the revised appreciation and partly from a recalculation of wastage rates. The scale of reserves in Appendix XXII held good until the end of the war. They were reviewed in 1943 but it was not considered advisable to modify them.

In the United Kingdom static units including those of the Anti-Aircraft Command had no specific allotment of 2nd or 3rd line reserves. Local stocks in depots, increased to meet the commitment, would have been drawn on for this purpose.

Additional Scales for P.A.D. Purposes

These were allotted on a station basis for use in gas cleansing centres, gas defence centres, and aid posts in accordance with local Passive Air Defence schemes at home and in rear areas overseas. A detailed allotment of anti-gas equipment for each of these centres and posts was laid down in " Protection against Gas and Air Raids—Pamphlet No. 3—Passive Air Defence, 1939 ". The changes in the anti-gas organization for Passive Air Defence have been referred to in Chapter II, Section 4, and the method of allocation of equipment at the end of the war will be found in " Regulations for P.A.D. Organization, 1944 ".

Scales of Training Equipment

Prior to and in the early part of the war separate scales of equipment were authorized for training. These are referred to in Chapter VII.

War Office Scales for the War against Japan

In 1945 revised War Office scales for troops proceeding eastwards were in preparation when the war ended. These did not vary fundamentally from the scales in the Appendix XXII with the following exceptions :—

(*a*) It was intended to provide a " Poncho " instead of a Cape.

(*b*) When the poncho became available it was to be issued to M.T. drivers, etc., in lieu of a suit, anti-gas, light.

(*c*) Only motor cyclists were to receive both a suit, anti-gas, light and a poncho.

(*d*) Scales of signs, warning, gas were reduced by 50 per cent.

(*e*) Sets of clothing for decontamination tasks consisted of :—

3 pairs gloves, anti-gas, oilskin, *or*
2 pairs gloves, anti-gas, rubber.
2 pairs overmittens, anti-gas.
1 pair overboots, anti-gas.

plus either { 1 jacket, light, anti-gas.
1 pair trousers, light, anti-gas.

or .. { 2 jackets, jungle, A/G, impregnated.
2 pairs trousers, jungle, A/G impregnated.
2 sets underclothing, impregnated.
3 pairs socks, impregnated.
1 hood, impregnated.

India

Up to December, 1941, War Office scales were applied to all formations proceeding overseas from India ; special scales were authorized for troops allotted to the defence of the North West Frontier, but troops located in any other part of India received a training scale only. After the outbreak of war with Japan, Field Army scales on the War Office model were introduced. Individual issues of personal equipment were similar to the War Office scales, but some unit reserves were reduced and certain items of unit equipment were issued on a restricted basis. This was partly because the scale of attack from Japan was likely to be less intensive than that which was to be anticipated in a European theatre, partly to ease the strain on difficult supply lines, and partly, in the case of oilskin clothing, owing to the unsuitability of such clothing for wide scale use in the tropics. Allotments of sets of oilskin clothing for decontamination tasks were cut by 50 per cent, and issues for special purposes were eliminated or reduced. Reconditioning reserves were correspondingly adjusted, and the allocation of these to echelons in rear of 2nd line were modified to suit the Indian system of supply in the field.

As the threat to India receded, the issue of Field Army scales was restricted accordingly, a special India scale on a percentage of establishment basis being authorized for most of India, except those areas within range of Japanese air attack. Ultimately, in 1945, the field force scales promulgated by S.E.A. Command were on a percentage establishment basis.

Carriage of Anti-Gas Equipment

During the early part of the war there was great insistence that a man should invariably carry his personal anti-gas equipment and never be parted from it under any circumstances. The object of this was to maintain a constant state of readiness against the first use of gas as well as to ensure the complete familiarity of the individual with his equipment, which is so essential under gas warfare conditions. It also facilitates the inclusion of gas as part of normal training.

The principle underlying the allocation of reserves was that replacement of damaged, expended or contaminated anti-gas equipment, as also of contaminated clothing, should be automatic in a similar manner to ammunition.

In 1942 the order in the United Kingdom that personal anti-gas equipment should be carried at all times was relaxed (*see* Chapter I, Section 5) and, as time passed and gas warfare continued to hang fire, it was ultimately accepted in the interests of mobility that personal equipment could be relegated to 1st line transport or dumped during phases of operations when the risk was acceptable to the commander concerned(1).

In April, 1943, 2nd and 3rd line reserves in the United Kingdom were withdrawn to Ordnance charge as a measure of economy, transport thereby released continuing to be earmarked for their carriage should the necessity arise. This practice was conformed to in all overseas theatres.

In 1940 loading trials were held to determine the best methods of loading unit, 2nd and 3rd line reserves. Consideration was given to special racking for vehicles carrying these reserves to enable individual items to be readily extracted. No racking, however, was ultimately provided, although in the early part of the war some formations improvised their own schemes. Most anti-gas equipment was packed in cartons which could be loaded in such a way that they would be within easy reach.

The unit reserves of an infantry battalion, including decontamination reserves of clothing, could be carried in one 30-cwt. lorry. The 2nd line reserves for an infantry division, *i.e.*, the quantities shown in Appendix XXII, Part VII, column (*i*) and Part VIII, column (*k*), required six 3-ton lorries, and additional transport was allotted accordingly to the divisional petrol company.

Ultimately this reserve was carried proportionately in each brigade, etc. company, corps troops petrol companies carrying this reserve for corps troops. Details regarding carriage, distribution and replacement of 2nd and 3rd line reserves will be found in R.A.S.C. Training Pamphlet No. 7, Part III—" Distribution of Blankets, Anti-Gas Equipment and Reserve Clothing in Advance of Railhead, 1942 ".

Note (1).—*See* Chapter I, Sections 5 and 6 for S.H.A.E.F. and 21 Army Group Orders regarding the carriage of personal equipment, unit equipment and 1st line reserves, both for the " Overlord " battle and subsequently. In all overseas theatres similar practices were adopted, varying in detail according to the particular circumstances in each theatre.

SECTION 5.—DEFENCE OF A.F.Vs. AGAINST CHEMICAL ATTACK

Development of gas weapons for attack against A.F.Vs. demonstrated the vulnerability of A.F.V. crews to certain forms of attack (*see* Chapter III), in particular, attack with the jet, anti-tank, and the 65-lb. air bomb. A direct hit or very near miss with the latter led to gross contamination of the tank and considerable ingress of liquid through gun mantlets, hatch doors, turret rings, periscopes, revolver ports, ventilation cowls and louvres. In addition to the dangers of contamination to personnel, subsequent decontamination of the tank presented a serious problem. In the case of the jet, anti-tank, the action of prussic acid was so rapid that, in a surprise attack, timely adjustment of the respirator was improbable.

Tear gas grenades were also very effective causing severe weeping in unprotected crews which continued for a few minutes after respirators were adjusted. (*See* Chapter IV, Section 1 for development of anti-tank gas weapons.)

Early in 1942 consideration was given to special means of protection of tanks over and above that afforded by the use of anti-gas equipment provided for the Army as a whole.

Drastic modification in the design of tanks in order to prevent ingress of liquid was out of the question, since this could not have been done without adversely affecting production, and even if attempted seemed incapable of achievement in time to be of use during the war.

The perspective with which the subject was approached was that the tank already provided the crew with a considerable degree of mechanical protection which was denied to the ordinary soldier, and that, important as the tank was as a weapon of war, the crew could not expect to be afforded 100 per cent. safety nor should any attempt be made to provide this, as it was likely to involve elaborate precautionary or protective measures. Experience indicated that the wearing of respirators, particularly the light respirator, did not, with practice, constitute an unacceptable handicap under European climatic conditions, and that tanks could be adequately fought with the crew wearing respirators.

Drastic re-design of tanks being impracticable, the only method of approach lay in the development of equipment which was readily applicable to existing designs of tanks. Various methods were tried out on the Churchill tank Mark 1. A system of canvas or fabric covers, for the outside and inside of the turret and driver compartment, and sponge rubber gaskets for hatch covers, revolver ports, etc., were fitted, as were cowls for roof ventilator covers and baffles for ventilation louvres in the base of the turret. All possible points of ingress were also heavily greased. Similar trials were carried out with Valentine tanks.

The liquid still got in, but in much reduced quantity, so that personnel could have protected themselves by normal methods and the decontamination problem would have been greatly lessened. Unfortunately, apart from the possibility of damage to covers by gunfire and other causes, the fitting of the equipment led to such a marked reduction in ventilation efficiency that the gain in protection from liquid gas was achieved only with considerable increase in the risk of poisoning by carbon monoxide.

" Lee " and " Sherman," " Cavalier " and " Centaur " tanks, and the Canadian " Ram " were later subjected to attacks and found to be equally vulnerable to attack with the 65-lb. bomb and with the jet, anti-tank, although the Sherman III and Sherman V were less vulnerable to liquid ingress when attacked from the front with the 65-lb. bomb.

Since the above methods for the prevention of ingress of liquid had proved unacceptable, little else could be done except from the angle of developing protection which could be afforded to crews by clothing, etc. To cater for the prussic acid risk, a light respirator container with improved protection against this gas (the " white dot " container) was introduced for A.F.V. crews.

A form of oilskin apron was also developed, because the cape, anti-gas was not well suited for wear inside a tank. This apron was not put into production since in any case it only provided partial protection, and it was ultimately decided that, as far as the European theatre of war was concerned, the threat to personnel in tanks did not warrant their being afforded special protection above that provided for other troops.

In the climatic conditions of the Far East the wearing of protective clothing inside tanks, as also the wearing of the respirator for long periods, was proved by trials in the theatres concerned to be too exhausting to be practicable. Furthermore, the prussic acid danger was the more immediate as the Japanese possessed a glass grenade with this charging. As a short-term project for the war against Japan a requirement was stated for the development of a breathing apparatus of a design originating in Canada, and which was to be worn in conjunction with an air-conditioned suit.

Development of the Canadian design of filtration apparatus (Individual Protective Unit—I.P.U.) was carried out in England. The final design consisted of an electric blower drawing air through a dust filter from existing intakes in the roof of the vehicle : the blower delivered the air to two, or three, boxes each containing four Light Type containers, Mk. II arranged in parallel, and thence to a loosely fitting face-piece supplied for each member of the crew. The air flow, and hence air pressure, was sufficient to prevent ingress of gas to the facepiece, and also provided cooling ventilation at all times. The delivery of air in the facepiece was arranged to overcome discomfort from jet effects, and the facepiece was fitted with the same speech transmitter as the standard light respirator for tank crews. Sets were available for trial when the war with Japan ended.

Superficially, the use of air-conditioned clothing appeared highly desirable to lessen fatigue in hot climates, and anti-gas oilskin clothing for protection against liquid gas should have been wearable over this without ill effect. The space available in existing A.F.Vs. was, however, insufficient to accommodate a filtration unit large enough to remove mustard gas vapour from the volume of air required for the suit, and the development of the system as an anti-gas device was abandoned. Air-conditioned clothing *without* gas protection was developed and manufactured in small quantity towards the end of the war.

It was realized from the beginning that the only really satisfactory answer to the related problems of gas, dust, ventilation and flash-back in fully enclosed A.F.Vs. was a system of positive pressure ventilation with one central intake connected with a purifier, and that this was the ideal to be aimed at, although incapable of achievement during the war. Until A.F.V. design has reached a stage when the sealing of the fighting chamber has been greatly improved and space has been provided for a suitable filtration unit, this is obviously impracticable ; reliance must be placed either on individual protection (the respirator) or on some form of breathing device as referred to above.

Decontamination of Tanks

In most circumstances in which tanks would have become contaminated either directly by weapons or through crossing contaminated ground, decontamination would not normally have presented a problem which could not have been overcome by field decontamination processes applied by crews or in the unit.

In the worst case, namely, a direct hit or very near miss by a mustard gas bomb, such as our 65-lb. bomb, resulting in gross external contamination and considerable ingress of liquid into the interior, decontamination would have been a recovery job. The tank could have been made reasonably safe to fight after about one hour spent in field decontamination processes followed by weathering for 24 hours, but there would still have been a vapour danger from liquid remaining in inaccessible places, and the replacement crew would have had to take precautions. To do the job properly would have involved removal to base and dismantling.

The results of a direct hit with a mustard gas bomb should be viewed in their proper perspective *vis-à-vis* a direct hit with H.E. or incendiary. The chances of it occurring are no greater ; but whereas in the two latter cases the tank is an almost certain " write-off," in the former, the tank could, if necessary, be back in action in 24 hours.

Teams toured all R.A.C. units in 1942–43 to demonstrate the methods of decontaminating tanks.

CHAPTER VII

GAS TRAINING—PRINCIPLES AND DEVELOPMENT

Introductory

Before 1941, defensive gas measures had always been taught under the heading of "Anti-Gas Training." The linking of anti-gas measures with passive air defence, which has been referred to in Chapter II, Section 4, led to the purely passive aspects of defence against gas becoming overstressed, and this was in some degree perpetuated in the title "Anti-Gas Training." There was a tendency amongst a large portion of the Army to feel that if gas were used there was little more that could be done about it than adjust the respirator, clean up from blister gas, sit tight and hope for the best. This passive outlook was particularly in evidence in relation to gas attack from the air—and in fact, in relation to air attack as a whole. It was to assist in counteracting this trend that training in protection against gas was given the more comprehensive name of "Gas Training." This was more indicative of the measures required to combat the enemy's use of gas as a whole. It was essential that the soldier should continue to fight or carry out his allotted duties despite the presence of gas.

The details of gas training are given in the pamphlets "Gas Training, 1942" and "Advanced Gas Training, 1943"—the latter pamphlet being supplementary to the former. They replaced the series of handbooks introduced in 1939—"Protection against Gas and Air Raids"—Pamphlet No. 1 (Protection against Gas in the Field), Pamphlet No. 2 (Respirators) and also—in so far as it referred to protection against gas—Pamphlet No. 3 (Passive Air Defence).

This book is not a textbook. The details of training will be found in the textbooks referred to above. In the following paragraphs only the general principles of training are referred to in order to show the major developments which occurred during the war period and the reasons for them.

The same principles applied and similar developments occurred in all theatres of war and throughout Empire Forces. Details of application varied in some cases according to special conditions prevailing in the theatre. Liaison between United Kingdom and overseas schools was lacking in the early days of the war, but great improvements were effected.

Training Policy—General

The fundamentals underlying training and protection against gas did not themselves change in any major degree during the war. It was rather the methods of training which changed.

The most important development was the carrying to its logical conclusion of the trend, which had started before the war, towards the simplification of the subject matter as taught to the soldier, both by the deletion of all technical detail and by the introduction of more practical methods of presentation.

Next in importance was the practice of training with actual war gases, the establishment of gas compounds and the introduction into individual training of exercises in which real mustard gas was used.

Other changes in training policy included the basing of defensive training on enemy methods of weapon development, as they became known, rather than on a wide range of possible methods of gas attack, some of which were largely of historical interest only ; the wide issue of a gas training film ; the results of our own equipment development ; modifications in teaching in conformity with the change in viewpoint on the efficacy of high spray attacks, and also on the relative danger of mustard gas in its liquid and vapour form, particularly in tropical climates ; and the inclusion of the passing of the Tests of Elementary Training for gas as a qualification for proficiency pay.

Simplification of Training

During the early and middle 1930s a very considerable amount of scientific and technical detail was taught to officers and N.C.Os. attending gas schools, in particular in respect of the war gases themselves and their possible means of discharge. Training was on theoretical lines and involved considerable numbers of lectures, practical periods being devoted almost entirely to respirator drill and care of the respirator, although there were some demonstrations of gas reconnaissance, and duties of decontamination squads. There were also some exposures to war gases. Although students were warned that a large amount of the technical detail taught was to give them the necessary background as regimental instructors, and that it was both undesirable and unnecessary to attempt to impart their full knowledge to every soldier, no precise instructions were given about the exact details to be taught to junior leaders, trained soldiers, and recruits respectively. Some tests of elementary training were laid down in 1935, when a new gas manual was introduced, as also were respirator lessons for recruits and a suggested syllabus for unit cadres.

At the time there was some justification, particularly when training officers, for attempting to train them to a standard above that essential solely as the unit instructor. The organization in the Army for defence against gas was only in a formative stage, and there were no staff officers in formations to handle gas matters. Knowledge throughout the Army as a whole was negligible, and the desire to have as many officers as possible with a wide knowledge and a solid background of technical detail to enable them to propagate "gas" in its correct perspective was excusable.

Between 1937 and 1939 the text book was re-written and more detailed syllabi for unit cadres etc. were included, as well as a few standard lessons for certain items which lent themselves most readily to practical methods of instruction, *e.g.*, decontamination of a rifle, duties of sentries. At the Gas School the subject matter as a whole began to be presented in a more straightforward and less technical form.

With the introduction of conscription in 1939 followed by the outbreak of war, the amount of time which could be economically allotted in units to gas training *vis-à-vis* other forms of training became a matter of great importance. Another factor of considerable import, particularly in the case of N.C.O. instructors, was the necessity for eradicating the desire of some instructors to dazzle their audiences with pseudo-scientific detail—in most cases inaccurately absorbed, insufficiently digested and largely irrelevant.

Although gas is fundamentally a scientific subject, it is totally unnecessary for every soldier to understand the scientific aspects. It became vital for the subject to be reduced to its essentials. As with all other subjects the standard to be attained by the recruit, soldier, junior leader and superior commander respectively, was that required by his rank.

The above factors accelerated the trend towards simplification, and led to the complete standardization of the detail to be taught in all units to the varying ranks. This was laid down in a new pamphlet " Gas Training, 1942 ", which took the form of a series of lessons which were to be taught to each soldier followed by additional detailed lessons for junior leaders. To give practice to leaders and men in certain aspects of gas warfare which were most likely to occur, a series of outdoor exercises adapted to arms of the service were also included in detail. These lessons and exercises were to be strictly adhered to in units, and no extraneous information of a scientific or technical nature was to be taught.

The fundamentals behind this development, as far as the ordinary soldier was concerned, were that he should be able to recognize the presence of gas either as a cloud or as a liquid, or by any suspicious smell or by any feeling of choking or irritation to the eyes or nose, but without identifying the exact type. It was unnecessary to train every man to attempt to recognize the many individual gases that might be used. Had gas warfare started, the range of gases commonly used by the enemy would have been rapidly learnt by experience. Only the most salient characteristics of groups of gases, *e.g.*, choking gases, blister gases, were taught.(1)

Having recognized or suspected the presence of gas the man had to be able to do one of two things—or in some circumstances, both :—

(*a*) Adjust the respirator quickly.

(*b*) Carry out personal decontamination.

A third requirement was for an almost automatic reaction by all ranks to the unit gas alarm system.

Junior leaders had to be taught the minimum essentials not only to enable them to act as instructors to their men but that they might be capable of protecting their commands while continuing with the task in hand in the general run of situations likely to be encountered had gas warfare started.

Officers and N.C.Os. attending courses at the gas schools (unit gas officers and N.C.Os.) were taught how to train junior leaders to instruct their men in each of the lessons and exercises laid down for soldiers in " Gas Training, 1942 ". In addition, to enable them to act as advisers to their applicable commanders, to inspect anti-gas equipment, and to assist in the gas training of their unit as a whole and in its gas protection in the field, they were given certain additional instruction. This was incorporated in a pamphlet "Advanced Gas Training, 1943 " which contained sufficient information for the above purposes and, in addition, gave them the necessary background required as a unit instructor. Technical and scientific detail was omitted, and the whole subject reduced to straightforward essentials.

Note.—(1) Compare this with U.S. policy referred to in Chapter II, Section 9. It is not untrue to say that whereas British practice was directed to creating confidence to meet the threat of gas, the United States relied on the fear of consequences for propagating an incentive to training.

As has already been mentioned in Chapter II, Section 4, the principle was increasingly emphasized that it was the duty of every unwounded man to prevent himself from becoming a blister gas casualty by the use of his personal equipment.

By 1942, the drill known as " Personal Decontamination " had become firmly established. This drill was to be carried out by every individual on becoming contaminated with liquid blister gas by any means. It was divided into two parts—Part I being the removal of any free liquid on the exposed skin, the application of ointment, and the changing of contaminated eyeshields. This was to be an automatic reaction and could be carried out on the move and wherever tactical conditions permitted. Part II had to be done at the halt and involved the application of ointment to contaminated clothing, the decontamination of weapons, and the changing of sleeve detectors (if necessary). The drill varied according to whether or not the cape (or light suit) was being worn at the time of contamination, and it was to be carried out under all circumstances unless tactical considerations prohibited it. For ease of remembrance the drill was based on the mnemonic " COECDO ".

The points noted above are the major points connected with the simplification of gas training and its application to the requirements of varying ranks. Details of the above and other points of interest will be found in " Gas Training, 1942 ".

Training with real war gases

Up to 1940 the use of real war gases for training in units had been restricted to tear gas (CAP) and nose gas (DM) in ampoule form for use in gas chambers during individual training ; a set of smelling bottles containing mustard gas, lewisite, phosgene, and liquid tear gas (BBC), for use during lectures on war gases ; and tear gas generators for use out of doors. From 1940 onwards a wider range of war gases was introduced into training. The use of real war gases for training was of the utmost value and did more than almost anything else towards banishing the fear of gas, creating confidence, and establishing a correct perspective in the minds of all ranks on the real dangers involved.

To begin with, the policy was that every man should experience nose gas in the gas chamber at least once in his service and the fact that he had done so be recorded in his paybook. As pointed out in Chapter III, Section 1, it is almost impossible to convey the true effects of these gases by means of verbal description. The next step was for every man to have a spot of mustard gas placed on his arm and then to apply anti-gas ointment. This was to create confidence in the use of ointment and prove that by its early application the effects of liquid mustard gas could be counteracted. The use of phosgene out of doors from small cylinders began concurrently with training with real mustard gas.

By 1941 gas compounds were established on a wide scale in all commands in the United Kingdom so that they should be readily accessible to all units. A gas compound consisted of a gas chamber where nose gases and tear gases could be demonstrated, a concrete surface on which actual mustard gas could be used, and a selection of typical road and other surfaces on which blister gas training substitutes could be placed to show what mustard gas looked like on varying surfaces. In addition phosgene and liquid tear gas (BBC) were also available. Mustard gas and BBC were initially provided in pint pots, and later in bombs, ground, 6-lb., from which the explosive charge had been removed. Units could also draw any of the stores mentioned from these compounds and use them in improvised gas compounds established in unit areas.

The final development in training with real mustard gas was the introduction in 1942 of a series of exercises in which contaminated weapons were decontaminated, contaminated ground was traversed, and men were exposed to mustard gas vapour for periods up to 30 minutes during which handling of weapons was practised. Their purpose was to demonstrate that contaminated ground does not present a major obstacle to traversing, and to prove that short exposures to mustard gas vapour were ineffective provided that the respirator was worn. In the case of specialist and ancillary arms, the exercises were adapted to the normal tasks and situations likely to be met with by those arms. C.W. staff and technical officers were available to advise as to the extent of any risks run, and the implementation of the whole policy of training with real war gases was a special responsibility of these officers in their own formations. Although this form of training was adopted throughout all Empire Forces, both in the United Kingdom and elsewhere, only a few accidents occurred, mostly not of a serious nature ; the isolated fatalities that occurred were due to foolish acts on the part of inexperienced individuals. Bombs, ground 6-lb., were exploded for contaminating ground for these exercises.

Nose gas generators were introduced for collective training but real mustard gas was not used on such exercises, owing to the obvious difficulty in control. From 1939 onwards, training ground bombs charged mustard gas substitute had been available to all arms and training chemical mines to R.E. units.

The policy was that all ranks, including the A.T.S., should undergo the above training with real war gases and, although this ideal was not fulfilled, a very large proportion of the Army was so exercised. This method of training unquestionably paid a high dividend and proved that, with the war gases as known during the war, the use of real gas in training is essential if a correct perspective is to be attained. At the time of writing our knowledge of nerve gases is incomplete and their nature is more insidious than any other gas hitherto known. They are lethal in very low concentrations and it is not yet possible to say whether their use in training will ever be a possibility.

The Basing of Defensive Training on Enemy Methods of Projection

Until 1941 the tactical background on which our defensive training was based was largely that of our own offensive doctrine (*see* Chapter V) combined with methods of projection which had been current during the 1914–18 war, although certain of these methods were regarded as being unlikely. As the war progressed, our knowledge of enemy methods became more complete through captured documents and equipment, and it was essential that known enemy weapons and doctrine should be high-lighted in defensive training. This was particularly necessary in relation to the German development of the H.E./chemical shell which introduced problems of recognition and protection of a somewhat different nature from other means of gas projection regarded as common practice, both by ourselves and our enemies.

In addition to the fact that for security reasons any wide dissemination throughout the Army of our own gas policy was impracticable, it was sound from a defensive point of view to paint the picture of gas warfare as it was likely to be actually delivered on us by the enemy. With the publication of " Gas Training, 1942 ", it was this viewpoint that formed the background for defensive training throughout the Army, rather than the general means of projection of gas which had formed the basis of the 1939 pamphlets.

Gas Training Films

In 1941 a gas training film was issued on a sufficiently wide basis to enable it to form part of the normal gas training of all recruits. It showed the action to be taken against various types of gas attack and demonstrated the methods of teaching the major lessons which had to be instilled into all ranks. A justifiable criticism of this film is that it was over-detailed and attempted to teach too much in a short space of time. It was largely based on the dangers of aerial spray, particularly high altitude spray, and it soon became out of date as a true picture of the relative dangers from spray and other forms of attack. In spite of these failings it was of great value. It gave the man a good general picture of gas warfare and emphasized one vital point, namely that if attacked by gas it is essential to get on with the job in hand. It played its part in counteracting the passive outlook which had become prevalent in the early years of the war owing to the linking of anti-gas training with passive air defence which has been referred to above.

Our Own Equipment Development

In 1939, the issue of an ointment which could be used prophylactically had only just begun and impregnated uniforms were not introduced until after the beginning of the war. Until 1943, when the pocket vapour detector became available, there was no detector for mustard gas vapour, nor was there any means of assessing the degree of danger present when mustard gas vapour was encountered, in relation either to tactical hazards or to defensive methods. When these items were introduced into the Service the training in measures to be adopted against mustard gas vapour had to be brought up to date. In addition, the simplified means of decontaminating ground (referred to when discussing bleaching powder in Chapter VI, Section 2) were incorporated in training, as were easier and more standardized methods of decontaminating vehicles. Respirator drill, also, was considerably simplified, and adjusted to the light type respirator when introduced.

Modification of Training in Relation to the Relative Dangers of Mustard Gas Liquid and Vapour

As the spray hazard came to be seen in better perspective (*see* Chapter V) the background for training correspondingly changed. Initially, too great an emphasis had been placed on the spray hazard. Although high spray attack continued to be taught as a possible means of attack, only passing reference was made to it during instruction as being a method which was unlikely to have any serious effects beyond a nuisance value provided that the normal drills for protection against spray attack were observed.

The enhanced value of mustard gas vapour particularly under tropical conditions came to be emphasized, and from the end of 1944 onwards, when it was visualized that numerous formations would be proceeding to the Far East on conclusion of the European war, the tropical aspects of chemical warfare were increasingly taken as the background for instruction at the Army School of Chemical Warfare, and in the formations concerned. In fact, during the various phases of the war the background for instruction at the School had been the conditions prevailing in the theatre or theatres of immediate importance.

The Inclusion of Tests of Elementary Training as a Qualification for Proficiency Pay

In 1943 the Army Council laid down that all ranks must qualify in the tests of elementary training for gas before being entitled to proficiency pay. This assisted in maintaining the minimum standard of training throughout the Army and at least ensured that gas training was not entirely overlooked.

General Standard of Gas Training Attained throughout the Army

The fact that gas warfare never eventuated was a factor which created great difficulties in maintaining a proper balance throughout the Army in relation to gas training. The soldier had to learn many new weapons and apply them to the many conditions of the war as they developed. Time was short and it was natural that commanders should get down to the immediate essentials and that training against a weapon which had not been used, and which, as the war progressed, many considered probably never would be used, should take a back place. The value of senior officers' courses in assisting in maintaining a correct perspective has been referred to in Chapter II, Section 5. The danger was that so long as gas warfare hung fire the element of surprise in the first use of gas was at a premium, and it was essential that a minimum standard in gas training should be maintained to ensure that the casualties occurring from the initial surprise attack should not be so large as to cause, at the worst, more than a local tactical reverse of short duration. Policy, therefore, was to maintain training at an essential minimum standard to ensure that no serious reverse occurred.

Owing to the element of surprise and the fact that the initiation of gas warfare was in the hands of the enemy and not of ourselves, the first use of gas, even against the best trained troops, would almost certainly have caused considerable casualties and afforded a temporary gas supremacy to the initiator.

Although the standard of training reached by many units in the Army was not very high and, in particular, the details of the drills against mustard gas liquid were not universally well applied, it may be said that in general the Army was reasonably well trained to meet the risk in so far as the war gases then known were concerned. The standard varied at different stages in the war according to the probabilities of the enemy using gas at various strategic phases. The standard of training was higher in 1942 than in 1943, but rose again before "D" day.

Had the Germans used nerve gas against us the initial results would have been serious. The respirator and eyeshield, however, afforded adequate protection, and it is fair to say that our system of training was readily adaptable, except in so far as the question of detection was concerned. This latter is dependent on scientific development.

Our policy of continuing to maintain training, as well as our " set-up " at the gas schools, was known to the enemy, and after the war it transpired that the Germans had assessed our standard of training and general preparedness as being very high. This was one factor—and a very decisive one—in their attitude towards the initiation of gas warfare. It is true to say, therefore, that, although our general standard of gas training may not have been as high as would have been desired by those whose special responsibility it was, it contributed, with other factors, to preventing the outbreak of gas warfare, and, in this sense, adequately achieved its object.

Training Stores and Scales

Both before and during the early part of the war, it was the policy to issue special scales of training equipment, partly in order to conserve war stores, and partly, in the case of detectors, to have a store which would react with training substitutes for mustard gas with a reasonable similitude to the reaction between the genuine articles. In the former category came rubberized capes and suits of light clothing in lieu of oilskin articles, and in the latter category were Detectors Gas Ground (Training), Detectors Gas Spray (Training) and Detectors Gas Sleeve (Training). Powdered whiting was issued in place of bleaching powder, as was vaseline for use in empty ointment tins.

Liquid mustard gas penetrated rubber garments more quickly than it penetrated oilskin garments, and initially great insistence was placed on the order that the rubberized garments should not be used as war equipment. In July, 1940, rubberized clothing was recognized as adequate for war use and training issues were merged with G 1098 equipment, the rubberized articles wasting out as replaced by oilskin. Henceforward units used G 1098 oilskin clothing for training purposes.

In 1941 the issue of powdered whiting and vaseline was discontinued and in 1942 training detectors were abolished. The reaction between the training substitutes when compared to that between the real articles was not sufficiently lifelike to warrant retention of the substitutes, and they caused confusion by their apparent likeness to the real articles. In any case by then they had served their purpose, and the availability of real mustard gas to demonstrate the true reaction on war detectors made them redundant. Henceforward no special scales of training equipment were issued in addition to war equipment.

There was initially a great insistence by most instructors on the need for smelling sets. In actual fact they were more in the nature of an instructor's box of tricks than of real value in training in recognition. They unduly highlighted detection by smell which was an unlikely means of recognizing a most probable type of attack, *viz.* with H.E./chemical shell. With the wider use of real war gases in training, these sets were no longer required.

CHAPTER VIII

DEVELOPMENT IN ENEMY COUNTRIES

SECTION 1.—GERMANY

A history of German gas warfare is being written by the Chemical Defence Department of the Ministry of Supply.

In the preceding chapters reference has been made to relevant aspects of German development, and their notable achievement, the discovery of the nerve gases, has been dealt with in detail in Chapter III, Section 4. In this chapter a general survey only is undertaken, largely with the object of comparing the German gas effort with our own.

The Germans expended a very considerable effort in gas warfare preparations. Even before the war had started they had set up research and training establishments on a lavish scale and had put several factories into commission for the manufacture of gas and gas warfare appliances. Other new factories and filling stations were constructed during the war, and much research effort was put into development work. Altogether about 70,000 tons of gases were manufactured, the greater part of which was charged into munitions of various types, totalling some 8 million rounds. Details will be found at Appendix XXIV.

Their defensive measures were elaborate and equipment was provided in abundance to cater for almost any imaginable contingency that might arise.

As far as can be ascertained, apart from Hitler's aversion to gas, the major factors which governed the German decision not to initiate gas warfare were that in the earlier phases of the war the successes of German arms rendered the use of gas unnecessary, the more so as it was not suited to the tactics of a war of rapid movement. In the later phases, shortage of raw materials, dislocation of their chemical industry through Allied bombing, loss of air superiority, and appreciation of the great chemical potential of the Allies induced the Germans to give up any idea of initiating gas warfare and to take all possible measures to avoid any incident which might precipitate the Allied use of gas. They were very impressed with the reported scale of United States preparedness, and anxiety about the intentions of the Allies in relation to gas warfare forced them to devote considerable production effort towards maintaining defensive preparations and sustaining as far as possible their offensive potential for counter gas warfare. Finally, they did not succeed in developing their newest gases (the nerve gases) according to programme while they still possessed an adequate bomber force. The fact that they assumed that the Allies also had these gases has been referred to in Chapter I, Section 6.

The organization for the offensive use of gas in the German Army hinged on the Artillery, and on the smoke troops—the " Werfer Brigades." A short history of these Brigades is given in Appendix XXV. The types of gas favoured by the Germans, their means of discharge, and the allocation to weapons can be seen from Appendix XXIV.

Defensively all units had a number of officers and N.C.Os. trained at Army Schools. Each headquarters and sub-unit down to companies or equivalent organized gas scout sections of one N.C.O. and three men who were specially equipped with detection equipment and protective clothing and received special training. The organization for large-scale decontamination—" Decontamination Batteries "—is referred to in Appendix XXV.

The High Command organization for the direction of chemical warfare and for research, supply and development is outlined at Appendix XXVI, which also shows the training and research establishments which it controlled.

There were two research establishments—Spandau (Berlin), and Raubkammer, near Munster, north-east of Hanover. The former consisted of laboratories where research of a more fundamental nature was carried out, and the latter both of laboratories and of extensive, well-equipped and highly-organized experimental grounds and ranges, where applied research and field trials were undertaken.

There was a parallel staff organization in both the Navy and the Air Force, which also had laboratory and experimental facilities of their own.

Army and Air Force experimental stations and gas munition depots were located in the Munster area. During the months immediately preceding the German collapse most of the Spandau laboratories had been evacuated to Munster as well as most of the office of Wa Prüff 9, thus concentrating in the area most of the people of consequence in official C.W. circles.

In some respects the Germans exhibited a curiously unrealistic approach to many of the problems of gas warfare. For example, they made no attempt to determine the casualty producing dosage for mustard gas vapour on the skin. The significance of drop size of mustard gas spray was also not appreciated and the true importance of thickening agents to reduce shatter was, therefore, missed. They did use thickeners but with the object of increasing ground contamination performance. Also, in their research work no real effort was taken to extrapolate toxicity from animals to man, with the result that the dosages underlying their munition expenditure figures were considerably lower than our own.

In the case of ground contamination with mustard gas, judged from the standpoint of their whole defensive set-up and their elaborate procedure for dealing with such areas, which was not dissimilar to that for a river crossing, they appear to have over appreciated both the dangers involved and the tactical effects of such contamination. In their doctrine, as compared to the British there was less emphasis on the need for improvisation and the acceptance of normal risks.

In assessing the German effort, however, their weaknesses must not be allowed to obscure their undoubted achievements—in particular their discovery of the nerve gases—which more than offset any deficiencies.

Comparison between Wartime and Postwar Knowledge of German Gas Warfare

During the war intelligence contained frequent and numerous references to the movement of stocks of gas weapons by the Germans both to their Eastern and Western fronts. These references were mostly of a somewhat indefinite nature, but both their frequency and the sources from which they came caused attention to be given to them, although to a lesser degree during the latter stages of the war. They were, in fact, all false, since not one single round of gas munitions was ever moved outside the borders of the Reich. In 1944 the Germans did consider forming dumps behind the coastal area in France : orders to do so were actually issued, but subsequently cancelled to avoid the possible occurrence of C.W. incidents.

Our knowledge of their defensive equipment and organization was almost complete through the capture of equipment and documents, and we had long known of the gas school of Celle. With the exception of the nerve gases,(1) we had a good insight into their offensive methods through the capture of the document " Gasscheisen aller waffen " by the Russians. This document also gave us an indication of their system of marking gas munitions with coloured rings.

In our efforts to discover the location of experimental stations, etc., we had accumulated enough evidence to show that Munsterlager was the centre of the area where experimental work was conducted and that large stocks of gas munitions existed there, but the size of the research establishment at Raubkammer,(2) and the lavish scale of its equipment, far exceeded anything that we had anticipated. We knew of Spandau before the war and we obtained further information of its activities during the war.

We had little knowledge of German manufacturing establishments or methods. Their system of manufacturing C.W. products only at specially built factories and their elaborate schemes of code names for intermediary and final products hampered the work of agents.

Note (1).—The lack of tangible evidence regarding these gases has been referred to in Chapter III, Section 4.

Note (2).—It is curious that after one mention in 1935 the name of Raubkammer itself did not appear in any report until 1945. Breloh, where experimental work was conducted during the 1914–18 war, is the immediately contiguous village and we had assumed the presence of a research establishment near Munsterlager. Reports of the reorganization of laboratories in this neighbourhood and of field experiments, troop exercises with gas, etc., had come in from 1937 onwards.

SECTION 2.—JAPAN

The gas warfare activities of the Japanese have been investigated mainly by the U.S. authorities.

No clearly defined C.W. service existed in the Japanese armed forces, but for research, development, and training, organizations specifically identified with C.W. were encountered. There was an antagonistic spirit between the Army and the Navy, and the lack of technical co-operation between these Services is exemplified in the field of chemical warfare, each Service carrying out its own independent research and manufacture of war gases and anti-gas equipment. There was no separate air force, each service having its independent air arm.

Postwar information shows that the Japanese were fully aware that they were in no position to wage aggressive gas warfare on a modern scale, as had been indicated by captured documents before the end of hostilities.

Very shortly after the surrender the Imperial General Staff issued orders for the destruction of records of research and development, and although countermanded within a few days, a pretty complete job seems to have been done in the meantime, since few documents of importance on C.W. activities were located.

Both offensive and defensive development were on stereotyped lines and not indicative of any great ingenuity. Although both the Army, and to a lesser extent, the Navy examined numbers of substances in the search for a new gas, they found nothing superior to the established war gases (mustard gas, lewisite, phosgene, prussic acid, and the tear and nose gases). The Japanese maintain that the Germans refused them information on new German developments, but did tell them that the Russians might employ nitrogen mustard gas.

During the war we, if anything, overestimated the Japanese gas warfare capacity. Their total production of war gases was only some 10 per cent. of the German effort.

In the Japanese defensive equipment the Army material for light protective clothing, comprising rubberized silk and cellophane, is the item of greatest interest, owing to its very high resistance to penetration by mustard gas. A fairly wide range of detector equipment was also developed.

With regard to the repeated Chinese claims from 1938 onwards that the Japanese used gas against them, interrogation of leading personalities—including General Hideki Tojo—supported by a few isolated documents, established that tear and nose gases were used, and that this use was authorized by the Imperial General Staff at the discretion of commanders in the field. All those interrogated, however, drew a distinction between the use of these harassing gases—which they sometimes referred to as " light " gases—and of gases having a lethal effect, and maintained that the use of the latter types was not permitted by the Imperial General Staff, and that they were not used. It must be remembered, however, that interrogations were carried out in the shadow of war crime trials ; it seems probable that earlier in the Chinese " incident " a limited use of mustard gas, or mustard gas/lewisite mixtures, occurred on a few isolated occasions, but the extent to which this had the approval of the Imperial General Staff is not known. After the outbreak of war with the Allies, in view of their awareness of the weakness of their position and their fear of retaliation, there is more reason for believing the denials of the Imperial General Staff.

APPENDIX I

(*See* pages 40 and 118.)

Extract from Prime Minister's Broadcast Speech on 10th May, 1942.

" There is however one serious matter which I must mention to you. The Soviet Government have expressed to us the view that the Germans in the desperation of their assault may make use of poison gas against the Armies and people of Russia. We are ourselves firmly resolved not to use this odious weapon unless it is used first by the Germans. Knowing our Hun, however, we have not neglected to make preparations on a formidable scale. I wish now to make it plain that we shall treat the unprovoked use of poison gas against our Russian ally exactly as if it were used against ourselves, and if we are satisfied that this new outrage has been committed by Hitler, we will use our great and growing Air superiority in the West to carry gas warfare on the largest possible scale far and wide upon the towns and cities of Germany. It is thus for Hitler to choose whether he wishes to add this additional horror to aerial warfare. We have for some time past been bringing our defensive and precautionary arrangements up to date, and I now give public warning so that there may be no carelessness or neglect. Of one thing I am sure—that the British people, who have entered into the full comradeship of war with our Russian ally, will not shrink from any sacrifice or trial which that comradeship may require."

Serial No. (3rd September, 1939)	Item	Type of Charging: Mustard Gas (Y). Tear Gas (B). Phosgene (G).	Stocks: Charged.	Stocks: Empty.	Remarks.	Serial No. (19th April, 1940)	Stocks: Charged.	Stocks: Empty.	Rem[arks]
	ARMY WEAPONS								
1	4·5-in. How. Shell (1914–18 war type).	Y	—	18,000	Charging completed by December, 1939.	1	18,000	—	In France.
2	6-in. How. Shell (1914–18 war type).	Y	—	28,000	Charging completed by November, 1939.	2	28,000	—	In France.
3	25-pdr. Base Ejection Shell ..	Y	—	—	In development stage.	3	—	—	Design clear—ch and gas ava production fo
3A	25-pdr. Base Ejection Shell ..	B	—	—	In development stage.	3A	—	—	Design clear—ch not available.
4	5·5-in. Gun How. B.E. Shell ..	Y	—	—	In development stage.	4	—	—	Design not clear
4A	5·5-in. Gun How. B.E. Shell ..	B	—	—	In development stage.	4A	—	—	Design not clear
5	6-in. How. B.E. Shell	Y	—	—	In development stage.	5	—	—	Design not clear
6	Livens Projector Drums ..	Y	—	—	Some drums available for charging, but bursters only just cleared for production.	6	4,700	—	–
6A	Livens Projector Drums ..	G	—	—	Under consideration—no requirements stated.	6A	—	—	Charging begu 1940.
7	4·2-in. Mortar Bombs	Y & G	—	—	No requirement for the 4·2-in. Mortar was stated until 1941.	7	—	—	As for Septembe
8	5-in. U.P. Rocket	G	—	—	In early development stage.	8	—	—	Still in developm
9	Bombs, Ground, 6-lb.	Y	—	—	Requirements stated—no production forecast available.	9	—	—	–
10	Mines, Chemical	Y	—	—	Requirements stated—no production forecast available.	10	—	—	No production f
11	Bulk Contamination Vehicle (Installation Type A/LP).	Y	—	—	Design clear.	11	—	—	No production f
	AIR WEAPONS								
12	Bombs, Aircraft, 30-lb., Mk. I..	Y	—	?	Charging begun October, 1939.	12	42,000(*a*)	18,900	(*a*) Only 10,000 fuze productio ready for desp Total charge France (24,5 available end
13	Bombs, Aircraft, 30-lb., Mk. II	Y	—	—	—	13	—	4,000	Stored in Un uncharged o storage life To be charged
14	Bombs, Aircraft, L.C., 65-lb. ..	Y	—	—	Not introduced until 1941.	14	—	—	–
15	Bombs, Aircraft, 250-lb. ..	Y	?	?	Charging begun August, 1939.	15	10,000	500	–
15A	Bombs, Aircraft, 250-lb. ..	G	—	—	No R.A.F. requirement for Phosgene weapons at this date.	15A	—	—	As for Septembe
16	S.C.I., 250-lb.	Y	—	?	Charging begun September, 1939.	16	361	39	In France 261 ; 100.
17	S.C.I., 500-lb.	Y	—	?	Charging begun October, 1939.	17	815	85	In France 809.
18	S.C.I., 1,000-lb.	Y	—	—	—	18	—	—	Production begi
19	S.C., 400-lb., Type S.L.3 ..	Y	—	—	Not developed until 1941.	19	—	—	As for Septembe
	WAR GASES								
20	Mustard Gas	Y	500 tons		Capacity 90 tons a week. Cabinet approval for 1,200 ton a week plant given on 13th October, 1939.	20	400 tons		Capacity 200 to
21	Tear Gas	B	5 tons		Capacity 5 tons a week. Cabinet approval for 30 ton a week plant for Army weapons given on 13th October, 1939. R.A.F. had no requirement.	21	4½ tons		Capacity 5 tons
22	Phosgene	G	—		25 ton a week plant available to meet demand for carbamite (intermediary of cordite). No capacity for C.W. purposes. R.A.F. had no requirement at this date. Cabinet approved 100 ton a week plant on 13th October, 1939.	22	—		As for Septemb ton a week p pated availab of 1941.

APPENDIX II

(3rd September, 1939 to 31st December, 1941)

	1st October, 1940.				19th December, 1941.			
...rks.	Serial No.	Stocks. Charged.	Stocks. Empty.	Remarks.	Serial No.	Stocks. Charged.	Stocks. Empty.	Remarks.
	1	17,000	—	—	1	—	—	Obsolete—Stocks held.
	2	25,000	—	—	2	—	—	Obsolete—Stocks held.
...arging machine ...ilable but no ...ecast.	3	200	2,000	—	3	201,000	—	
...arging machine	3A	—	—	—	3A	73,000	—	
.	4	—	—	Design not clear.	4	—	—	
.	4A	—	—	Design not clear.	4A	—	—	
.	5	—	—	Design not clear.	5	22,500	—	
-	6	16,000	5,000 (6 and 6A)	—	6	60,000	30,000 (6 and 6A)	
...n in August,	6A	9,000		—	6A	34,000		
...r, 1939.	7	—	—	As for September, 1939.	7	—	—	
...ent stage.	8	—	—	Specified June, 1940; Production initiated October, 1940, but owing to technical difficulties no operational rounds available until 1942; production discontinued December, 1942.	8	—	—	See remarks under October, 1940.
-	9	73,300	92,000	—	9	132,000	330,000	Charged and empty reserves complete.
...recast available.	10	4,500	525	Production capacity temporarily reduced by bombing.	10	13,000	27,000	Charged and empty reserves complete.
...recast available.	11	65	109	—	11	127	240	Charged and empty reserves complete.
...fuzed (owing to ...n hold-up) and ...atch to France. ...d reserve for ...0) anticipated ...May, 1941.	12	59,592	408	In Middle East 9,100—Target December, 1940, 17,000.	12	173,000 (12 and 13)	—	United Kingdom 156,000; Northern Ireland 2,200; Egypt 10,000; Kenya 3,000; Sudan 2,500.
...ited Kingdom ...wing to short ...when charged. ...when required.	13	—	32,000	As for April, 1940.	13			Storage difficulties overcome and stored charged.
-	14	—	—	—	14	48,000	—	Plus 5,000 empties and charging drums *en route* to Middle East.
-	15	14,483	5,517	In Middle East 2,300—Target December, 1940, 8,000.	15	28,000	—	United Kingdom 20,000; Northern Ireland 288; Egypt 4,500; Kenya 1,500; Sudan 2,000; Far East 500.
...r, 1939.	15A	—	—	Charging of small quantities beginning.	15A	27,000	—	All in United Kingdom.
In Middle East	16	416	24	In Middle East 100—Target December, 1940, 1,200.	16	1,800	—	United Kingdom 1,000; Egypt 400; Sudan 300.
	17	928	27	In Middle East, Nil—Target December, 1940, 900.	17	6,400	—	United Kingdom 4,500; Northern Ireland 260; Egypt 1,250; Sudan 400; in transit to Far East 300.
...nning.	18	301	6	All in United Kingdom.	18	2,900	—	All in United Kingdom.
...r, 1939.	19	—	—	—	19	—	—	Introduced and production beginning.
...s a week.	20	1,040 tons		Capacity 350 tons a week.	20	4,000 tons		Capacity 900 tons a week. This met foreseen requirements.
a week.	21	71 tons		Capacity 5 tons a week.	21	76 tons		Output 30 tons a week.
...er, 1939; 100 ...lant not antici- ...le before spring	22	Nil		Capacity 45 tons a week; including original 25 ton a week plant.	22	Nil		Charged into weapons as available. Output 160 tons a week of which 50 tons a week required for carbamite.

APPENDIX III

Stock Position—Chemical Weapons and Gas—"D" day, 1944

Serial No.	Item.	Type of Charging. Mustard (Y). Tear Gas (B). Phosgene (G).	Theatre.									Total all Theatres.	Remarks.
			United Kingdom.	Central Mediterranean Forces.	Middle East.	Persia and Iraq Command.	India and South East Asia.	Australia.	New Zealand.	South Africa.	Canada.		
	ARMY WEAPONS												(a) Includes War Office reserves. (b) Including War Office reserve and reserve for Central Mediterranean Forces. (c) Including Central Mediterranean reserve in Middle East.
1	25-pr. Base Ejection Shell.	Y	1,171,000 (a)	51,000	266,000 (b)	143,000	336,000	260,000	93,000	44,000	—	2,364,000	
1A	25-pr. Base Ejection Shell.	B	430,000 (a)	17,000	93,000 (b)	43,000	93,000	86,000	31,000	15,000	—	808,000	
2	5·5-in. Gun How. Base Ejection Shell.	Y	229,000 (a)	9,000	13,000 (c)	1,000	9,000	—	—	3,000	—	264,000	
2A	5·5-in. Gun How. Base Ejection Shell.	B	81,000 (a)	3,000	3,000 (c)	—	2,000	—	—	1,000	—	90,000	
3	6-in. How. Base Ejection Shell.	Y	115,000 (a)	—	—	1,000	15,000	27,000	—	—	—	155,000	
4	4·2-in. Mortar Bombs, Chemical.	Y	428,000 (a)	84,000	13,000 (c)	—	15,000	40,000	—	—	6,000	586,000	
4A	4·2-in. Mortar Bombs, Chemical.	G	119,000 (a)	33,000	2,000	—	5,000	8,000	—	—	3,000	170,000	

APPENDIX III—*continued*

Stock Position—Chemical Weapons and Gas—" D " day, 1944—*continued*

Serial No.	Item.	Type of Charging. Mustard (Y). Tear Gas (B). Phosgene (G).	Theatre. United Kingdom.	Central Mediterranean Forces.	Middle East.	Persia and Iraq Command.	India and South East Asia.	Australia.	New Zealand.	South Africa.	Canada.	Total all Theatres.	Remarks.
	ARMY WEAPONS —*continued*												
5	Bombs, Ground (*d*).	Y	122,000	4,000	—	—	3,000	20,000	5,000	—	500	154,500	(*d*) No operational requirement and production discontinued. Arrangements existed for emergency production of a simple drum for use as a chemical mine, to be burst with an external burster, should a requirement for a direct contamination weapon have arisen. Stocks of Livens drums charged " Y " could have been similarly used. The Bomb, ground, 6-lb., was being used as a training store. (*e*) Manufactured in South Africa.
6	Mines, Chemical (*d*).	Y	—	5,000	—	—	1,000	5,000	—	—	—	11,000	
7	Bulk Contamination Vehicles (*d*).	Y	66 charged. 18 empty.	—	—	—	—	—	—	—	—	84	
	AIR WEAPONS												
8	Bombs, Aircraft, 30-lb., Mk. I.	Y	150,000	2,600	14,000	—	—	8,000	—	—	—	174,600	
9	Bombs, Aircraft, L.C., 65-lb.	Y	340,000	4,400	19,000	—	6,400	1,200	—	45,000 (*e*)	—	416,000	
10	Bombs, Aircraft, 250-lb.	Y	14,000	600	7,500	—	—	4,800	—	—	—	26,900	
10A	Bombs, Aircraft, 250-lb.	G	41,700	440	4,900	—	2,500	3,600	—	—	—	53,140	

APPENDIX III—*continued*

Stock Position—Chemical Weapons and Gas—" D " day, 1944—*continued*

Serial No.	Item.	Type of Charging. Mustard (Y). Tear Gas (B). Phosgene (G).	Theatre. United Kingdom.	Central Mediterranean Forces.	Middle East.	Persia and Iraq Command.	India and South East Asia.	Australia.	New Zealand.	South Africa.	Canada.	Total all Theatres.	Remarks.
	AIR WEAPONS —*continued*												
11	Bombs, Aircraft, 500-lb.	G	41,000	—	—	—	—	—	—	—	—	41,000	(*f*) Stock figure at end of June, 1944, was 2,500.
12	Bombs, Aircraft, 500-lb.	Y	coming forward. (*f*)	—	—	—	—	—	—	—	—	(*f*)	
13	S.C.I., 250-lb. ..	Y	940	140	680	—	—	—	—	—	—	1,760	
14	S.C.I., 500-lb. ..	Y	4,100	140	700	—	400	850	—	—	—	6,190	
15	S.C.I., 1,000-lb. ..	Y	2,500	—	—	—	—	—	—	—	—	2,500	
16	S.C., 400-lb., Type S.L.3.	Y	28,000	125	—	—	—	—	—	—	—	28,125	

(C29058) K**

APPENDIX III—*continued*

Stock Position—Chemical Weapons and Gas—" D " day, 1944—*continued*

Serial No.	Item.	Type of Charging. Mustard (Y). Tear Gas (B). Phosgene (G).	Theatre.			Total all Theatres.	Remarks.
			United Kingdom.	South Africa.	Canada.		
	War Gases						
17	Mustard Gas ..	Y	*Stock* :—7,700 (*g*). *Capacity* :— Existing—900 tons a month. Within 6 months of the outbreak of gas warfare—3,680 tons a month.	*Stock* :—300 tons. *Capacity* :— Small current production. Maximum output 450 tons a month.	Rated output of 1,200 tons a month should have been achievable in four months.	*Stock* :—8,000 tons. Maximum ultimate Commonwealth capacity —5,330 tons a month.	(*g*) Includes stocks at R.A.F. Forward Filling Depots.
18	Phosgene ..	G	*Stock* :—Nil. *Capacity* :— 400 tons a month expanding to 1,480 tons a month within 6 months of gas warfare starting.	Maximum output 100 tons a month.	—	Maximum ultimate Commonwealth capacity —1,580 tons a month.	
19	Tear Gas (B.B.C.)	B	*Stock* :—111 tons. *Capacity* :— 100 tons a month.	—	—	—	

APPENDIX IV

(*See* page 43.)

Types of Target Suitable for Gas Attack in Relation to the " Overlord " Battle

Serial No.	Type of Target.	Type of Attack.	Remarks.
1	Rail communications and bottlenecks such as marshalling yards, railway installations, detraining stations.	Mustard gas and H.E. bombs	Suitable for day or night attack.
2	Defiles such as bridges, road centres	Mustard gas and H.E. bombs	It was recommended that all H.E. attacks should have a proportion of mustard gas bombs, since contamination would greatly delay repair.
3	Higher H.Q., troops in built-up areas, gun areas, Battalion or Company localities.	Phosgene and/or mustard gas with H.E.	The persistence of mustard gas and its effect on our own plans and troops as the battle moved forward was noted. This interference was regarded as an inconvenience rather than a danger.
4	Supply Depots and Dumps ..	Mustard gas and H.E. ..	Regarded as profitable targets during the later phases of " Overlord."
5	Troop movements and concentrations by day ..	Low spray and fighter attack	Initially limited to gas bombing as S.C.I. aircraft were not immediately available.
6	Enemy forming up areas and O.Ps. ..	Mustard gas bombs	Contamination of small areas within range of our own troops might have been effective in delaying movement and preparations for attack.

As mustard gas persists for some days under European weather conditions and the effects are often not apparent for from 12 to 24 hours, it was therefore regarded as more of a harassing agent. Phosgene, being a killer, was especially likely to get results on tired troops at night. The decision of Supreme Headquarters was required for all gas attacks outside the immediate tactical area.

(C29058) K**2

APPENDIX V

Total Production of Gas and Gas Weapons in the United Kingdom, 1939 to 1945 (*a*)

Serial No.	Items.	Charging.	Quantity.	Cost (*c*). (£ millions)
1	25-pdr. B.E. Shell	Mustard Gas (Y)	3,394,093	7·212 (Shell only).
1A	25-pdr. B.E. Shell	Tear Gas B	942,960	1·838 (Shell only)
2	5·5-in. Gun Howitzer B.E. Shell	Y	300,766	2·752 (Shell only)
2A	5·5-in. Gun Howitzer B.E. Shell	B	101,195	·632 (Shell only)
3	6-in. Howitzer B.E. Shell	Y	155,448	1·367 (Shell only)
5	Livens Drum	Y	60,003	·168
5A	Livens Drum	Phosgene G	39,867	·105
6	5-in. U.P. Rocket	G	743,884	1·859 (Head only)
7	4·2-in. Mortar Bomb Chemical	Y	635,278	1·079 (without Fuze)
7A	4·2-in. Mortar Bomb Chemical	G	210,330	·365 (without Fuze)
8	Bomb, Ground, 6-lb.	Y	307,761	·272
9	Mine, Chemical	Y	21,417	·097
10	Installation, Type A/LP	Y	195	·015
				Total £17·768
11	Bomb, A/C, L.C., 30-lb., Mks. I and II.	Y	179,218	·331
12	Bomb, A/C, L.C., 65-lb.	Y	399,322	1·597
13	Bomb, A/C, L.C., 250-lb.	Y	28,349	·184
13A	Bomb, A/C, L.C., 250-lb.	G	54,647	·355
14	Bomb, A/C, L.C., 500-lb.	G	55,638	·723
15	Bomb, A/C, L.C., 500-lb.	Y	13 738	·178
16	S.C.I., 250-lb.	Y	1,831	·053
17	S.C.I., 500-lb.	Y	6,464	·387
18	S.C.I., 1,000-lb.	Y	2,915	·320
19	S.C.I., 400-lb.	Y	48,202	2·241
				Total £6·373
				Grand Total £24·141
20	Bulk stocks, Mustard Gas (exclusive of that charged into weapons) (*b*).		7,774 tons	(*d*)

(*a*) This table should not be read as showing stocks in hand at the end of the war. Production tailed off from the end of 1943 onwards, and stocks in May, 1945, were not materially greater in most cases than those shown in Appendix III, although there was some increase in gas shell and in certain air gas weapons. The discrepancies between Appendices V and III are in the main due to wastage, but the reasons vary with the individual weapons, *e.g*, losses at sea, deterioration in chargings, leakers, out of date marks, use in trials or in training, transfer to the United States.

(*b*) Total Gas production (including gas charged into weapons) was :—

	Mustard Gas. (tons)	*Phosgene, Tear, and Nose Gas.* (tons)
Army Weapons	6,711	5,618
R.A.F. Weapons	23,242	8,023
Bulk Stocks	6,266	401
United Kingdom (R.A.F. filling station)	1,508	—
Bulk (to United States)	2,992	—
	40,719	14,042
		54,761 tons

(*c*) The figures represent the normal costs relating to manufacture, *i.e.*, material, labour, etc., or estimated values, but do not include any element for departmental expenses.

(*d*) Figures for the quantity of each type of mustard gas included in total stocks are not now available. As the cost of each type varies considerably, it is impracticable to assess total value.

APPENDIX VI

(*See* page 51 (footnote 2).)

The Bari Disaster

The ship " John Harvey " loaded with 100 tons of mustard gas in United States 100-lb. bombs, as well as a large quantity of munitions, was berthed at a mole in Bari Harbour when enemy planes raided the harbour on the night of 2nd December, 1943. The " John Harvey " was not hit during the raid, but shortly after the raid a petrol ship exploded and started to burn. There were two big explosions, the second of which blew up the " Harvey " which sank with no survivors. The main explosion of the evening was of tremendous violence—windows seven miles away being smashed. An order was dispatched to scuttle the " Harvey " when fires began, but it was not known whether this message was ever received.

In the maelstrom of fire and explosion 16 ships were sunk and four partially destroyed. Many survivors were blown into the water or forced to swim at some time in the evening. The surface of the water was thickly covered with oil at some points in the harbour. In the huge fires which were raging much of the mustard gas was burnt, empty bomb cases on the mole accounting for the release of a ton or so of mustard gas just above it. Such mustard gas as was not burnt either sank to the bottom or dissolved in the oil, and it was this solution of mustard in oil which produced most of the severe casualties and subsequent deaths (*see* Note (1)).

More than 1,000 men were killed or missing after the disaster, and 800 cases were admitted to hospital of which 628 were suffering from mustard gas exposure and 67 deaths were wholly or partly caused by this gas.

It seems remarkable that there was no general gas alarm. In the resulting confusion the smell was either not detected or escaped recognition. Someone on one ship did call " Gas " and many of the crew adjusted respirators for a short period but did not themselves recognize it. No information about the possibility of gas was circulated to the hospitals, and neither rescue squads nor hospital personnel had any idea that casualties might have been exposed to it. Many casualties were covered with crude oil with which the mustard gas was in solution and, under the impression that they were suffering from immersion and exposure, were wrapped in blankets and given warm tea. Surgical cases received priority, but others were left in their blankets for as long as 12 or 24 hours. There was no attempt at decontamination nor at removal of clothing, and in the rush no odours were detected. Only a few medical attendants sustained burns although several had eye irritation and this, as well as the development of symptoms in the mustard gas casualties, pointed to the mustard gas in oil solution being dilute.

As a result of the confusion following the raid and the fact that the existence of a gas cargo was not common knowledge, the presence of gas was not immediately realized. The medical services were severely handicapped because they had not been given previous information that the cargo was expected and therefore no special preparations had been made.

On another occasion at Algiers, a fire occurred in a ship as a result of the leaking of R.A.F. white phosphorous smoke bombs. H.E. munitions were also on board but the stowage plan was not sufficiently well-known either to the Master or to the port officials to enable correct action to be taken to isolate the fire from these munitions or to dump them overboard. The ship was being towed out of harbour by a destroyer and fire fighting was proceeding when she blew up, with very heavy loss of life which included many port officials and fire fighters on board as well as the ship's company and that of the destroyer.

A detailed knowledge of the stowage plan should at least have led to reduction in the loss of life had it not enabled the explosion to be prevented.

It is essential that the most meticulous care should be taken in the stowage of gas and smoke munitions in ships and in their receipt at and passage through ports. All relevant authorities in a port should be informed in advance of the arrival of such cargoes and details of stowage should be available. Such measures should not only be co-ordinated between all services in the port but arrangements must also be made between allies to ensure that the necessary details are provided before the arrival of ships.

Note (1).—Pure mustard gas, being heavier than water, sinks to the bottom. It might be thought that there is a potential value in a floating mustard gas/oil mixture in defence against beach landings. The whole question was extensively examined and the following facts emerged.

The most concentrated mustard gas/oil solution which will float contains 30 to 40 per cent. mustard gas. If unrestricted, such mixtures spread rapidly on the surface of sea water and travel down wind very slowly, tending to pile up on beaches. The mustard gas content decreases rapidly by evaporation, solution, and hydrolysis, and after a quarter to half an hour, under temperate conditions, the concentration is too small to be of practical casualty producing significance. Army service boots and gaiters, although completely water-logged, give adequate protection against the surface film, and the amount picked up by serge battle-dress is less than would be thought, and unlikely to produce significant skin reaction. The vapour hazard from areas of water contaminated with a mustard gas/oil film would not, in general, be significant. The film might be more effective in tropical conditions against troops wearing shorts, but the difficulties in using it would be greatly increased owing to the much shorter persistence of the film.

APPENDIX VII

(*See* page 62.)

Inter-Service Chemical Warfare Committee

Terms of Reference

(i) To relate practical requirements to the resources available.

(ii) To make recommendations to the Chiefs of Staff as to the best method of employing the available resources for offensive purposes, the study of defensive aspects, however, not being precluded. To advise upon the allotment of resources between the Services.

(iii) To bring to the notice of the Chiefs of Staff the advantages and disadvantages of any particular method of using gas.

(iv) To keep the supply and research situation under constant review ; to advise upon the desirability of putting new weapons and new gases into production ; and to place before the Chiefs of Staff recommendations to ensure that supplies keep pace with operational requirements.

(v) To keep a check on the anti-gas arrangements made by the Service Departments,

and from November, 1942, (vi) to be the agency for implementing the joint U.S.–British supply and procurement programme (*see* Chapter II, Section 9).

APPENDIX VIII
(See Chapter II, Section I)

DEVELOPMENT OF THE GENERAL STAFF AT THE WAR OFFICE 1937 TO 1945

(Branches or appointments wholly concerned with Chemical Warfare are encircled with thick black lines)

1937

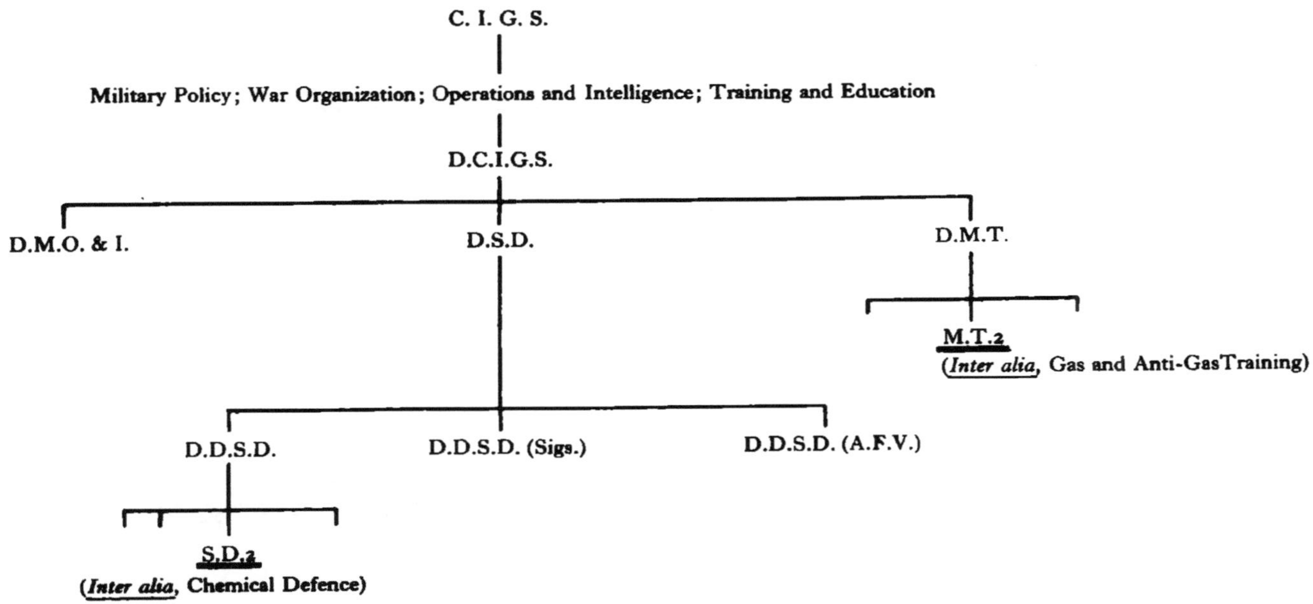

APPENDIX VIII (cont.)

DEVELOPMENT OF THE GENERAL STAFF AT THE WAR OFFICE 1937 TO 1945

(Branches or appointments wholly concerned with Chemical Warfare are encircled with thick black lines)

1938

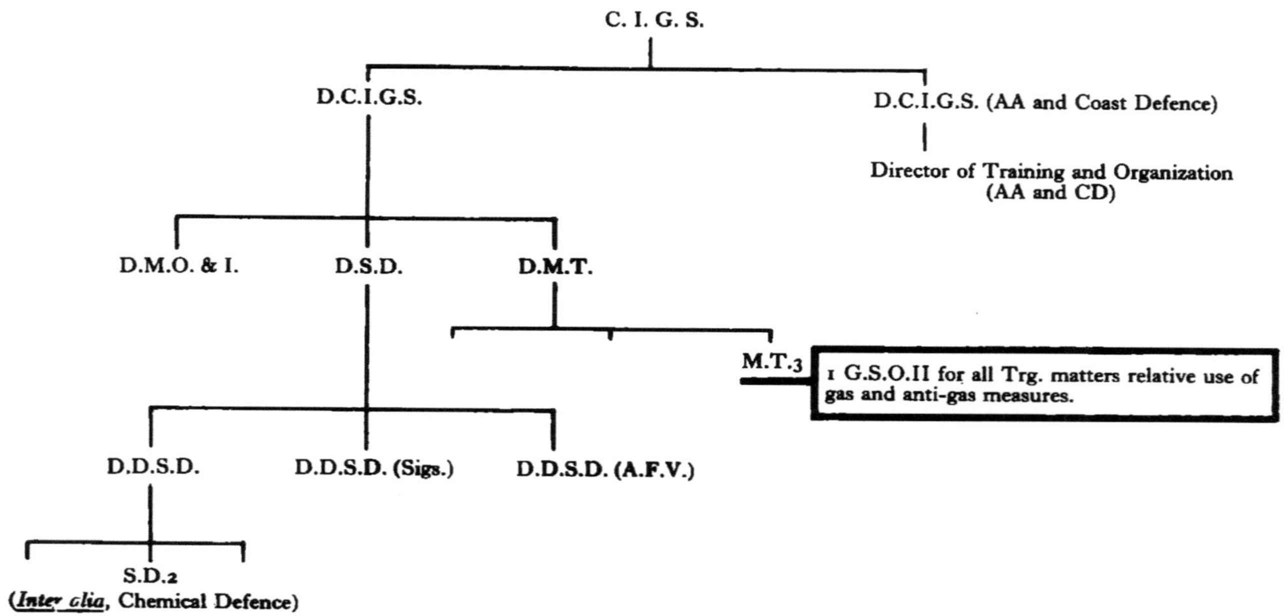

DEVELOPMENT OF THE GENERAL STAFF AT THE WAR OFFICE 1937 TO 1945

(Branches or appointments wholly concerned with Chemical Warfare are encircled with thick black lines)

1939/40

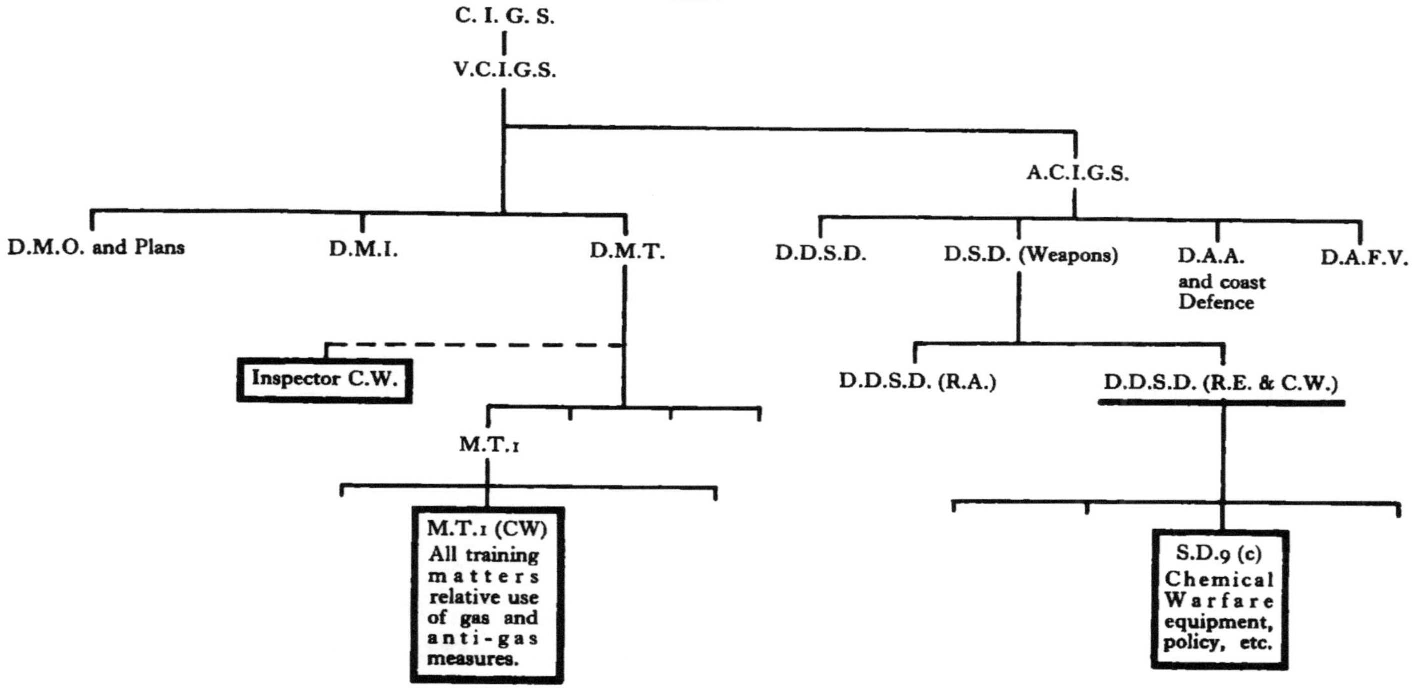

DEVELOPMENT OF THE GENERAL STAFF AT THE WAR OFFICE 1937 TO 1945

(Branches or appointments wholly concerned with Chemical Warfare are encircled with thick black lines)

1941

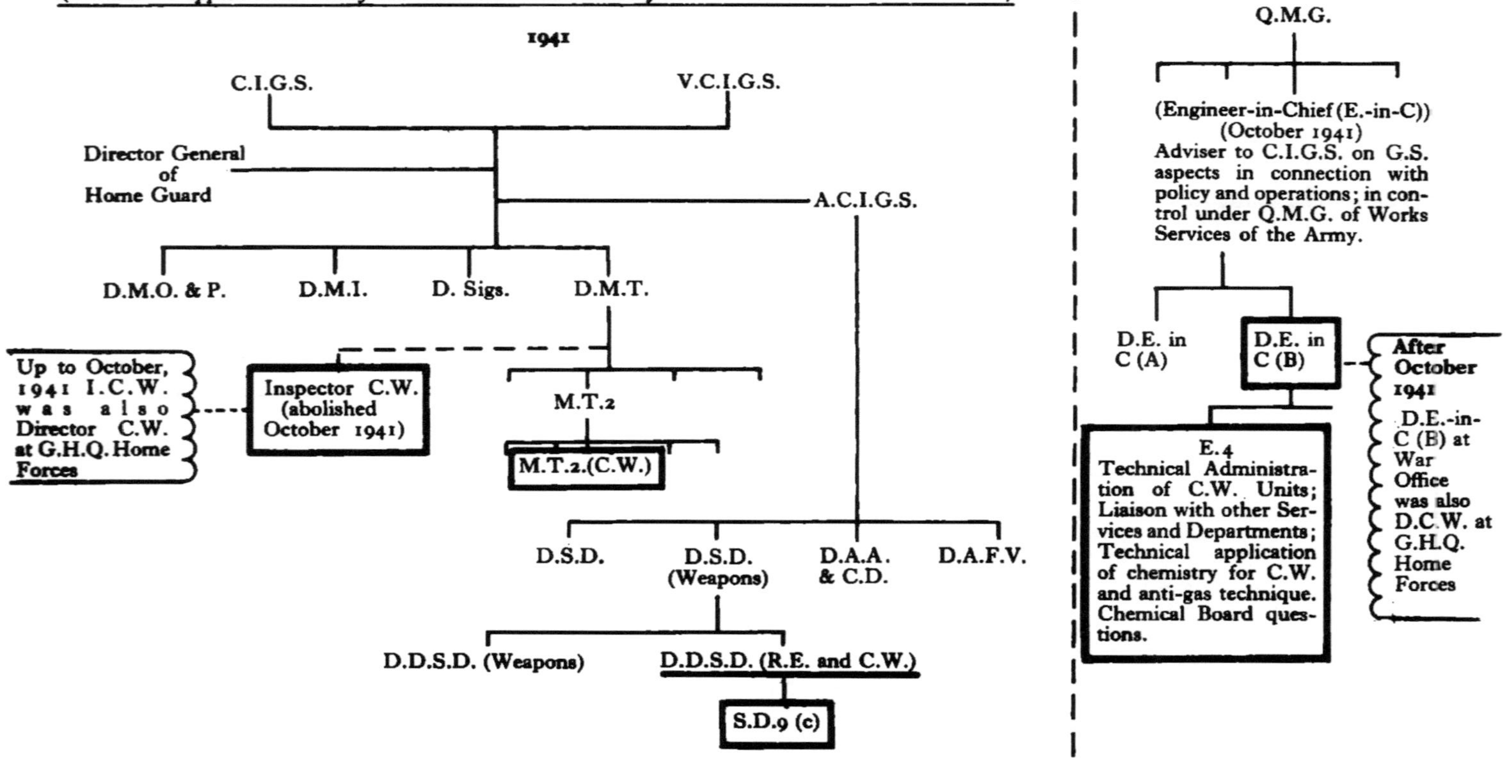

DEVELOPMENT OF THE GENERAL STAFF AT THE WAR OFFICE 1937 TO 1945

(Branches or appointments wholly concerned with Chemical Warfare are encircled with thick black lines)

1942/43

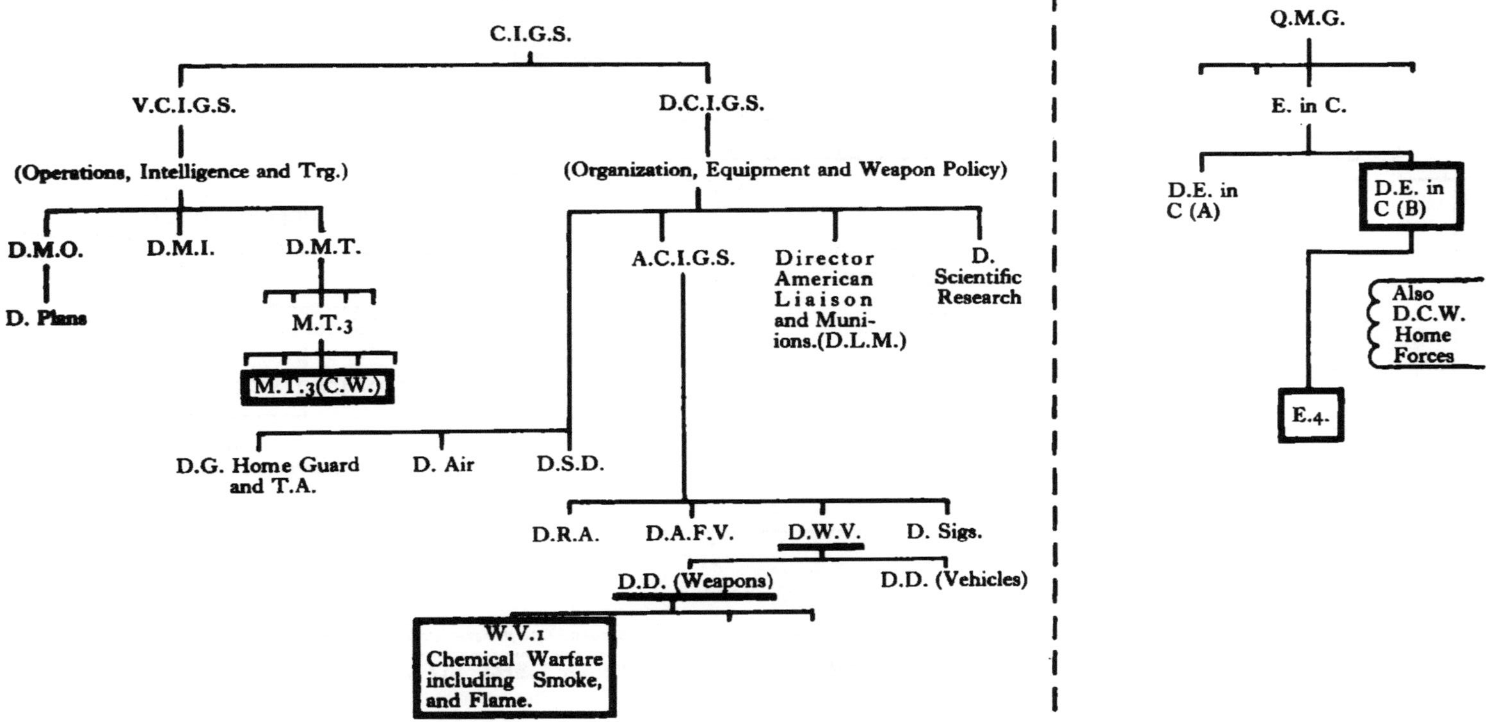

APPENDIX VIII (cont.)

DEVELOPMENT OF THE GENERAL STAFF AT THE WAR OFFICE 1937 TO 1945

(Branches or appointments wholly concerned with Chemical Warfare are encircled with thick black lines)

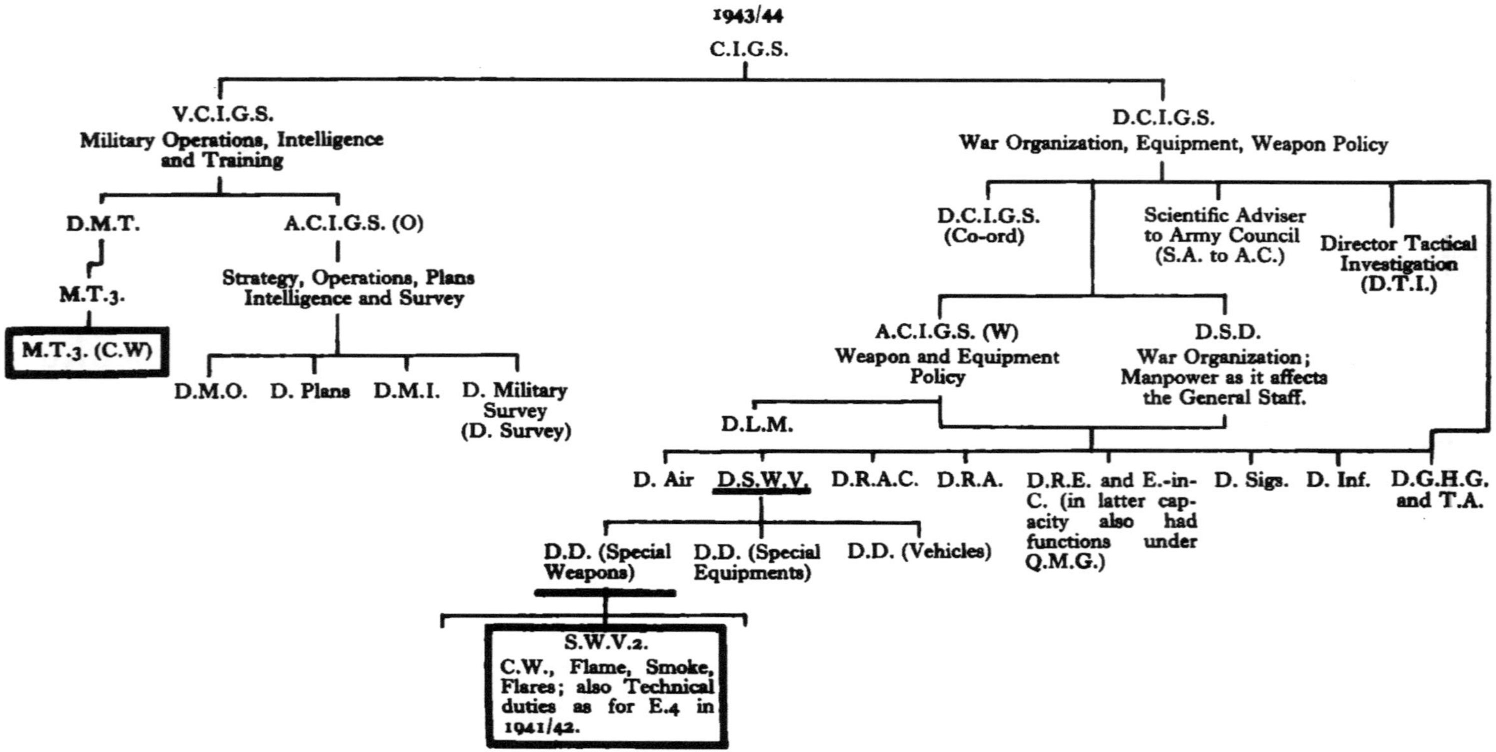

APPENDIX VIII (cont.)

DEVELOPMENT OF THE GENERAL STAFF AT THE WAR OFFICE 1937 TO 1945

(Branches or appointments wholly concerned with Chemical Warfare are encircled with thick black lines)

1944/45

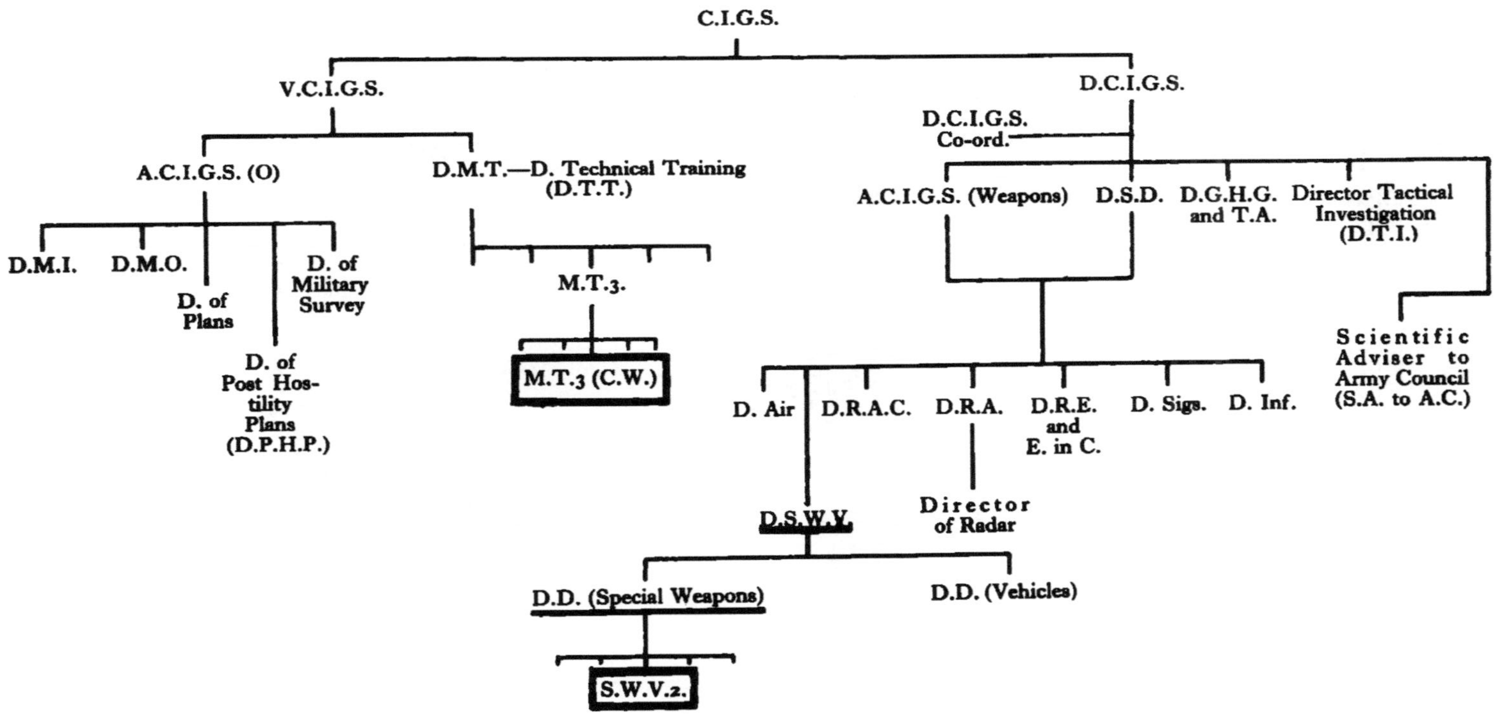

APPENDIX IX

(See Chapter II, Section I)

ORGANIZATION AND FUNCTIONS W.V. DIRECTORATE AND S.W.V. DIRECTORATE—JULY 1942 TO END OF WAR

W.V. Directorate — July, 1942

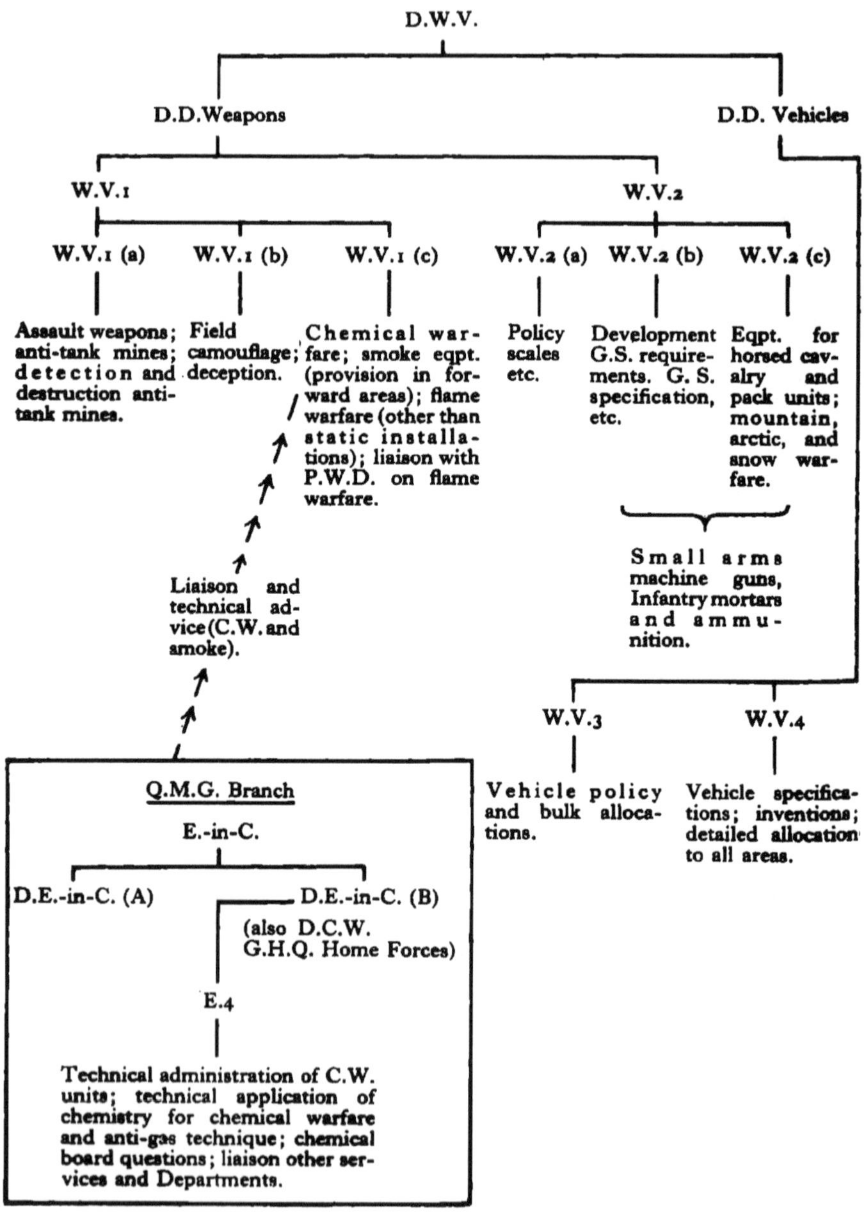

APPENDIX IX (cont.)

W.V. DIRECTORATE—MARCH, 1943

- D.W.V.
  - D.D. Weapons
    - W.V.1
      - Smoke, chemical warfare and flame warfare (including technical aspects and liaison with appropriate technical authorities of other departments)
    - W.V.2
      - Camouflage (field and static); deception; development of special eqpt. not appropriate to particular arms; development of common user personal eqpt.
  - D.D. Vehicles
    - W.V.3
      - Co-ordination of production and allocation of vehicles *ex* U.K., Canada and U.S.A.
    - W.V.4
      - Development; Co-ordination of all W. E. questions concerning vehicles; vehicle lifting inventions.

APPENDIX IX (cont.)

ORGANIZATION AND RESPONSIBILITIES OF THE "SPECIAL WEAPONS AND VEHICLES DIRECTORATE"—31st JULY, 1943

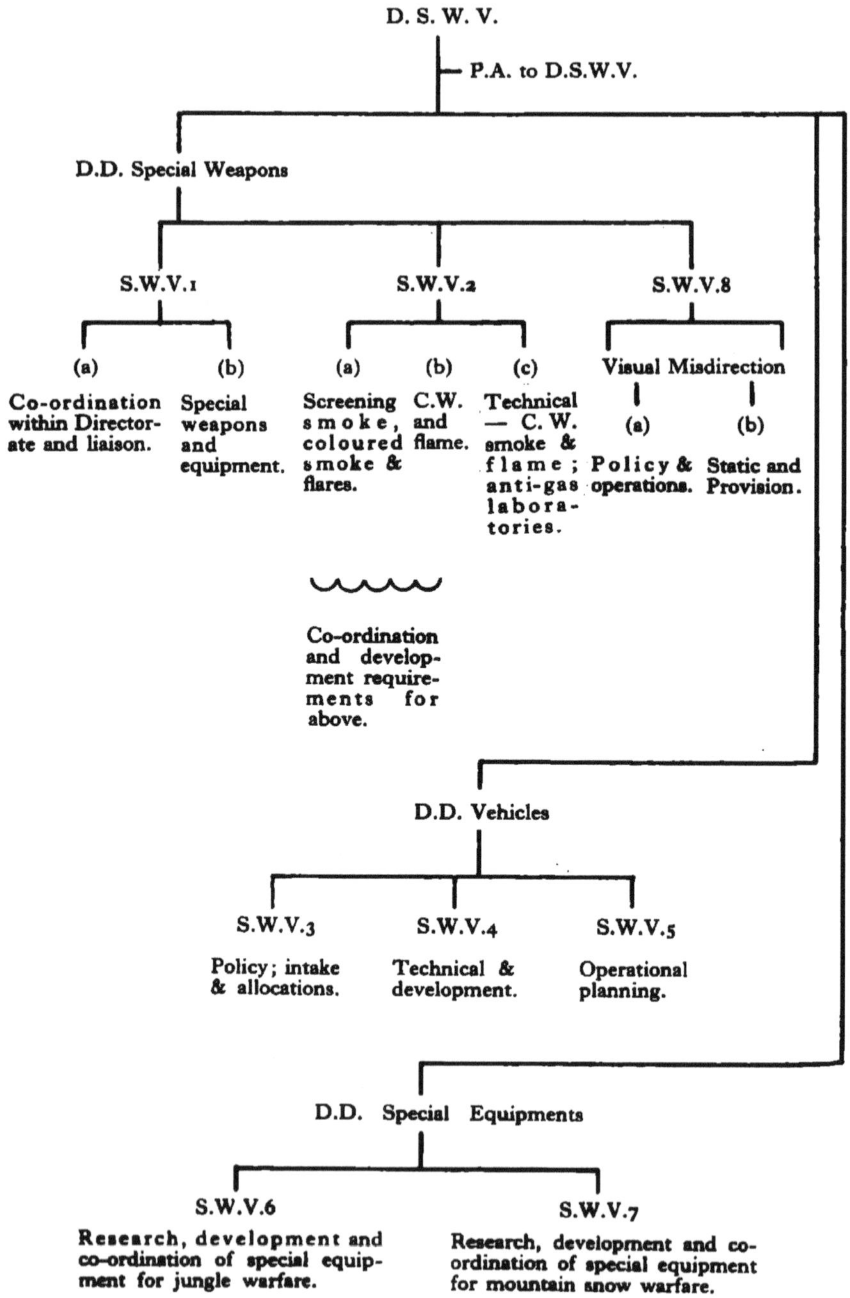

ORGANIZATION AND RESPONSIBILITIES OF THE "SPECIAL WEAPONS AND VEHICLES DIRECTORATE"—17th FEBRUARY, 1944

- **D. S. W. V.**
  - P.A. to D.S.W.V.
  - S.W.V.1 (a) (Co-ordination within Directorate and liaison with overseas Theatres)
  - **D.D. Special Weapons**
    - S.W.V.1 (b)
      - Development policy and control of special eqpt. not appropriate to particular arms; development of common user personal equipment.
    - S.W.V.2
      - S.W.V.2 (a)
        - **Smoke**
        - Co-ordination of development and requirements for all screening smoke, coloured smoke and flares; development requirements and allocation of all A. A. screening smoke stores.
      - S.W.V.2 (b)
        - **C.W. and Flame**
        - C.W. policy (offensive and defensive). Flame warfare policy. Development requirements and allocation of C.W. and flame eqpt.
      - S.W.V.2 (c)
        - Technical and chemical matters in connection with chemical and flame warfare, smoke and flares; technical matters relating to technical officers (C.W.) and anti-gas laboratories.
    - S.W.V.6
      - Research, development and provision of special equipment for jungle warfare.
    - S.W.V.8
      - S.W.V.8 (a)
        - Camouflage policy, operational questions, scales and equipment; devices and flash simulators.
      - S.W.V.8 (b)
        - Camouflage, W. Es. personnel, static and special devices.
  - **D.D. Vehicles**
    - S.W.V.3
    - S.W.V.4
    - S.W.V.5
    - No change

APPENDIX X

(See page 71)

MINISTRY OF SUPPLY—WAR-TIME ORGANIZATION OF C.W. RESEARCH AND PRODUCTION

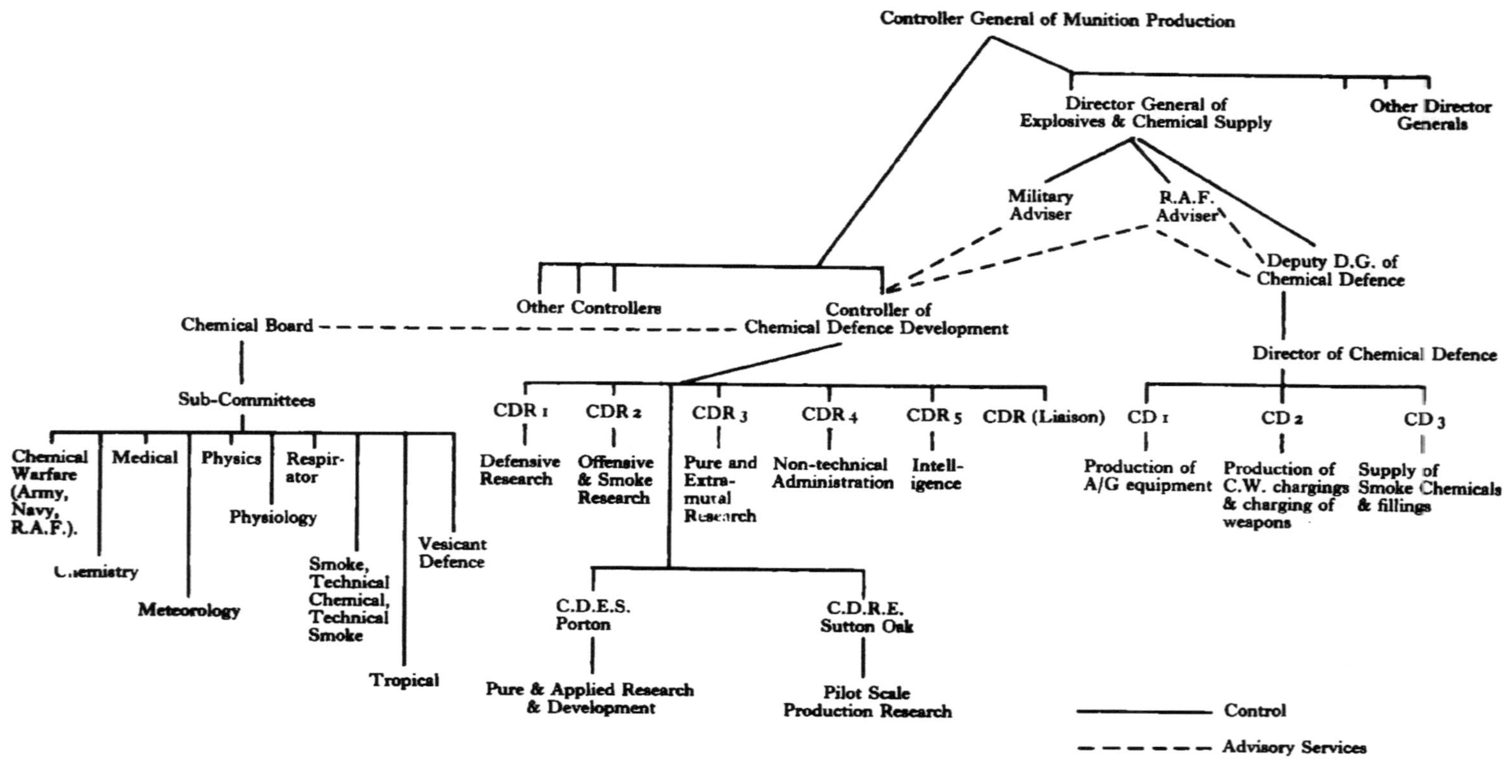

APPENDIX XI

(*See* pages 78 and 109.)

Duties of Technical Officers (C.W.) at Headquarters of Formations

1. To assist the General Staff officers (trained in C.W.) of the formation concerned in considering the general technical aspects of chemical warfare (including gas, smoke, and flame) and in maintaining a high standard of training throughout their formation, and to advise them on all technical matters particularly in relation to the following :—

(*a*) The condition of gas, smoke, and flame equipment and accessories in use or awaiting issue.

(*b*) The investigation (in connection with those services and adviser branches which may also be concerned) of technical problems arising in connection with the storage and use of gas, smoke, and flame weapons and accessories.

(*c*) The sifting of intelligence reports which deal with gas, smoke, or flame warfare, to ensure that no technical implications are overlooked.

(*d*) The examination of captured enemy gas, smoke, and flame equipment and accessories in accordance with the standing instructions of the formation.

2. To maintain close liaison with the Medical Services, and to provide medical officers, laundry technical officers, and chemists of all categories in the Army with information and advice on chemical matters.

3. To assist, on request, in operational research, when this can be done without prejudice to their other duties in the field.

4. The following channels of communication are recognized :—

(*a*) The normal channel of communication on purely technical matters is between the technical officers (C.W.) of formations.

(*b*) For duties under para. 2 above, technical officers (C.W.) will communicate directly with the officer or unit concerned.

APPENDIX XII

(*See* pages 77, 100 and 109.)

Duties of G.S.Os. III (Ops.) Trained in C.W.

The duties of the G.S.Os. III trained in Chemical Warfare on the H.Q. of Armoured and Infantry Divisions are as follows :—

1. To advise the commander on all C.W. matters.
2. To advise the commander on all flame warfare matters.
3. In conjunction with the C.R.A., to advise the commander on the use of smoke.
4. To assist in maintaining a high standard of anti-gas training throughout the division.
5. To work in close liaison with the Intelligence branch, advising them on the detailed information required for smoke, flame, and gas warfare. He will be responsible for ensuring that such information is brought to the notice of the commander.
6. To receive and examine all specimens of enemy gas munitions and equipment of technical interest, and to deal with these items in accordance with the instructions in force in the theatre in which the division is operating.

APPENDIX XIII

(*See* page 82.)

Order of Battle—Chemical Warfare Troops

Serial No.	Unit.	Location.	Remarks.
	3RD SEPTEMBER, 1939.		
1	No. 58 Chemical Defence Coy. ..	WINTERBOURNE DAUNTSEY (near SALISBURY).	Redesignated No. 58 Chemical Warfare Coy., 12th October, 1939.
2	No. 11 Chemical Warfare Training Bn., R.E.	WINTERBOURNE DAUNTSEY (near SALISBURY).	Cadre staff under training.
	15TH APRIL, 1940.		
1	No. 1 Chemical Warfare Group (Nos. 58, 61, 62, C.W. Coys.).	B.E.F., France.	
2	No. 63 C.W. Coy.	BULFORD.	
3	No. 11 C.W.T.B., R.E. ..	WINTERBOURNE DAUNTSEY.	Joined Serial 4 in July, 1940.
4	Chemical Warfare Training Centre (Officer Cadet Training Wing, No. 24 C.W. Depot Bn. R.E., No. 12 C.W. T.B., R.E., No. 13 C.W.T.B., R.E.).	BARTON STACEY.	Forming.
	15TH JULY, 1941. (On completion of raising of C.W. Units.)		
1	Commander C.W. Troops and H.Q. ..	WINTERBOURNE DAUNTSEY.	
2	No. 1 C.W. Group (58, 61, 62, 67 C.W. Coys.).	Various Locations United Kingdom.	
3	No. 2 C.W. Group (64, 65, 66, 74 C.W. Coys.).		
4	No. 3 C.W. Group (68, 69, 70, 75 C.W. Coys.).		
5	No. 4 C.W. Group (71, 72, 73, 76 C.W. Coys.).		
6	No. 5 C.W. Group (77, 78, 79, 86 C.W. Coys.).		
7	No. 6 C.W. Group (80, 81, 82, 87 C.W. Coys.).		
8	No. 7 C.W. Group (83, 84, 85, 88 C.W. Coys.).		
9	No. 8 C.W. Group (89, 90, 91 C.W. Coys.)		
10	No. 9 C.W. Group (92, 93, 94 C.W. Coys.)		
11	C.W.T.C. (No. 11 C.W.T.B., R.E.) ..	BARTON STACEY.	No. 12 C.W.T.B. (less H.Q. and two Training Coys.) disbanded 15/4/41. No. 13 C.W.T.B. and H.Q. and two Trg. Coys., No. 12 C.W.T.B. disbanded 15/6/41. No. 24 C.W. Depot Bn. disbanded 13/7/41.
12	Chemical Warfare Training and Experimental School.	SOUTH MOLTON (N. DEVON).	Raised 16/5/41.
13	Petroleum Warfare Experimental Section		

APPENDIX XIII—*continued*

Order of Battle—Chemical Warfare Troops—*continued*

Serial No.	Unit.	Location.	Remarks.
	15TH JANUARY, 1943. (Prior to abolition of C.W. Troops).		
1	Commander C.W. Troops and H.Q. ..	Stanhope Gate, London, W.1.	
2	No. 1 C.W. Group (Nos. 58, 67 C.W. Coys. and No. 62 C.W. (Mortar) Coy.	INDIA.	
3	No. 2 C.W. Group (Nos. 65 and 74 C.W. Coys. and No. 66 C.W. (M) Coy.).	MIDEAST.	
4	No. 3 C.W. Group (Nos. 69, 70 C.W. Coys.).	Various Locations in United Kingdom.	
5	No. 4 C.W. Group (Nos. 71, 73 C.W. Coys. and No. 72 C.W. (M) Coy.).		
6	No. 5 C.W. Group (Nos. 77, 79 C.W. Coys. and No. 80 C.W. (M) Coy.).		
7	No. 6 C.W. Group (Nos. 81, 82 C.W. Coys. and No. 87 C.W. (M) Coy.).		
8	No. 7 C.W. Group (Nos. 84, 85 C.W. Coys.).		
9	No. 8 C.W. Group (Nos. 90, 91 C.W. Coys. and No. 89 (C.W. (M) Coy.).		
10	No. 9 C.W. Group (Nos. 92, 93 C.W. Coys. and No. 94 C.W. (M) (Coy.).		
11	No. 10 C.W. Group (Nos. 61, 75 C.W. Coys. and No. 64 C.W. (M) (Coy.).		
12	No. 11 C.W. Group (Nos. 78, 88 C.W. Coys.).		
13	No. 68 C.W. (Mortar) Coy. .. No. 86 C.W. (Mortar) Coy. ..	1st Army, North Africa.	In infantry support role using H.E. and smoke.
14	No. 83 C.W. (Mortar) Coy. ..	United Kingdom —earmarked for North Africa.	Proceeded North Africa in infantry support role 1/4/43.
15	No. 1 Chemical Warfare Training Coy., R.E.	WINTERBOURNE DAUNTSEY.	Administered by No. 2 L. of C. Training Battalion, R.E.
16	Chemical Warfare Training and Experimental School.	SOUTH MOLTON (N. DEVON).	

Conversion to R.E. role or disbandment.

H.Q. C.W. Troops disbanded 31st March, 1943.

H.Qs., C.W. Groups in United Kingdom converted to H.Q. G.H.Q., Army or Corps Troops Engineers (9th March, 1943).

C.W. Coys. and C.W. (Mortar) Coys. in United Kingdom converted to Field Coys., R.E. (27th February, 1943).

No. 1 C.W. Group (H.Q. and the C.W. Coys.) in India similarly converted in May, 1943.

No. 2 C.W. Group in Middle-East. H.Q. disbanded ; Coys. reverted to Field Coys., R.E.

Three C.W. Coys. in North Africa were disbanded. (No. 86, April, 1943 ; No. 83 in May, 1943 ; No. 68 in August, 1943.)

Chemical Warfare Training and Experimental School—Reduced to Cadre strength 15th April, 1943 ; disbanded 23rd June, 1943.

No. 1 Chemical Warfare Training Coy., R.E.—Disbanded in July, 1943.

ORGANIZATION OF A CHEMICAL WARFARE GROUP

(See pages 82 and 164)

C.W. GROUP
(W.E.111/14B/2).

Headquarters...... Lieut.-Colonel
Adjutant
Intelligence Officer (Subaltern)
Survey Officer (Subaltern)
Regimental Serjeant-Major
2 Serjeants
39 rank and file
4 bicycles
7 motor cycles
1 car, 4 str.
7 trucks (15-cwt.)
2 lorries (3-ton)
1 water trailer (15-cwt.)
2 No. 11 wireless sets

Attached
1 Medical Officer
1 L.A.D. Type A
3 A.C.C. cooks

C.W. (Rocket) Coy.
(W.E.111/1931/15/2).

H.Q. Major
Captain
C.S.M.
C.Q.M.S.
3 Serjeants
(1 Fitter)
(1 Tpt.)
(1 Meteor)
29 rank and file
3 motor cycles
4 trucks (1 fitted for wireless)
5 lorries
2 water trailers (15-cwt.)
1 No. 11 wireless set

Sec — Sec — Sec

C.W. (Rocket) Coy.

C.W. (Mortar) Coy.
(W.E.111/145/1).

H.Q. Major
Captain
Subaltern
C.S.M.
C.Q.M.S.
3 Serjeants
(1 fitter)
(1 Tpt.)
(1 Meteor)
54 rank and file
6 motor cycles
10 trucks (1 fitted for wireless)
7 lorries
2 No. 18
1 No. 11
wireless sets

Sec — Sec — Sec

To face page 274

APPENDIX XIV

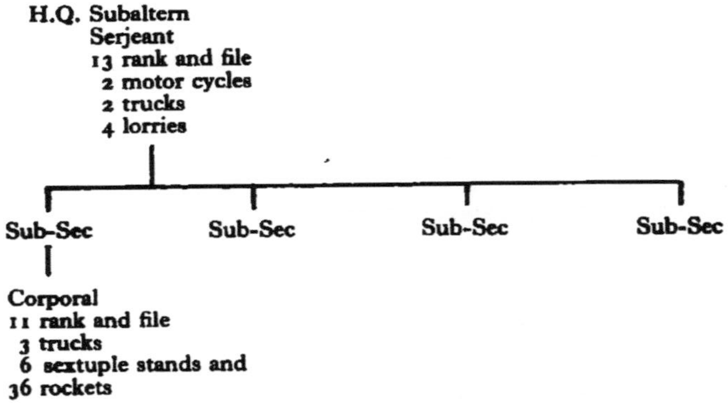

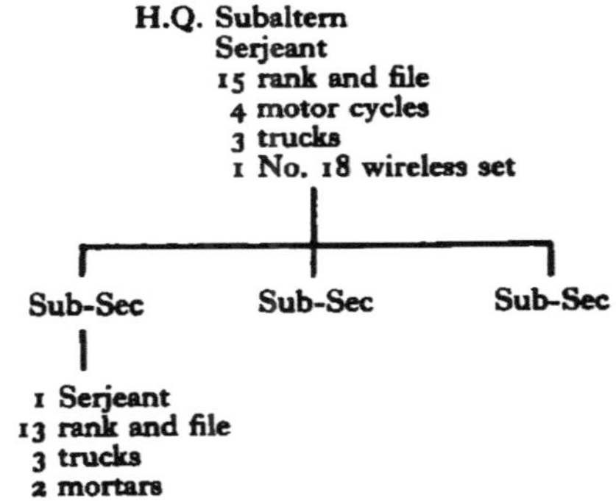

SUMMARY—C.W. GROUP	Officers	O.Rs.	M.C.s	Trucks	Lorries	Rockets	Mortars
Group H.Q. (including M.O. and L.A.D.)	5	59	8	9	5	–	–
One C.W. (Mortar) Coy.	6	240	18	46	7	–	18
Two C.W. (Rocket) Coys. (each).	5	216	12	46	9	432	–
Total C.W. Group, R.E.	21	731	50	147	30	864	18

APPENDIX XIVA

(See pages 85 and 86)

METEOROLOGICAL ORGANIZATION IN THE FIELD

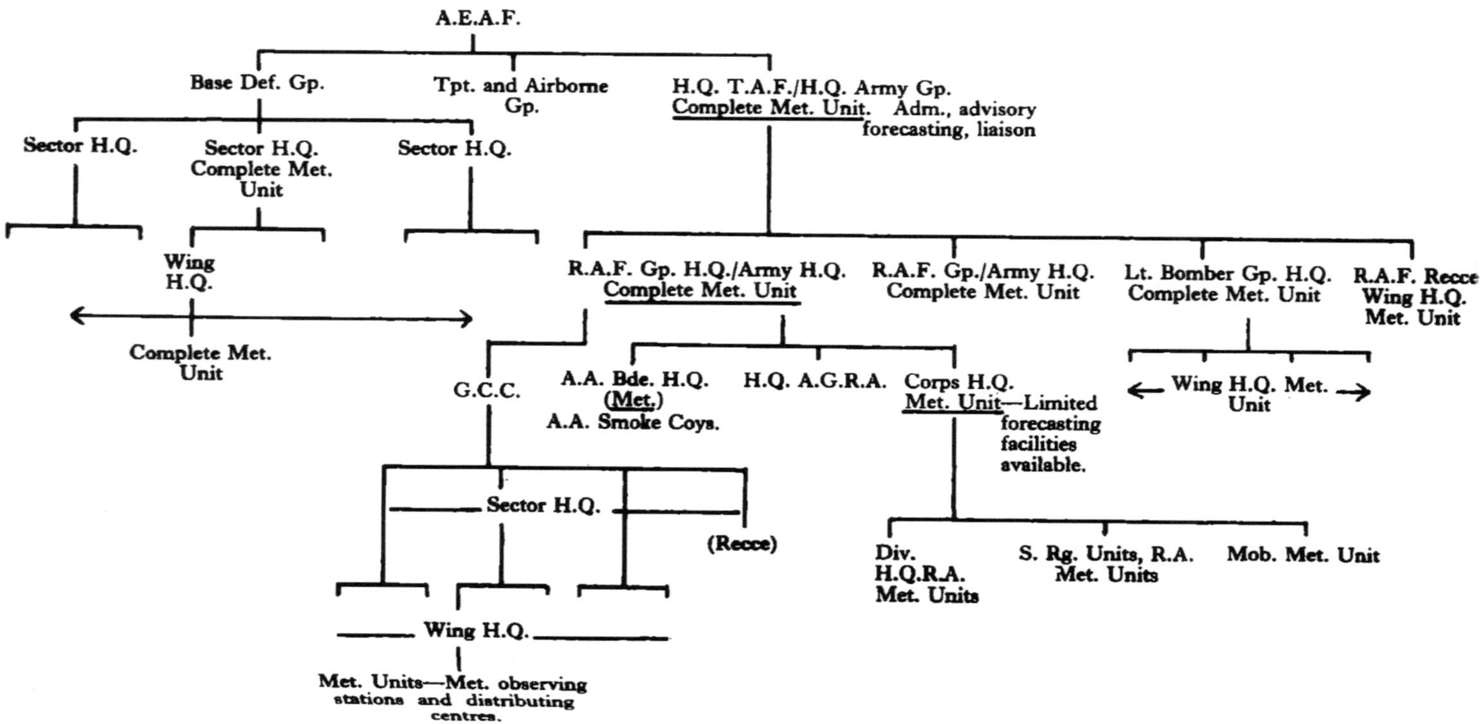

APPENDIX XV

(*See* pages 103 and 105.)

War Establishment XIV/959/1

Notified in A.C.Is. 23rd May, 1945. Effective date 23rd April, 1945.
(*Incorporating Amendment No.* 1 *of 7th July*, 1945.)

No. 3 CHEMICAL DEFENCE LABORATORY, R.E., 21 ARMY GROUP WAR ESTABLISHMENT

Detail.	Main Laboratory.	Mobile Detachments (each).	Total, No. 3 C.D. Laboratory, R.E., 21 Army Group.
(i) Personnel			
Officer commanding (major) (chemist)	1	—	1
Chemists	—	(*a*) 1	2
Total, officers	1	1	3
Company serjeant-major	1	—	1
Total, warrant officers	1	—	1
Serjeants (chemists)	—	1	2
Serjeant (technical weapons)	1	—	1
Total, serjeants	1	1	3
Corporals	5	—	5
Sappers (includes 1 lance-corporal)	5	1	7
Drivers (includes 1 driver-mechanic)	3	1	5
Total, rank and file	13	2	17
Total, other ranks	15	3	21
Total, all ranks	16	4	24
Attached :—			
R.A.M.C.—Serjeant (for D.A.D. (physiologist))	1	—	1
A.C.C.—Cook	1	—	1
Total, No. 3 Chemical Defence Laboratory R.E., 21 Army Group (including attached).	18	4	26
(ii) Distribution of Rank and File by Trades and Duties			
Tradesmen :—			
Chemical laboratory assistant (corporal)	1	—	1
Clerk, general duties (corporal)	1	—	1
Carpenter and joiner (corporal)	1	—	1
Fitter (corporal)	1	—	1
Photographer (corporal)	1	—	1
Pioneers, R.E., for duty as :—			
Motor-cycle orderly	1	—	1
General dutymen	2	1	4
Orderly	1	—	1
Storeman	1	—	1
Total, tradesmen	10	1	12

(*a*) Captain in one detachment, subaltern in the other.

Detail.	Main Laboratory.	Mobile Detachments (each).	Totals, No. 3 D.C. Laboratory, R.E., 21 Army Group.
(ii) Distribution of Rank and File by Trades and Duties—*continued*			
Non-tradesmen :—			
Drivers, I.C., for duty as :—			
Drivers of vehicles	1	1	3
Batman	1	—	1
Batman-driver	1	—	1
Total, non-tradesmen ..	3	1	5
Total, rank and file ..	13	2	17

(iii) Transport

Motor-cycle, solo	1
Car, 5-cwt., 4 × 4	1
Car, 4-seater, 4 × 2	1
Truck, 15-cwt., 4 × 4, G.S.	1
Lorries, 3-ton, 4 × 2, G.S.	2
Trailer, 10-cwt. 2-wheeled, lightweight, G.S.	1

(iv) Weapons

L.M.Gs., ·303-in.	2

Note on First Reinforcements

First reinforcements—	
Subaltern ..	1
Corporal ..	1

APPENDIX XVI

(*See* pages 44 and 111.)

PROCEDURE FOR REPORTING AND CONFIRMING THE USE OF GAS BY THE ENEMY AGAINST THE UNITED KINGDOM

Part I.—Method of Reporting

1. OBJECT

The object is to transmit reasonably substantiated reports of the use of gas with the greatest possible speed to one of the following :—

Ministry of Home Security.
Admiralty.
War Office.
Air Ministry.
Ministry of Aircraft Production.

2. REPORTING

Reports may reach the above by the following channels :—

(*a*) CIVIL

G.I.Os. will investigate incidents and reports through Civil Defence channels to Regional H.Q. Senior Gas Advisers will examine reports and samples and will report to Ministry of Home Security. The Regional Hospital Officer will investigate any suspected gas casualties.

(*b*) NAVAL ESTABLISHMENTS

Suspected gas incidents will be investigated on request by Civil Organization. Reports will pass through Civil channels to Ministry of Home Security and through Naval channels to the Admiralty for information.

(*c*) MILITARY

(i) ***British***

Home Forces.—Suspected gas incidents will be investigated by the unit concerned and reported through the usual channels to G.H.Q., Home Forces ; who will inform the War Office (S.W.V.2).

A.A. Command.—Suspected gas incidents will be investigated, by Civil Organization. Reports will pass through Civil channels to Ministry of Home Security and through the usual channels to A.A. Command for information.

Isolated British units or detachments in an area administered by a U.S. formation will report through U.S. channels.

(ii) *U.S. Forces* (including U.S. Air Force)

U.S. formations will investigate gas incidents occurring in their area and report through the usual channels to H.Q. E.T.O.U.S.A., who will inform the War Office (S.W.V.2).

Isolated U.S. units or detachments in an area administered by a British formation will report through British channels.

(*d*) Air Force

Suspected gas incidents will be investigated by the Station concerned, in accordance with instructions issued by the Air Ministry. R.D. Arm 6 M.A.P. will be responsible for informing the Air Ministry (Duty Gp. Captain or D.D. A.R.P.).

3. Priority of Messages

(*a*) Reports on the following subjects will be sent with the highest degree of priority :—

(i) First use of gas by the enemy.

(ii) Suspected new gas.

(iii) New weapon or new method of use.

(iv) Heavy casualties due to suspected failure of our defence measures.

Speed in reporting the above should be regarded as *more important than secrecy.*

(*b*) Routine reports will be forwarded with the degree of priority and security appropriate to normal intelligence matters.

4. Liaison

Liaison will be established between the Headquarters of Corps, Commands, and A.A. Groups, U.S. Corps and Base Sections, R.A.F. Groups, and the Civil Region, whereby all information will be exchanged. Warning of suspected gas incidents should be passed between Military and Civil Defence personnel at all levels, but exchange of technical information will not take place below the level of Corps/Region, except where G.I.Os. are called in to investigate incidents, which in the case of A.A. Command will be the normal procedure.

5. Facilities for Investigation

(*a*) H.Q., E.T.O., U.S. Army will be given facilities for investigating any incident in the United Kingdom on request.

(*b*) E.M.S. Specialists and C.D.E.S., PORTON will be given facilities for investigating any incident or examining any casualties as ordered.

Part II.—Final Confirmation of Reports

6. Object

The following procedure is designed to ensure that identical technical advice is given simultaneously to Ministry of Home Security, the Admiralty, the War Office and the Air Ministry.

7. First Reports

(*a*) The first reports of the use of gas, or other high priority reports, will come, by the channels laid down in Part I, to one or more of the following :—

The Ministry of Home Security The Admiralty The War Office The Air Ministry	For phone numbers, etc., *see* Appendix A*

* Appendix A to this directive has been omitted.

(*b*) The first recipient of an apparently authentic report will immediately inform the remainder and also C.C.D.D. Ministry of Supply.

(*c*) The War Office will ensure that S.H.A.E.F., H.Q. E.T.O.U.S.A., H.Q. 21 Army Group, G.H.Q. Home Forces and H.Q. A.A. Command are informed of the reports received. The Air Ministry will be similarly responsible for informing RD. Arm 6 M.A.P., War Office (S.W.V.2) will also inform War Office (M.I.10), Main Building, Ext. 86.

(*d*) The technical branches of the Ministry of Home Security, the Ministry of Supply, the War Office, E.T.O.U.S.A. and M.A.P. will exchange all technical information at this stage.

8. Technical Meetings

(*a*) When any one of the technical branches in para. 7 (*d*) considers that sufficient evidence is available, they will ask the Scientific Adviser to the Ministry of Home Security to convene a meeting between representatives of all.

(*b*) Technical representatives from other H.Qs. concerned may be asked to attend as necessary.

(*c*) The Ministry of Home Security War Room will send official notification of this meeting, to, *inter alia*, The Cabinet War Room, The Admiralty (Duty Captain), The War Office (M.O.2), and the Air Ministry (Duty Gp. Captain).

9. Promulgation of Conclusions

After complete technical discussion the conclusions of the meeting will be promulgated as follows :—

(*a*) If the validity of the report is disproved The Ministry of Home Security War Room will inform officially the parties mentioned in para. 8 (*c*).

(*b*) If the report is confirmed the Minister will decide whether to :—

(i) Release the broadcast messages to the public, etc.

(ii) Instruct the Ministry of Home Security War Room to make the signal " The use of gas by the enemy in this country/Northern Ireland has been confirmed. Press and broadcast notice released " to (*inter alia*) :—

The Admiralty (Duty Captain).

The War Office (M.O.2) who will inform :—

The War Office
D.M.O.—who passes information to the Chiefs of Staff.
D.M.I.
V.Q.M.G.
Q (Ops.).
C.-in-C., 21 Army Group.
C.-in-C., Home Forces.
G.O.C.-in-C., A.A. Command.
C.G., E.T.O.U.S.A.

The Air Ministry (Duty Gp. Captain).

PROCEDURE FOR REPORTING AND CONFIRMING THE USE OF GAS BY THE ENEMY ON THE CONTINENT OF EUROPE

(To be read in conjunction with the procedure for reporting in the United Kingdom)

1. It must be considered unlikely that the use of gas solely against the United Kingdom will occur. Two contingencies must therefore be considered :–

(*a*) The use of gas against our invading forces on enemy-occupied territory.

(*b*) The use as in (*a*) above with simultaneous use against the United Kingdom.

In the case of (*a*) standard military procedure for reporting the situation would operate but the procedure requires amplification and definition to deal with the specific case of the reporting of the first use of gas. With alternative (*b*) the standard military and civil procedures would have to operate together.

2. The following procedure is set out to show how the purely Service channels used in 1 (*a*) will be linked with the existing arrangements under the Ministry of Home Security.

Initial Reports of the Use of Gas

3. Reports on the use of gas on the Continent will pass both by " I " (G.2) and C.W. channels through the highest headquarters, and thence to S.H.A.E.F. and to the War Office (M.I.10). S.H.A.E.F. will keep E.T.O.U.S.A. informed ; the War Office will notify the Ministry of Supply, who will be responsible for advising other civilian ministries. The reports will be screened at the Headquarters through which they pass by the joint action of " I " (G2) and C.W. officers.

Gas and Pathological Samples

4. Samples of gas and equipment of urgent operational importance will be sent by fastest possible means under the orders issued by Field Formations, in the case of British Forces to Porton, and U.S. to 44 Chemical Laboratory Coy.

Pathological samples will likewise be sent to Porton by British Forces and to No. 1 General Medical Laboratory by U.S. Forces.

Whenever samples are sent direct to any of the above laboratories, 21 Army Group will inform S.H.A.E.F. and the War Office (M.I.10). The War Office will inform the Ministry of Supply.

Reports on Samples

5. Immediate reports on samples sent to them direct will be made by Porton direct to 21 Army Group. These reports will be sent also by Porton by the quickest means to C.D.R.D., to the War Office, and to S.H.A.E.F. To these reports Porton will add any general suggestions for immediate defensive measures to enable D.C.W. 21 Army Group to issue such instructions as he deems requisite.

The same procedure will be followed by the two U.S. Laboratory Coys. There should be the closest agreement and co-ordination between Porton and the U.S. Laboratories, and wherever possible agreed reports will be submitted.

Final Confirmed Report

6. The formation of a " final confirmed report " is unlikely to be difficult, because it will probably be obvious if gas really has been used.

A suggested mechanism for its formation follows from the fact that the presence of a suspected sample at Porton or a U.S. Laboratory will be known to 21 Army Group, S.H.A.E.F., War Office, E.T.O.U.S.A., and Ministry of

APPENDIX XVII
(See page 129)

ANGLO-AMERICAN ORGANIZATION

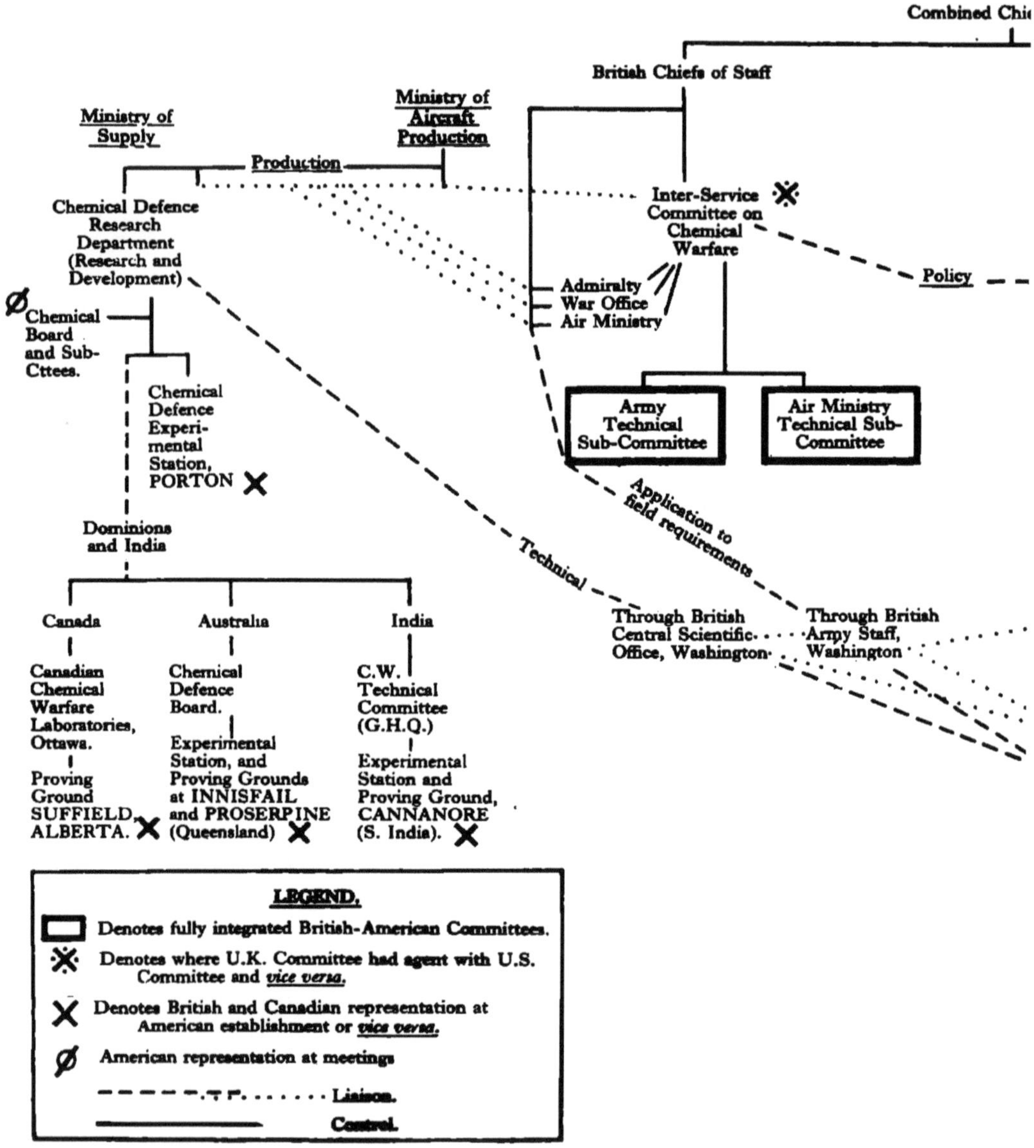

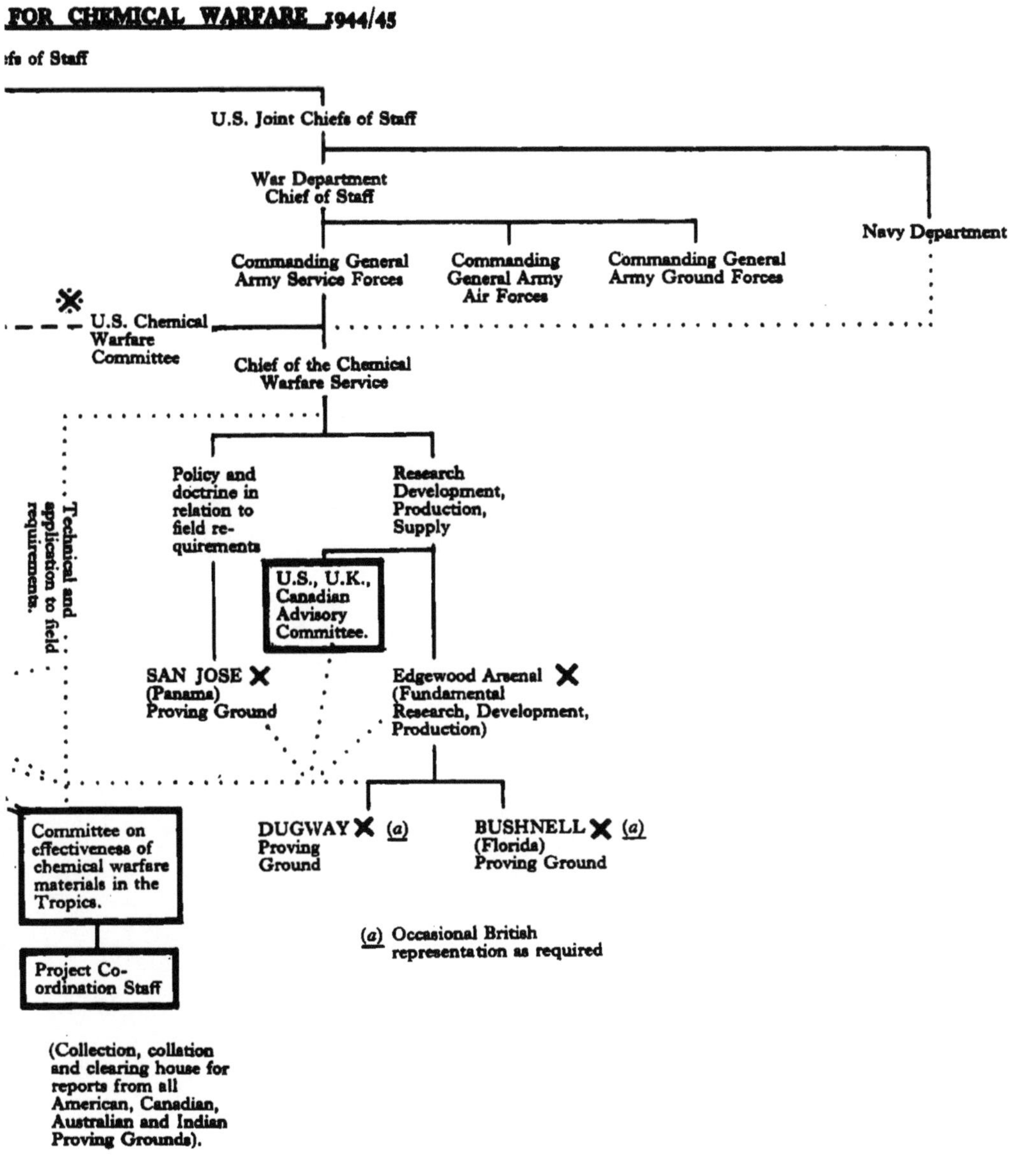

To face page 281

Supply. Further, this " confirmed report " would be needed for immediate action by the War Office (M.O., M.I., and S.W.V.). Information will therefore be received and action taken by the War Office, which will undertake the responsibility of obtaining before issue of a " confirmed report " the full concurrence to this report, including its wording, of C.C.D.D. Ministry of Supply, E.T.O.U.S.A., and, if obtainable, S.H.A.E.F. and 21 Army Group. Subsequent to the issue of this report the same authorities will make agreed recommendations on the immediate defensive measures to be adopted.

This report will then be sent to the British Chiefs of Staff, whereby the other Services are automatically informed, and to the Cabinet War Room. The War Office will inform the Ministry of Home Security. Reports will reach the Combined Chiefs of Staff through the British Chiefs of Staff and through S.H.A.E.F.

WARNING TO MINISTRIES OF POTENTIAL ACTION REQUIRED

7. The general principle will be that all operational matters of general concern that are within the cognizance of the War Office M.O. branch concerned are automatically known to the corresponding branches in the other Service Ministries. The technical information required will however be exchanged between the various technical branches which are :—

Admiralty	D.T.S.D., D.T.M. and M.D.G.
War Office	D.M.I., D.S.W.V. and D.Path.
Air Ministry	D.D.A.R.P. and D.Hyg.
Ministry of Aircraft Production	D. Arm D.
Ministry of Home Security	Scientific Adviser
Ministry of Supply	C.C.D.D.
E.T.O.U.S.A.	C.W.S.

APPENDIX XVII—*see* pull-out sheet.

APPENDIX XVIII

(*See* page 137.)

Dosages of War Gases

The effects on men of persistent gas vapour and of non-persistent gases depend not only on the amount of vapour or gas present in the air but also on the time during which men are exposed to it. Both of these factors must be taken into account in defining the effectiveness of any gas. Short exposures to high concentrations may produce the same effects as long exposures to low concentrations and for practical purposes the effects caused are in proportion to the magnitude of the product of the concentration (C) multiplied by the time of exposure (T). This product is known as the Dosage (D).

For example, exposure for 20 minutes to a concentration of say 100 milligrammes per cubic meter would be a dosage of 2,000. Forty minutes' exposure to a concentration of 50 mg./cu.m. would also be a dosage of 2,000. The unit of dosage is in terms of milligrammes minutes per cubic metre (mg.min./cu.m.).

Both dosages although attained by different combinations of C and T will produce similar effects. This example is illustrative of the values of C and T which might be encountered in mustard gas exposures.

For each gas there is a minimum dosage which must be attained in order to produce effects which are sufficiently severe to be incapacitating. Conversely, there is a maximum dosage to which our own troops can safely be exposed if fighting efficiency is not to be reduced by minor injuries from gas. This is of particular import in relation to mustard gas owing to its long persistence.

Dosage therefore readily shows the comparative toxicity of various substances which might be used as war gases. The table below is given with this object. In the case of mustard gas it also shows the enhanced effectiveness of mustard gas vapour under hot humid climatic conditions.

Serial No.	Type of Gas.	Dosage required to attain worth-while propotion of casualties. (In mg.min./cu.m.)	Remarks.
(a)	(b)	(c)	(d)
1.	Phosgene ..	3,200 (lethal).	Smaller dosages will necessitate hospital treatment.
2.	Prussic Acid ..	5,000 (lethal).	Small dosages ineffective.
3.	Cyanogen Chloride..	11,000 (lethal). (200,000 for penetration of respirator container).	Small dosages non-lethal but irritant–intolerable at once.
4.	Arsine ..	5,000 (lethal).	
5.	Chloropicrin.. ..	6,750 (lethal).	
6.	Nerve Gas—"Tabun"	300 (lethal).	Other compounds lethal in even lower dosages.
7.	Mustard Gas Vapour	200 (no respirator—will produce temporary blindness in any climate).	
8.	Mustard Gas Vapour	750 (respirator but no protective clothing—temperature 80° F.—sweating skin —will produce severe generalized skin blistering and total disablement).	
9.	Mustard Gas Vapour	4,000–10,000 (respirator but no protective clothing—temperature 40° to 60° F. —cool dry skin—severe generalized blistering—total disablement).	
10.	Nitrogen Mustard Gas Vapour (HN2).	70 for eye effects—otherwise ⅓rd as effective as mustard gas.	
11.	Lewisite vapour ..	Disabling dosages cannot be attained in the field.	

APPENDIX XVIIIA

(*See* page 173.)

Characteristics of Aircraft Spray

Extract from Military Training Pamphlet No. 5 " The Use of Gas in the Field "—1943.

" Liquid blister gas can be sprayed from aircraft at any height, and reaches the ground in drops, the size of which is virtually independent of the height of the aircraft. The size of the area sprayed, which will all be downwind, depends on the height and speed of the aircraft, the speed of the wind, the quantity of liquid released, and the course of the aircraft relative to the direction of the wind. The density will be greater the smaller the area.

Spray falls at about 1,000 ft. per minute, and is unaffected by the presence of cloud. The large drops will fall in the upwind part of the sprayed area, while the small drops will be carried farther downwind. Normally the upwind part of the area will also be the most heavily contaminated. Spray is classified as low or high, depending on whether the height of the spraying aircraft is below or above 1,000 ft.([1]).

Low spray attacks depend for their effectiveness upon the skill of the pilot, since only a relatively narrow area is showered with blister gas. The contamination produced can have a considerable casualty effect, but against this must be set the difficulty of allowing for wind in aiming, and the vulnerability of the aircraft.

High spray is suitable for harassing large areas. Such attacks should be prepared in advance, since their accuracy depends upon information on the wind direction and speed. The use of high spray might produce widespread casualties among badly trained personnel. Note that several minutes will elapse between the passage of the aircraft and the falling of the spray.

Spray attacks are less effective in very cold weather, when freezing of the spray is possible and the wearing of thick clothing by the sprayed troops will reduce its casualty-producing value.

In general, it may be said that spray produces no immediate casualties (except eye casualties), and any form of overhead cover gives a measure of protection."

---

(1) Although this definition was never officially amended low spray attacks were ultimately regarded as only being useful if carried out from very low heights, *viz.*, 50 feet and upwards.

APPENDIX XIX

(*See* Chapter IV, Section I)

LIVENS PROJECTOR

Projectors emplaced.

Drum, Projector barrel and shoe, (improvised) stand, sheer wire pistol and burster for drum, charge box (one open to show cordite charges), firing cable.

5-in. U.P. ROCKET

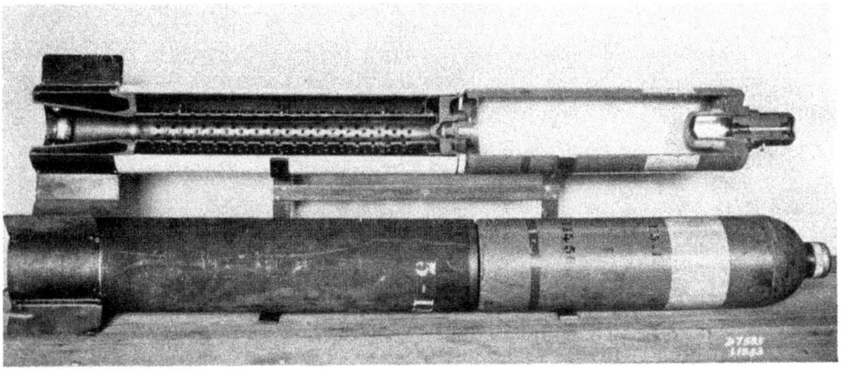

Bomb and rocket tail.

Sextuple stands.

4·2-in. MORTAR (BRITISH)

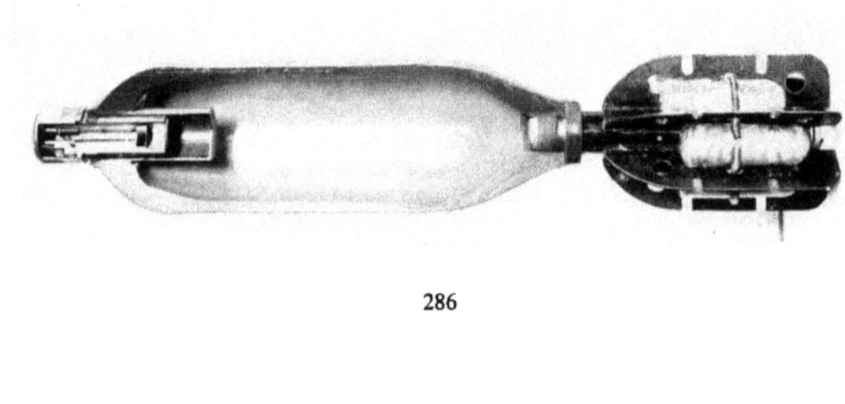

Bomb, chemical, Mk. I.

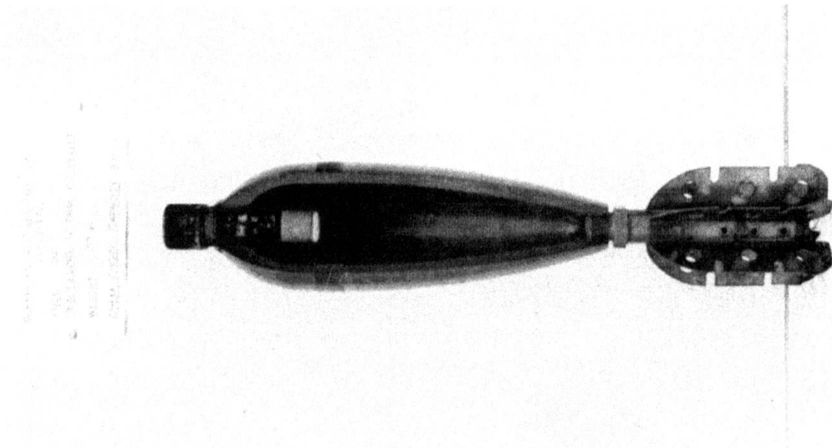

Bomb, chemical, streamlined.

4·2-in. MORTAR (UNITED STATES)

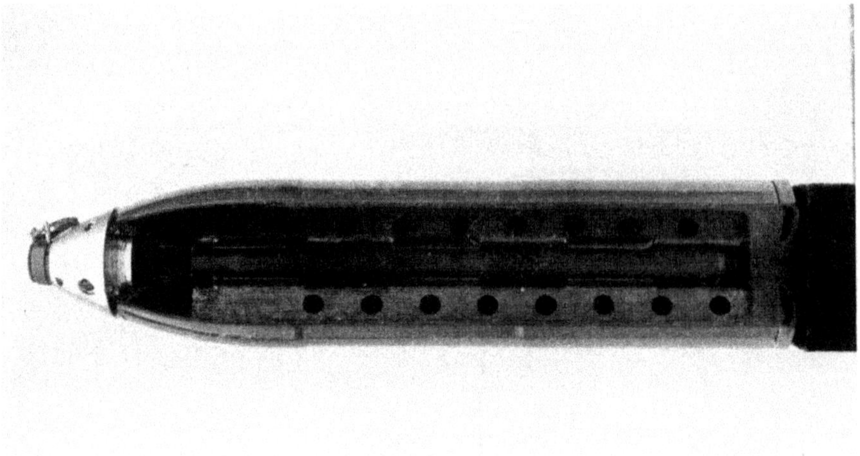

BULK CONTAMINATION VEHICLE
(Installation, Type A/L 3)

JET, ANTI-TANK, PORTABLE, Mk. 1

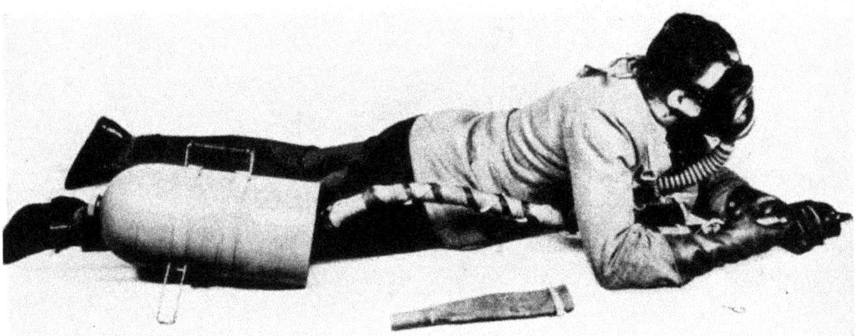

APPENDIX XIX—*continued*

JET, CHEMICAL, No. 1, Mk. I

GRENADE, A.C., 600 c.c.

APPENDIX XX

(*See* Chapter IV, Section 2)

BOMB, AIRCRAFT, L.C. 500-lb.

Chargings—Phosgene (ground burst). Mustard gas (air burst, but not finally cleared when war ended).

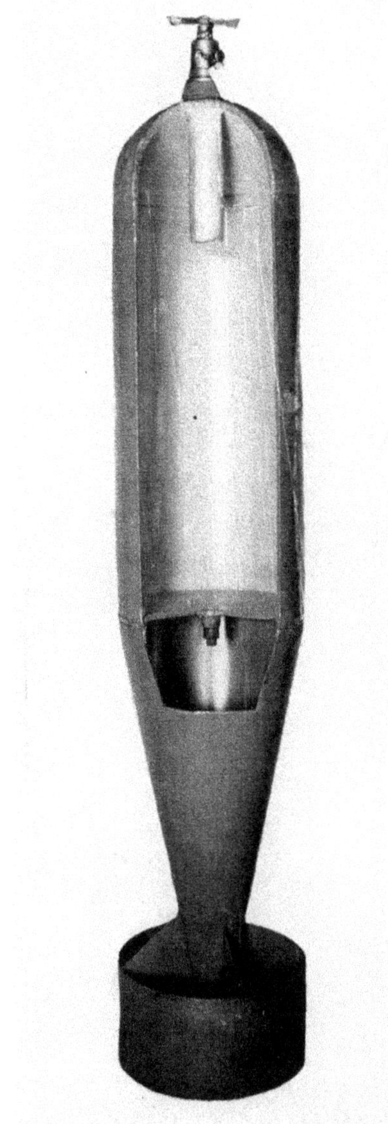

BOMB, AIRCRAFT, L.C. 250-lb., Mk. II

Chargings—Phosgene or Mustard gas.
Charged mustard gas—*Area* of
Heavy contamination, 170 sq. yds.
Light contamination, 550 sq. yds.
Anti-personnel effect, 2,300 sq. yds.

When charged mustard gas and air burst the size of the area covered by droplets varied according to the height of burst and the strength of the wind. At height wind products of 5,000 to 15,000 (feet × miles/hour) the area of light contamination (10 g/m^2) was in the order of 1,500 to 2,000 square yards, an area of 13,000 square yards being covered by a density greater than 1 gram per square metre.

BOMB, AIRCRAFT, L.C. 30-lb., Mk. I

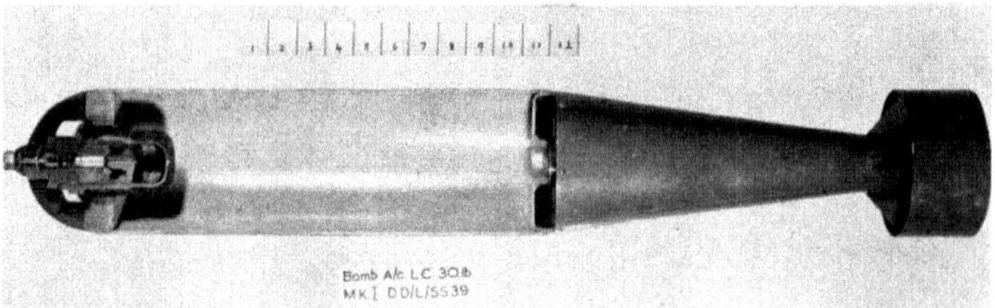

Charging—Mustard gas

Areas of

Heavy contamination (50 gms./m²)	10 square yards	on hard targets
Light contamination (10 gms./m²)	75 square yards	
Anti-personnel effect	400 square yards	

BOMB, AIRCRAFT, L.C., 30-lb., Mk. II

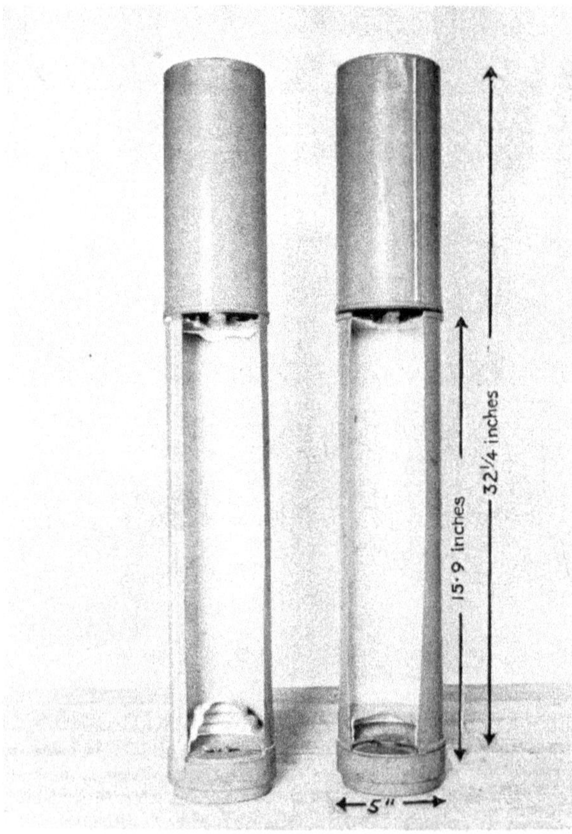

Charging—Mustard gas

Areas of

Heavy contamination	40 square yards	on hard targets
Light contamination	100 square yards	
Anti-personnel effect	250 square yards	

BOMB, AIRCRAFT, L.C., 65-lb., Mk. II (with Transit crate)

Charging—Mustard gas

Areas of	
Heavy contamination	.. 40 square yards
Light contamination	.. 300 square yards
Anti-personnel effect	.. 1,000 square yards

BOMB, AIRCRAFT, L.C., 50-lb. (CANADIAN)
Cluster of 5.

S.C.I, TYPE S/G, 250-lb., Mk. IV

S.C.I, Type S/G, 500-lb. FOR INTERNAL STOWAGE
Fairings could be fitted for external stowage, for which the emission pipe was also modified.

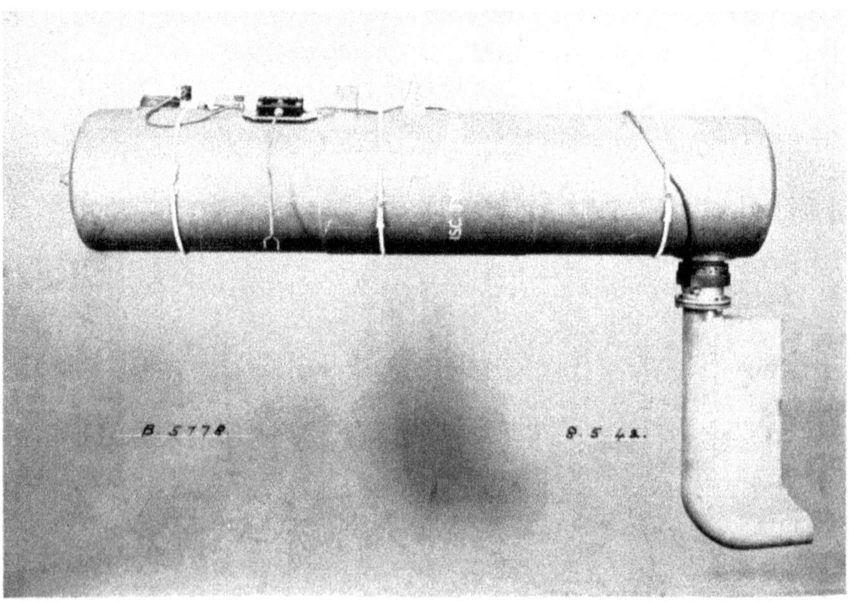

APPENDIX XX—*continued*

S.C.I, TYPE S/G, 1,000-lb.

UNITED STATES AIRPLANE SMOKE TANK—M.10,
with fairing over British suspension lug for carriage on British aircraft

I.S.C., TYPE S/L, 400-lb.
Spraying bomb " The Flying Cow "

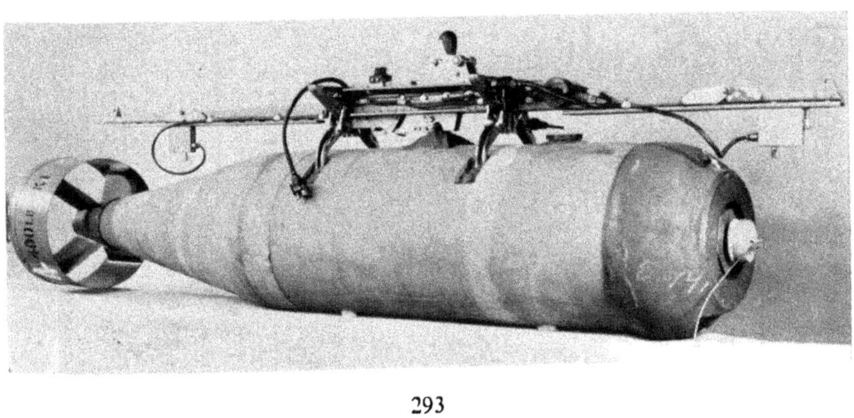

APPENDIX XXI

(*See* pages 47 and 192.)

United Kingdom Production Totals of Anti-Gas and Chemical Defence Stores from September, 1939 to September, 1945

Serial No.	Item.	Accounting Unit.	Number Produced.	Approx. Cost.
				£ millions
	PERSONAL EQUIPMENT			
1	Respirators, G.S.—Facepieces	singles	20 million	9
	Containers	singles	26 million	4·9
	Haversacks	singles	24 million	8·4
2	Respirators, Light Type—Facepieces	singles	5½ million	1·7
	Containers	singles	7½ million	0·4
	Haversacks	singles	5½ million	1·1
3	Outfits, Anti-dim	singles	40¾ million	0·25
4	Eyeshields, Anti-gas (plain and tinted)	singles	262½ million	4
5	Capes, Anti-gas	singles	24½ million	*
6	Curtains, Anti-gas	singles	16 million	*
7	Impregnated Battle Dress (A/V)	suits	13 million	0·36
8	Impregnated Socks (A/V)	pairs	2 million	0·009
9	Dubbin, Protective	tons	6,243	0·291
10	Wallets, Anti-gas	singles	13⅓ million	*
11	Ointments, Anti-gas (Nos. 1, 2, 3A, 5 and 6) ..	2-oz. outfits	75 million	1·27
	Ointments, Anti-gas (Nos. 1, 2, 3, 3A and 5 in 7-lb. jars).	lbs.	631,500	0·113
	Ointments, Anti-gas (Nos. 3 and 3A for animals)	8-oz. jars	375,000	0·037
12	Detectors, Gas, Sleeve	pairs	51 million	1·27
13	Detectors, Gas, Ground	booklets	7¼ million	0·03
	UNIT EQUIPMENT			
14	Light Oilskin, Jackets	singles	9¼ million	*
15	Light Oilskin, Trousers	pairs	8½ million	*
16	Light Oilskin, Hoods	singles	2¼ million	*
17	Gloves, Anti-gas (oilskin and rubber)	pairs	24 million	*
18	Overmittens	pairs	5¾ million	*
19	Overboots, Anti-gas	pairs	3½ million	*
20	Aprons, Anti-gas	singles	262,000	*
21	Gas Alarm—Local—Rattle	singles	293,281	0·07
22	Gas Alarm—General—Syren	singles	32,219	0·016
23	Gas Alarm—General—Strombos Horse ..	singles	3,040	0·007
24	Detector Paint	lbs.	3 million	0·08
25	Detectors, Gas, Spray (Mk. I and Mk. III) ..	singles	4 million	0·06
26	Detectors, Paper, Type A	sets of 25 papers	6½ million	0·002
27	Pocket Vapour Detector	sets	135,483	0·10
28	Signs, Warning, Gas	singles	5 million	0·03
29	Sacks, Paper, Gasproof	singles	3½ million	0·127
30	Pathway, Anti-gas	yards	885,000	0·03
31	Covers, Stretcher	singles	700,000	0·035
	MISCELLANEOUS			
32	Cases, War Gas, Testing	sets	95	0·0015
33	Cloth, Union, Anti-gas	yards	1½ million	0·26
34	Generators, Lachrymatory	singles	2 million	0·41
35	Capsules, Lachrymatory	singles	19½ million	0·06
36	Smelling Sets	sets	37,725	0·02
	ANIMAL EQUIPMENT			
37	Horse and Mule Respirators, Mk. I	singles	59,500	0·13
38	Eyeshields, Horse and Mule	singles	554,040	0·04

Total (less starred items)	36·282
Total (starred items)	35
Total (Civilian Respirators)	14
	85·282

REMARKS

Respirators.—In addition 90 million civilian, 4½ million civilian duty respirators, 3 million children's respirators, 2 million baby helmets and 86,000 hospital helmets were manufactured at a cost of approximately £14 million. The total of complete respirators (all types civilian and Service) was therefore over 125 million, in addition to which spare containers, haversacks, etc., were also manufactured.

Oilskin Clothing.—In addition the following anti-gas oilskin items were also produced : Anti-gas coats (1 million), cap covers (2 million), covers for tropical helmets (½ million), waders for assault troops (50,000), gas detector tippets (¾ million), heavy oilskin coats (3½ million), heavy oilskin jackets (2½ million), heavy oilskin trousers and leggings (3 million pairs). The three latter items include those provided for the Navy for anti-weather purposes.

The total cost of all oilskin items (including starred items) was approximately £35 million.

(C29058) M

APPENDIX XXII

(*See* Chapter VI)

Scales of Anti-Gas Clothing and Equipment

Extract from Document Security B.498

Part I

PERSONAL ANTI-GAS EQUIPMENT

for

(*a*) Field Force Units.
(*b*) Non-Field Force Units in Defended Ports Home and Abroad.
(*c*) Units of the Anti-Aircraft Command.

Serial No.	Item.	Scale of Issue.		1st line reserves.	Remarks.
		Every officer and other rank and every member of the Q.A.I.M.N.S. and of the V.A.D.	Every member of the A.T.S.		
(a)	(b)	(c)	(d)	(e)	(f)
1	Respirators	One.	One.	10 per cent.*	
2	Ointment, anti-gas ..	Two 2-oz. containers; one to be carried in pocket of respirator haversack the other in pocket of cape.	Those who are issued with cape A/G., as for column (c). Others: one tin to be carried in the pocket of the respirator haversack.	50 per cent.	
3	Eyeshields, anti-gas ..	One packet of six to be carried in pocket of respirator haversack.	As for column (c).	50 per cent.	

4	Capes, anti-gas ..	One.	1. One for every M.T. driver. 2. One for every officer and member employed with A.D.G.B. 3. One for all other A.T.S. personnel whose duties are carried out in the open, or, if otherwise, for whom no covered shelter accommodation exists.	50 per cent. (except units in column (*f*).)	(i) A.A. and S/L Btys., R.A., 1st line reserve of 10 per cent. only. (ii) A.T.S., no reserve, except those employed in an operational role with A.D.G.B. or as M.T. drivers 10 per cent., and those at A.A. Training Centres 20 per cent.
5	Detectors, gas sleeve (sometimes referred to as detector, gas, individual).	Two pairs.	For those employed as instrument numbers with A.D.G.B. on gun sites: two pairs. Others: two pairs when proceeding overseas.	100 per cent.	
6	Cotton waste	One ounce; half in the pocket of respirator haversack, half in pocket of cape.	Those issued with a cape: as for column (*c*), others: half an ounce in pocket of respirator haversack.	50 per cent.	
7	Curtains, anti-gas ..	This item will not be drawn until further notice, but will be held in ordnance depots.	One for each member of A.T.S. employed as instrument numbers with A.D.G.B. on gun sites.	50 per cent. of the quantities allowed under column (*d*).	
8	Detectors, gas, ground..	One pad of 25, to be carried by every officer, W.O. and N.C.O.	One pad of 25 to be carried only by officers and sub-officers.	50 per cent. (except units in column (*f*)).	(i) 1st line reserves for Fd. Sqn., Fd. Pk. Sqn., Fd. Coy., Fd. Pk. Coy. and A.T. Coy. R.E. is 200 per cent. (ii) 1st line reserves for units of A.A Command is 25 per cent.

Notes.

1st line reserves will be held in Unit Transport (if available) or in Unit Store.

*** In the case of the Light Respirator, 10 per cent. additional containers were also authorized.**

(C29058)

M*

APPENDIX XXII—*continued*

Part II

VEHICLE EQUIPMENT

for

(*a*) Field Force Units.

(*b*) Non-Field Force Units in Defended Ports Home and Abroad.

(*c*) Units of the Anti-Aircraft Command.

1. Detector Paint

Detector paint for vehicles will be provided on the scale of 4 oz. for each Armoured Fighting Vehicle, Dragon, Tractor, Ambulance, Lorry, Car, Van and Truck on the establishment.

2. Trays for Detector Paint (with brackets)

One will be provided for each forward controlled vehicle, on the establishment.

3. Sacks, Paper, Gas-proof (for contaminated clothing)

(*a*) For all vehicles except Tanks, Armoured Cars, Scout Cars and Carriers and Task vehicles of L. of C. Railhead Companies, R.A.S.C. One sack per vehicle.

(*b*) In addition, in 2nd and 3rd Line Ammunition and Petrol Units, R.A.S.C. Four sacks for each Task vehicle, making a total of five sacks for each Task vehicle.

4. Bleaching Powder

One 2-lb. tin per Tank, Armoured Car or Scout Car.

5. Sandbags

Two for each vehicle (all types) for decontamination purposes.

APPENDIX XXII—*continued*

Extracts from Part III. Table 1 (*a*)

UNIT EQUIPMENT FOR UNITS OF THE FIELD FORCE.

(Not applicable to African Colonial Forces)

Serial No.	Unit or Formation.	Set of Light Anti-gas Clothing. (*See* Note (*a*))	Set of Light Anti-gas Clothing, less Hoods and Overboots†.	Set of Light Anti-gas Clothing, less Hoods†.	Sets of Light Anti-gas Clothing, less Overboots, plus Boots, Rubber, Knee	Signs, Warning, Gas.	Detectors, Gas Spray. (*See* Note (*b*).)	Bleaching Powder (Super-Tropical). (*See* Note (*m*).)	Alarms, Local (Rattles).
(a)	(b)	(c)	(d)	(e)	(f)	(g)	(h)	(i)	(j)
7	H.Q., a Division	12	*See* Note (*h*)	*See* Note (*i*)	—	50	50	24—7 lb.	2
8	H.Q., Armoured Div. ..	12			—	50	50	20—7 lb.	2
23	Regt. H.Q., R.H.A. Field, Medium A/Tk., Heavy, Super Heavy, and Survey Regt.	4			—	20	20	24—17 lb.	1
33	An Armoured Regt. ..	48			*See* Note (*d*)	100	100	20—17 lb.	1 per Regt. H.Q., 1 per Sqn. H.Q., 1 per Troop.
71	Regt. H.Q. A.A., L.A.A. and Searchlight Regt.	8			—	30	20	24—7 lb.	2
64	Medium Bty.	24			—	120	120	24—7 lb.	2 per Bty. H.Q., 1 per Sec.
65	R.H.A. and Fd. Bty. ..	16			—	80	120	24—7 lb.	2 per Bty. H.Q., 1 per Sec.
97	H.Q. Rly Operating Group	2			—	5	20	5—2 lb.	1
103	A Field Company	24 (Note (*e*))			—	50	100	8–7 lb. per H.Q. 8–7 lb. per Sec.	—1 per H.Q. —1 per Sec.
128	A Docks Section	6			18	200	20	60—7 lb.	
196	An Infantry Bn.	24 (Note (*g*))			—	200	200	24—7 lb. per Coy.	1 per Bn. H.Q., 1 per Coy. H.Q., 1 per Pl.
263	Detail Issue Depot ..	4			—	10	20	20—2 lb.	—
261	Mob. Fd. Bakery	6			—	20	20	8—7 lb.	1
292	Div. Fd. Ambulance ..	24			—	50	40	48—7 lb.	1 per H.Q., 1 per Coy.
361	Mob. Laundry, R.A.S.C...	80			—	50	20	16—17 lb.	1

† Hoods were withdrawn in 1942.

(Notes (*a*)—(*n*) overleaf.)

APPENDIX XXII—*continued*

Notes

(*a*) A set of Light Anti-gas Clothing includes the following :—

(i) One light suit, *i.e.*, jacket and trousers.
(ii) One hood, anti-gas.†
(iii) Three pairs gloves, anti-gas, oilskin, or : two pairs anti-gas, rubber (*see also* Note (*n*)), two pairs overmittens.
(iv) One pair overboots, anti-gas.

(*b*) Detector Gas Spray frames will be replaced by an 18-in. metal sheet, and the scale for these sheets will then be 1/9 the scale of frames.

* * *

(*d*) To each Armoured Regt., Armoured Car Regt., and Army Tank Bn., 20 pairs boots, rubber, knee, for decontamination of vehicles.

(*e*) In addition, for reconnaissance purposes :—

(i) Two pairs overboots, and two pairs of shoes and overmittens per officer.
(ii) Five pairs overboots, and five pairs gloves and overmittens per H.Q. and each Troop or Section.

* * *

(*g*) In addition, 24 pairs of overboots for use in demarcating contaminated areas.

(*h*) Sets of light anti-gas clothing less hoods† and overboots will be issued to all units on a scale of one set each for the following :—

(i) Each motor cyclist.
(ii) Each transport driver (including Scout and Bren carrier).
(iii) Each R.A.S.C. lorry driver.
(iv) Each ambulance driver.
(v) Personnel of all units (except A.A. and S/L Btys. R.A.) who are specially employed in manning A.A./L.M.Gs.
(vi) A.A. and S/L Btys. R.A., for 90 per cent. of establishment after deducting those already provided under (i) to (iv) above.
(vii) Each tank, armoured car and scout car and each member of crews of Bren carriers.
(viii) Each transport driver of the A.T.S.

† Hoods were withdrawn in 1942.

APPENDIX XXII—*continued*

Notes—continued

(*i*) Sets of light anti-gas clothing less hoods† will be issued on a scale of one set per man to 50 per cent. of the signal establishment of Royal Signals and other units.

* * *

(*l*) Ten per cent. of items in columns (c) and (f) and in notes (*h*) and (*i*) above will be carried in 1st line reserves.

(*m*) Stocks of bleach should be decentralized on a basis of one 7-lb. tin for every four vehicles or one 2-lb. tin per vehicle on establishment when possible.

(*n*) Gloves, Anti-gas, will be issued in the following proportions of Rubber and Oilskin :—

Unit.	Rubber, at 2 pairs per suit.	Oilskin, at 3 pairs per suit.
Mobile Laundries Field Ambulances Casualty Clearing Stations Light Aid Detachments Anti-gas Laboratories	75 per cent.	25 per cent.
All Anti-Aircraft Units	20 per cent.	80 per cent.
All other units except Home Guard and P.A.D.	10 per cent.	90 per cent.
Home Guard and P.A.D.	Nil.	100 per cent.

† Hoods were withdrawn in 1942.

(C29058) M**

APPENDIX XXII—*continued*

Part III—Table 1 (*b*)

ADDITIONAL ITEMS

Unit Equipment for Units of the Field Force.

Serial No.	Item.	Scale of Issue.
1	Sacks, paper, gas-proof (for contaminated clothing).	10 per cent. of establishment all units.
2	Valises, anti-gas, clothing.	(i) One for every tank, armd. car and scout car and one for each member of the crew of Bren carriers.
3	Detector paint for painting capes, light suits and detectors, gas spray, frames. (In 2 lb. = 1 pint tins.)	(i) 1 lb. for each 32 capes or light suits. (ii) 1 lb. for each 250 detectors gas spray, frames, or 1 lb. for each 25 metal sheets, 18-inch, when issued.
4	Capsules, lachrymatory, for testing respirators.	10 per cent. of establishment of all units, quarterly.
5	Ointment, anti-gas, for decontaminating certain weapons.	8 oz. for each of the following weapons :— (i) Anti-tank rifle. (ii) Machine-gun (all types). (iii) 2-inch and 3-inch mortars. (iv) 10 per cent. reserve in unit transport.
6	Covers, stretcher.	One per stretcher on establishment plus 200 per cent. reserve.
7	Equipment for stretcher bearers.	(i) Gloves, anti-gas—1 pair per stretcher bearer. (ii) Overboots, anti-gas—1 pair per stretcher bearer. (iii) Capes, anti-gas—1 per every two stretcher bearers (for additional protection of patients).

Part III

Table 2

UNIT EQUIPMENT FOR UNITS OF THE ANTI-AIRCRAFT COMMAND

Scales of issue were higher than for Field Force Units.

* * * * *

Table 3

UNIT EQUIPMENT FOR NON-FIELD FORCE UNITS IN DEFENDED PORTS AT HOME AND ABROAD

Scales of issue were higher than for Field Force Units.

* * * * *

Part IV

UNITS NOT INCLUDED IN PART III. TABLES 1 TO 3

Training Units and Training Centres (All Arms).
Home Defence Battalions.
Holding Battalions.
Station Maintenance Companies.
Station Transport Companies.
Young Soldier Battalions.
All Schools of Instruction.

Personal Equipment .. One of each item in Part I of these Scales. Unit reserves 10 per cent. (capes 20 per cent.).

Vehicle Equipment .. One 4-oz. tin Detector Paint, one sack, paper gasproof and one light suit (with gloves) per M.T. vehicle.

Unit Equipment .. *Sets* of 30 detectors, gas, ground, 45 detectors gas spray, 10 light suits, 30 pairs gloves, 10 pairs over-boots, 12 2-oz. containers, two 7-lb. tins bleaching powder, 40 signs warning gas, 4 rattles, 4 sacks paper gasproof—per 100 men or less ; Unit reserves, $\frac{1}{2}$ set per 100 men.

APPENDIX XXII—*continued*

Part V

ANTI-GAS EQUIPMENT FOR ANIMALS

(all horse and veterinary units)

Item.	Scales.			
	Horsed Units (each).	Mobile Veterinary Sections (each).	Veterinary Evacuating Stations (each).	Veterinary Hospitals (each).
Pails, slop	1 per 40 animals.	2	4	6
Jugs, hot water ..	1 per 100 animals.	1	2	4
Brushes, horse	1 per 20 animals.	4	8	16
Scrapers, horse (flexible metal with two handles).	1 per 40 animals.	2	4	8
Bleaching powder (super-tropical) (2-lb. tin).	1 per animal for 50 per cent. of animal strength.			
Ointment, anti-gas (2-oz. containers)	6 per animal on establishment for each unit.			
Soft soap (*a*)	14 lb.	14 lb.	14 lb.	14 lb.
Soda (*a*)	14 lb.	14 lb.	14 lb.	14 lb.
Soda, bicarb. (*b*)	7 lb.	7 lb.	7 lb.	7 lb.
Respirators, horse or mule ..	1 per horse or mule on establishment plus 10 per cent. reserve.			
Eyeshields, horse or mule	2 pairs per horse or mule on the establishment.			

(*a*) For washing. (*b*) For eye irrigation.

APPENDIX XXII—*continued*

Part VI

ADDITIONAL ANTI-GAS CLOTHING AND EQUIPMENT FOR MEDICAL UNITS AND REGIMENTAL MEDICAL OFFICERS FOR DEALING WITH GAS CASUALTIES.

1. Medical Units

Unit.	Bleaching Powder, Super-tropical, 7-lb. tins.	Ointment, Anti-gas, in 2-oz. containers.	Aprons, Anti-gas.	Gloves, Anti-gas. (*a*) (*b*)
				pairs.
Advanced Depot of Medical Stores	143	—	500	500
Base Depot of Medical Stores ..	214	—	—	—
Light Field Ambulance ..	14	—	18	18
Casualty Clearing Station ..	43	400	50	50
Field Ambulance	14	—	24	24
Field Ambulance Ind. Bde. Gp. ..	18	—	30	30
General Hospital :—				
200 beds ..	4	—	6	—
600 beds ..	10	—	6	—
1,000 beds ..	21	—	6	—

(*a*) 75 per cent. of these quantities will be rubber.
(*b*) 25 per cent. of these quantities will be oilskin.

2. Regimental Medical Officers

Bleaching powder (super-tropical), 7-lb. tins	2
Aprons, anti-gas	6
Gloves, anti-gas, pairs, rubber	6
Overboots, anti-gas (Infantry Bns., Pioneer Bns. and Chemical Warfare Group only)	12
Overmittens, anti-gas, pairs	6

APPENDIX XXII—*continued*

Part VII

SCALE AND LOCATION OF RESERVES (FIELD ARMY)

Table I.—For an Infantry Division (not including Unit Reserves)

Note.—The strength of the Infantry Division has been taken as 15,000 all ranks. Where numbers are shown the numbers override percentages for exact determination as to what is to be carried.

Serial No.	Items of Anti-gas Stores and Clothing.	Expressed in Units.	Three Months Reserves expressed as a Percentage of Equipment in use.	Location of reserves carried (shown as percentage of equipment in use).				Numbers carried for an infantry division.				Total per Division.
				In 2nd line.	In 3rd line.	In L. of C. Rail-head Coy.	At the Base.	In 2nd line.	In 3rd line.	In L. of C. Rail-head Coy.	At the Base.	
(a)	(b)	(c)	(d)	(e)	(f)	(g)	(h)	(i)	(j)	(k)	(l)	(m)
			Per cent.									
1	Respirators :—											
	Facepieces	Nos.	30	—	—	—	30	—	—	—	—	4,500
	Haversacks..	Nos.	60	—	—	15	45	—	—	2,250	6,750	9,000
	Containers	Nos.	60	—	—	15	45	—	—	2,250	6,750	9,000
	Outfits, anti-dimming ..	Nos.	90	2	2	42	44	270	270	6,200	6,750	13,490
2(*a*)	Eyeshields	pkts. of 6	120	12	12	36	60	1,350	1,350	5,850	9,450	18,000
3	Ointment, anti-gas	2-oz. tins or jars.	150	15	15	45	75	4,464	4,464	13,392	22,320	44,640
4	Cotton waste	lbs.	75	10	10	17	38	112	112	168	336	728
5	Capes, anti-gas	Nos.	50	5	5	15	25	750	750	2,250	3,750	7,500
6	Wallets, anti-gas	Nos.	75	4	4	30	37	560	560	4,500	5,620	11,240
7(*b*)	Detectors, gas, ground ..	pads of 25	75	11	11	33	20	600	600	1,375	1,700	4,275
8(*c*)	Detectors, gas spray (frames) ..	Nos.	15	3	3	3	6	325	325	325	650	1,625
9	Detectors, gas spray (booklets)	Nos.	75	9	9	20	37	1,525	1,525	3,300	6,375	12,725

10	Detector, paper, type "A" ..	pkts.	75	11	11	15	38	2,000	2,000	2,680	6,680	13,360
11	Detector paint (for vehicles) ..	4-oz. tins	75	9	9	20	37	240	240	720	800	2,000
12	Detector paint (for capes and suits).	pint = 2-lb. tin	75	8	8	23	36	38	38	113	188	377
13	Suits, light, anti-gas	Nos.	75	9	9	20	37	450	450	990	1,890	3,780
14(*e*)	Gloves, anti-gas, oilskin ..	pairs	60	7	7	22	24	1,250	1,250	3,750	4,200	10,450
15	Overboots	pairs	75	8	8	23	36	180	180	520	870	1,750
16	Hoods, anti-gas	Nos.	75	8	8	23	36	100	100	300	500	1,000
17	Valises	Nos.	75	8	8	23	36	14	14	42	70	140
18	Ointment, anti-gas, for weapons	2-oz. tins or jars	150	15	15	45	75	720	720	2,160	3,600	7,200
19(*d*)	Bleaching powder (super-tropical)	cwts.	150	10	10	54	76	14	14	72	100	200
20	Signs, warning gas	Nos.	75	10	10	18	37	600	600	1,200	2,400	4,800
21	Alarms, general (sirens) ..	Nos.	12	2	2	4	4	3	3	4	9	19
22	Alarms, local (rattles)	Nos.	12	2	2	2	6	9	9	13	14	45
23	Detector, gas, sleeve	pairs	75	8	8	23	36	2,400	2,400	6,900	10,800	22,500

Notes

(*a*) A further 30 per cent. (to make a total of 150 per cent.) are included in the complete respirators held in decontamination reserve. See Part VIII.

(*b*) These numbers refer to the 6-in. metal frames. When existing stocks of these are used up the 18-in. sheet will be brought into use; the numbers will then be 1/9th of the present quantities.

(*c*) When booklets have become obsolete an increase in the amount of detector paint, of 1 lb. per 250 detectors, gas spray (frames) will be necessary.

(*d*) Five per cent. of the total quantity of bleaching powder will be in 2-lb. tins, the remainder in 7-lb. tins.

(*e*) Ten per cent. of these quantities will be rubber and 90 per cent. oilskin.

APPENDIX XXII—*continued*

Part VIII

SCALE AND LOCATION OF DECONTAMINATION RESERVES

Table I.—For an Infantry Division

Serial No.	Item of Clothing.	Scale of Reserves expressed as a Percentage of Establishment.	Location of Reserve. Numbers expressed as Percentage of Establishment.						Numbers applicable to an Infantry Division.					Total Reserve for a Division.
							At the Base.							
			In 1st line.	In 2nd line.	In 3rd line.	In L. of C. R/H Coy.	For troops in front R/H.	For remainder.	In 1st line.	In 2nd line.	In 3rd line.	In L. of C. R/H Coy.	At the Base.	
(a)	(b)	(c)	(d)	(e)	(f)	(g)	(h)	(i)	(j)	(k)	(l)	(m)	(n)	(o)
1	Respirators, complete (*a*)	30	—	6	6	6	12	18	—	900	900	900	1,800	4,500
2	Sacks, paper, gas-proof	60	—	3	3	24	30	54	—	525	525	4,150	5,200	10,400
		+ 12,000 per div. per month (*b*)	—	—	—	—	—	—	—	—	—	—	36,000	36,000
3	Web equipment for rifle armed soldier	30	2	1	1	2	24	26	300	150	150	300	3,600	4,500
	Additional for specialists	30	2	1	1	2	24	26	100	50	50	100	1,200	1,500
4	Ground Sheets	10	—	2	2	2	4	6	—	300	300	300	600	1,500
5	Blankets	10	—	—	—	4	6	10	—	—	—	600	900	1,500
6	Greatcoats	15	2	1	1	2	9	11	300	150	150	300	1,350	2,250
	Clothing, sets (*f*)	30	4	2	2	2	20	22	600	300	300	300	3,000	4,500
	Consisting of :—													
	B/D trousers (*c*)													
	Anklets (*d*)													
	Boots													
	Socks													
	B/D blouses (*e*)													
	Shirts													
	Vests													
	Drawers													
	Jerseys, pullover													
	Braces													

Notes

(*a*) Includes face-piece, haversack, container Type " E ", outfit anti-dimming, and a packet of eyeshields.
(*b*) These are general reserve but for convenience they have been grouped with the decontamination reserves of this item.
(*c*) Pantaloons in the case of horsed units. (*d*) Puttees in the case of horsed units. (*e*) S.D. jackets, in the case of horsed units.
(*f*) The appropriate article of clothing will be substituted in the case of units mobilized for service in tropical or semi-tropical climates.

APPENDIX XXIII

(*See* Chapter VI)

THE G.S. RESPIRATOR WITH ADDITIONAL TYPE EA CONTAINER

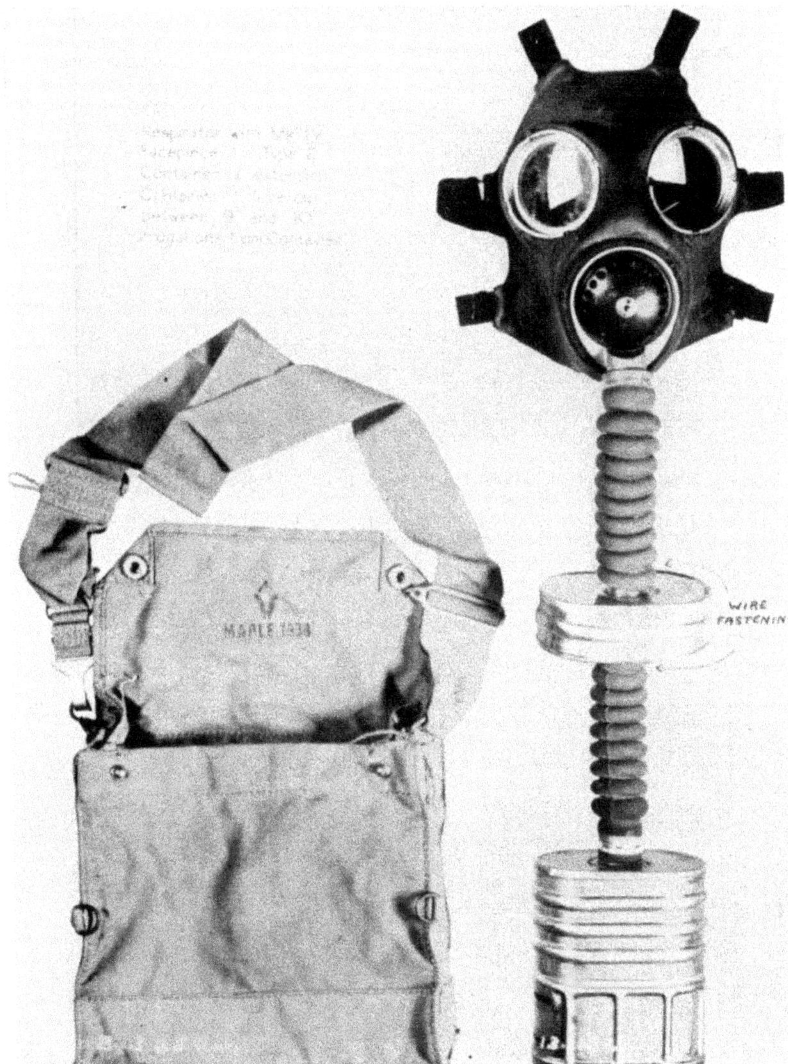

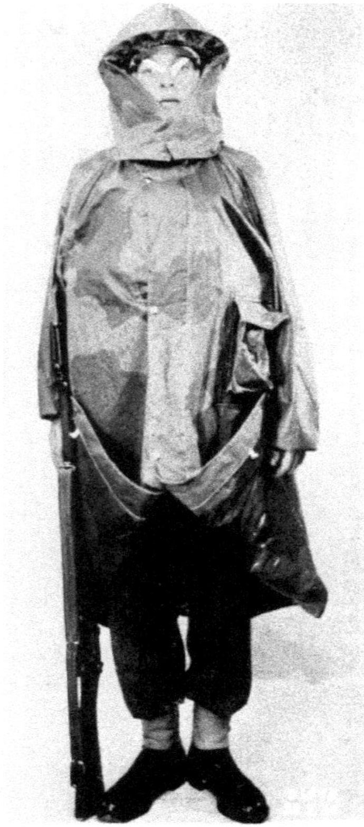

The British Gas Warfare Soldier—1940 model—wearing Eyeshield, Cape, and Curtain, Anti-gas.

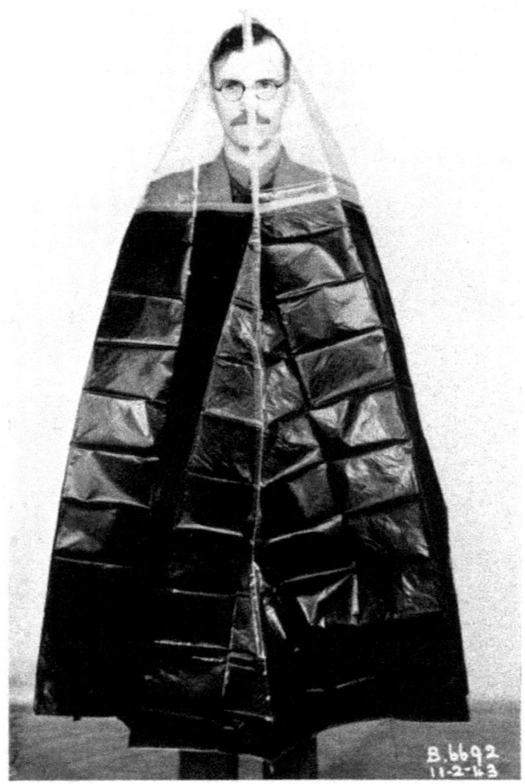

The United States " Cover, Protective, Individual."

The German " Gasplan "

DETECTOR, VAPOUR, POCKET, Mark I

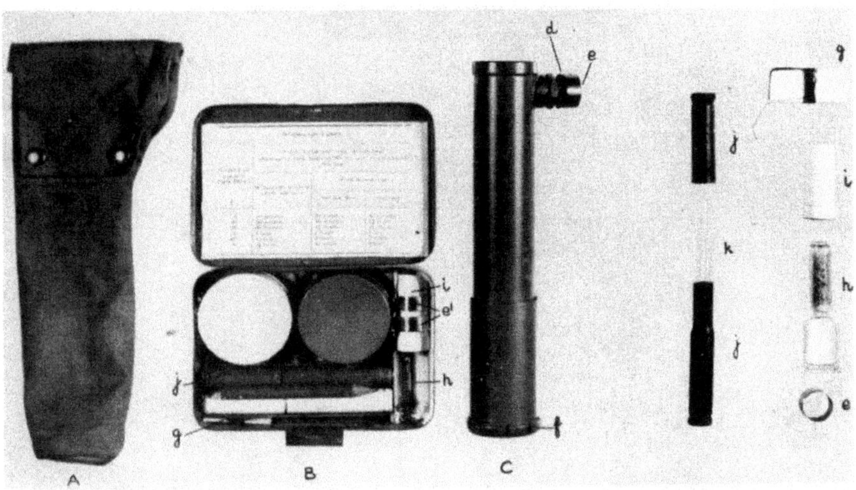

Holster—A Box—B Pump—C Components (withdrawn from box)

(d) Rubber Collar.
(e, e¹) Brass Rings.
(f) Press Cap.
(g) Thread and Oiled Silk.
(h) Guard Tube.
(i) Cover for Guard Tube.
(j) Cardboard Case.
(k) Glass Tubes.

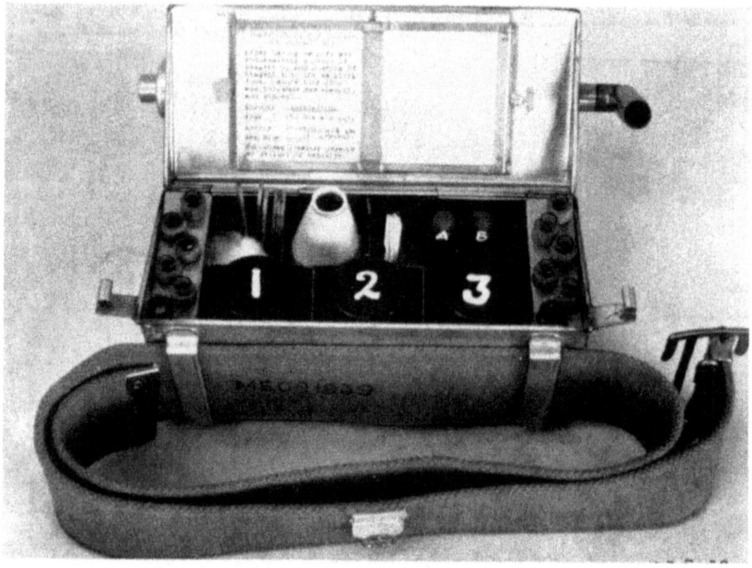

CIVIL GAS IDENTIFICATION OFFICER'S SET

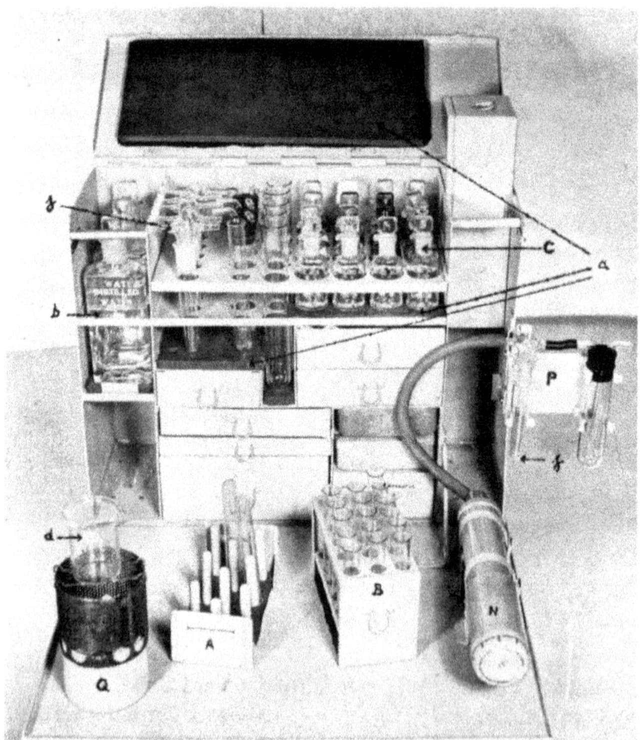

THE CASE, WAR GAS, TESTING
(Issued only to Technical Officers, C.W.)

THE Mk. I/L HORSE RESPIRATOR AND THE COVER, HORSE AND MULE

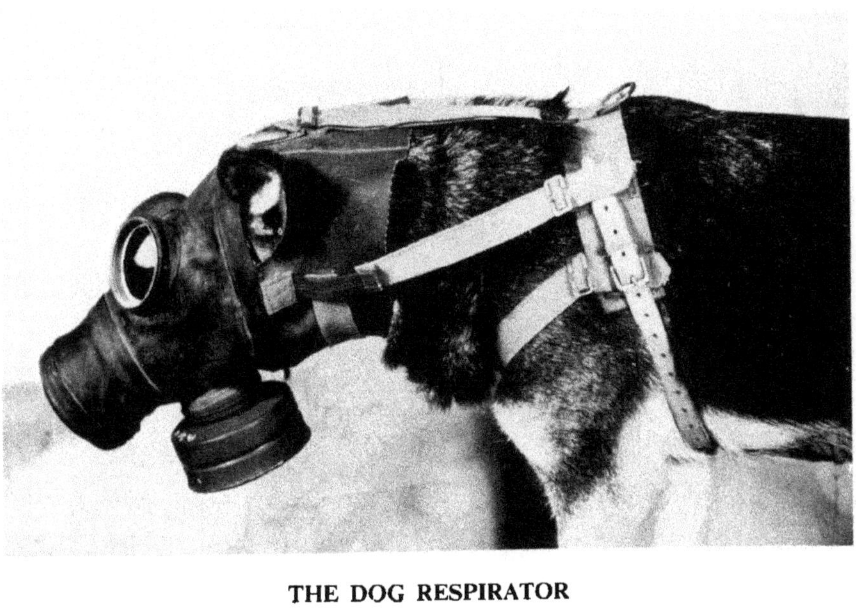

THE DOG RESPIRATOR

Developmental Horse Leggings, showing protection afforded to the heel, and method of fixing in position.

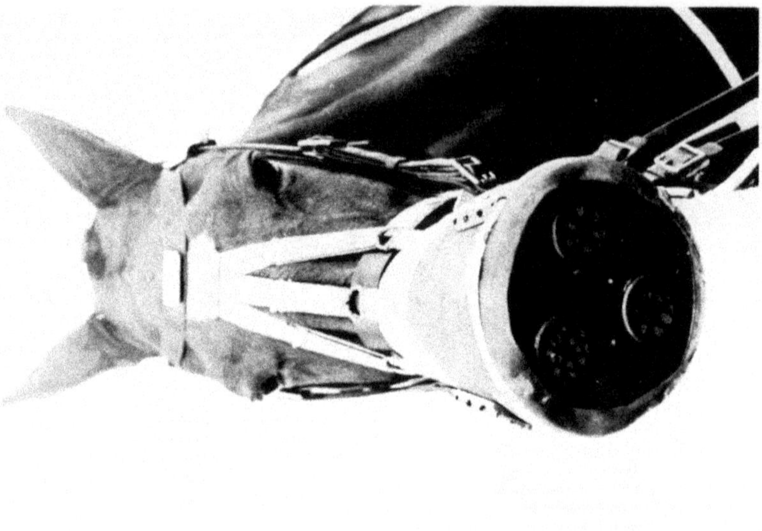

THE Mk. II HORSE OR MULE RESPIRATOR

APPENDIX XXIV

Notes on German Gas Weapons

1. The attached tables show the stocks of gas weapons and bulk stocks of gas uncovered at the end of hostilities. No records are available of stocks captured by the Russians, but there is evidence that large stocks were moved westwards as the Russians advanced and the quantities which fell into the hands of the Western Allies were almost certainly disproportionately great in relation to the territory occupied. Difficulties in making an accurate estimate of the large quantities of munitions involved and discrepancies in reported results were inevitable. The analysis was complicated by the fact that methods of reporting were not uniform, particularly in the case of gas shell where only about one-third of the stocks were classified. It has been assumed that the classified stocks were a reasonable cross section of the whole. That these assumptions were justifiable is indicated from documents recording production figures for war gases which showed a total production tonnage of a similar order to that given in the attached tables.

2. The use of mustard gas and nitrogen mustard gas in large burstered munitions should be noted. Used in this manner for initial cloud effect the skin blistering characteristics of these gases becomes secondary, and the primary effect is on the lungs and breathing tract through inhalation, and on the eyes. The properties of nitrogen mustard gas have been described in Chapter III, Section 3. The Germans considered this 20 to 30 per cent. superior to mustard gas for this purpose. Tabun was, of course, far superior to both in this role and would in time have superseded them.

Their views on the use of tear gas and nose gas were orthodox. The means of discharge was both in H.E./Chemical type and in generator loaded projectiles.

As with us, they regarded phosgene as a charging suitable only for aircraft bombs, but it was in any case superseded by Tabun.

3. Bulk stocks of gas were in proportion of 55 per cent. vesicant, 36 per cent. lung injurant (or, at least, gas employed tactically in a lung injurant role, including 19 per cent. Tabun), and 9 per cent. harassing agents. It is very probable that some proportion of the bulk stock of vesicant liquid was intended for use in munitions with high bursting charges to produce lung injurant effect, which would bring the ratio of (tactically) vesicant and lung injurant supplies nearer to equality.

Only a small but ill-defined proportion of green ring bombs contained phosgene. Apart from the 19 per cent. Tabun, therefore, blister gas forms the basis of the greater part of the tactically lung injurant munitions and, from the manufacturing viewpoint, about two-thirds of the total gas production would in fact possess blistering properties.

4. In addition to the weapons listed in the tables the Germans also possessed spray tanks which could be used for smoke or gas. These were both of the pressure type and the gravity flow type. They considered low spray only to be a useful weapon, but it was strongly opposed by the German Air Force owing to aircraft vulnerability. A development of interest was their experimental work with spraying prussic acid, originally based on captured Russian equipment. Such attacks were visualized from a very low height—25 ft. They concluded that lethal concentrations could be thus set up but their conclusions were based on an erroneously low lethal dosage.

Their bulk contamination vehicles were standard semi-tracked vehicles on which drums and discharge apparatus could be placed. They realized the disadvantages of laying contamination by such means and, with the object of laying it so that it would only become effective on the enemy entering the area, they introduced " Bodenkugeln ", for which large orders were placed as late as September, 1944. These were glass bulbs containing mustard gas which could be laid just under the surface of the ground and would be broken by foot pressure of troops passing over them.

5. V-Weapons. Except for the records of one static trial with a V.1 charged phosgene, no active steps were taken to develop these weapons as gas weapons.

APPENDIX XXIV—*continued*

German Air Gas Weapons

Serial No.	Bomb. (a)	Coloured Ring Marking. (b)	Charging. (c)	Classified Stocks (figures in brackets show percentage by types). (d)	Unclassified Stocks (e)
1	K.C. 250 Gb (*see* Note * below).	One yellow ring	Mustard or Mustard/Arsenol	146,020 (37%)	In addition to stocks in column (d) an unclassified tonnage of aircraft bombs was reported which was calculated to be the equivalent of approximately 13,800 bombs of all types, on the assumption that the classified tonnage was a representative cross-section of the whole stocks.
2	K.C. 250 II Gb	Two yellow rings.	Thickened Mustard or Mustard/Arsenol	36,737 (9·3%)	
3	K.C. 250 Gr (Grün)	One yellow ring and one green ring. Larger burster than Serial 1.	Mustard or Mustard/Arsenol	51,096 (12·9%) (Serials 3 and 4)	
4	K.C. 250 II Gr	One or two green rings	Phosgene		
5	K.C. 250 III Gr	Three green rings	Tabun	102,802 (26%)	
6	K.C. 50 II Bu (Blau)	Two blue rings	Nose gas generator type.	24,300 (6·15%)	
7	K.C. 250 W (Weiss)	Two white rings	Tear gas (CAP)	31,778 (8·05%)	
8	K.C. 500 W	Two white rings	Tear gas (CAP)	2,431 (0·6%)	
			Total	395,164	13,800
				408,964	

* Kampstoff Cylinderische 250 Kg. Gelbring.

APPENDIX XXIV—*continued*

German Gas Shell, Rockets, Mortar Bombs, Mines and Generators

Serial No	Shell.	Marking.		Charging and Type.	Classified Stocks.	Unclassified Stocks.
		Coloured Rings.	Letters and Arabic numerals in same Colour as Rings.			
	(a)	(b)	(c)	(d)	(e)	(f)
1	7·5 cm. (2·95-in. Light Field Gun) Weissring.	One white ring		Tear Gas (C.A.P.)—H.E./chemical type.	84,970 (3·5%)	Unclassified Tonnage calculated in a similar manner as for Air Gas Weapons.
2	10 cm. Gelbring	One yellow ring		Mustard Gas	506,400 (serials 2–4)	
3	10·5 cm. (4·14-in. Gun Howitzer) Gelbring.	One yellow ring	Gb GB	Mustard/Arsenöl, for ground contamination.		
4	10·5 cm. (4·14-in. Gun Howitzer) 39 Gelbring.	One yellow ring	B 39	Mustard/Arsenöl, for ground contamination.		
5	10·5 cm. (4·14-in. Gun Howitzer) Grünring.	One green ring	L/O	Mustard/Arsenöl } H.E./chemical with large burster for initial cloud effect. (serials 5–6)	257,600 (serials 5–6)	
6	10·5 cm. (4·14-in. Gun Howitzer) Grünring-Gelb.	One green and one yellow ring.	o 38 in green	Mustard/Arsenöl		
7	10·5 cm. (4·14-in. Gun Howitzer) 38 Grünring. 1.	One green ring	K 38	Nitrogen Mustard Gas (HN3)—H.E./chemical with large burster for initial cloud effect.	212,972	
8	10·5 cm. (4·14-in. Gun Howitzer) 39 Grünring. 3.	One green ring	3 39	Tabun—H.E./chemical with large burster for initial cloud effect.	443,350	

(70%) — serials 2–8, Classified Stocks

9	10·5 cm. (4·14-in. Gun Howitzer) Blauring. 1.	One blue ring	1	Nose Gas—H.E./chemical type	215,850 (Nos. 9–10)		
10	10·5 cm. (4·14-in. Gun Howitzer) 40 Blauring. 3.	One blue ring	3	Nose Gas—Base ejection generator type.			
11	10·5 cm. (4·14-in. Gun Howitzer) Weissring.	One white ring		Tear Gas (C.A.P.)—H.E./chemical type.	123,300		
12	15 cm. (5·91-in. Howitzer) Gelbring	One yellow ring	Gb GB	Mustard/Arsenöl, for ground contamination.	37,317 (Nos. 12–13)	(26·5%) (Nos. 12–18)	
13	15 cm. (5·91-in. Howitzer) 39 Gelbring.	One yellow ring	B 39	Mustard/Arsenöl, for ground contamination.			
14	15 cm. (5·91-in. Howitzer) Grünring.	One green ring	L/O	Mustard Gas—H.E./chemical type	238,400		
15	15 cm. (5·91-in. Howitzer) 38 Grünring. 1..	One green ring	K 38	Nitrogen Mustard Gas (HN3)—H.E./chemical type.	75,700		
16	15 cm. (5·91-in. Howitzer) 39 Grünring. 3.	One green ring	3 39	Tabun—H.E./chemical type	153,779		
17	15 cm. (5·91-in. Howitzer) Blauring 1.	One blue ring	1	Nose Gas—H.E./chemical type	76,261		
18	15 cm. (5·91-in. Howitzer) Weissring.	One white ring		Tear Gas (C.A.P.)—H.E./chemical type.	85,408		
				Total ..	2,511,207		4,800,000
					Approximately 7¼ millions.		
	15-cm. Rockets (*For Nebelwerfer 41—see Appendix XXV*)						
19	Gelbring	One yellow ring	B	Mustard/Arsenöl	28,800		
20	Grünring-gelb	One green and one yellow ring.	O 38 in green	Mustard Gas—H.E./chemical type	27,400		
21	Grünring. 1	One green ring	K 38	Nitrogen Mustard Gas—H.E./chemical type.	32,700		
				Total ..	88,900		

APPENDIX XXIV—*continued*

German Gas Shell, Rockets, Mortar Bombs, Mines and Generators—*continued*

Serial No.	Shell.	Marking.		Charging and Type.	Classified Stocks.	Unclassified Stocks.
		Coloured Rings.	Letters and Arabic numerals in same Colour as Rings			
	(a)	(b)	(c)	(d)	(e)	(f)
	10-cm. Mortar Bombs (*Standard Infantry Mortar*)					
22	Gelbring ..	One yellow ring	G/B	Mustard/Arsenöl	206,300	
	Chemical Mines					
23	10·1 Sp. Bü 37 Gelbring	One yellow ring		Two gallons Mustard Gas	38,700	
24	10·1 Sp. Bü 37 Doppel Gelbring ..	Two yellow rings	..	Thickened Mustard	140,450	
				Total ..	179,150	
	(Spruh Büchson—spraying mines—on being functioned projected an inner container into the air which burst 10 to 20 feet above the ground. They could also be ground burst.)					
25	Toxic Generators			Nose Gas	14,000	

APPENDIX XXIV—*continued*

Total Gas Tonnage including Gas charged into Weapons and Bulk Stocks

		Mustard and Mustard/Arsenöl.	Arsenöl.	Lung Injurant.	Nitrogen Mustard.	Tabun.	Nose Gas.	Tear Gas.	Total.
Aircraft Bombs ..	Tons	19,950	—	5,850	—	9,600	210	3,900	39,510
Shell	Tons	2,960	—	4,465	1,251	3,750	710	750	13,886
Mines	Tons	2,119	—	—	—	—	—	—	2,119
Mortar Bombs ..	Tons	310	—	—	—	—	—	—	310
Rockets	Tons	215	—	—	111	—	—	—	326
Toxic Generators ..	Tons	—	—	—	—	—	62	—	62
Bulk Stocks	Tons	12,150	480	—	50	—	741	—	13,421
		37,704	480	10,315	1,412	13,350	1,723	4,650	69,634

APPENDIX XXV

(*See* page 243.)

History of Development to "Werfer Brigades"

The following information on the development of the Smoke Arm (Nebe Truppe) was furnished by Oberst Hirsch of Wa.Prüf. 9.

1932–35.—In 1932 a " Smoke Squadron " was organized in Königsbrück, equipped with smoke spray apparatus and generators. This squadron was converted to a Battery (Nebel Abteilung) in 1934. The equipment remained the same but was considerably increased and the battery was " motorised ". In 1935, AHA/In.9 was organized, an Inspector of Smoke Troops was appointed and the Smoke Arm (nebel Truppe) was created, although initially it consisted only of the Nebel Abt. at Königsbrück and a detached instructional company.

1936–37.—The Smoke Arm was increased to two batteries, the first remaining in Königsbrück and the new battery being formed in Bremen. Each battery consisted of three platoons (troops), equipped with the new 10-cm. Smoke Mortar (Nebelwerfer 35), two sections each of four mortars being allotted to each platoon. The Nb.W. 35 had a maximum range of 3,500 metres with a maximum rate of fire for short periods of 15 rounds per minute. The charging for the mortar bombs was at first a mixture of SO_3 and pumice but this was later changed to SO_3 only. Results were excellent in humid weather conditions and still adequate in dry weather. In October, 1937, a third battery was formed at Celle.

1938–39.—In 1938 two further batteries were formed, at Munsingen and Kaiserslautern. Up to this time these Nebelwerfer batteries had, in addition to their mortar equipment, a decontamination and bulk contamination vehicle, so that they could be used as smoke troops, decontamination troops or gas contamination troops as required. Each battery consisted of an H.Q., a signal section, three platoons or troops, a light column for ammunition supply, and a heavy column of all types of vehicles for decontamination, bulk contamination, transport of decontaminants and gas supplies.

In manœuvres in 1938 this organization was shown to be too clumsy. Accordingly, in 1939, the smoke troops were reorganized into smoke mortar batteries and decontamination batteries, five of each being formed in the course of the year. The organization of these batteries was as follows :—

Nebelwerfer-abteilung (*Smoke Mortar Battery*)	*Entgiftungs-abteilung* (*Decontamination Battery*)
H.Q.	H.Q.
Signal Section (wireless and telephone on overland transport).	Signal Section (as in Mortar battery).
3 Platoons or troops, each of two sections of 4 (later 3) 10-cm. NbW. 35 mortars.	3 Platoons (equipped with medium and small decontamination vehicles and (tracked) gas identification vehicle).
Light column (for ammo. supply)	Light column (for transport of decontaminant).
	Heavy column (transport vehicles for gas supplies and bulk contamination vehicle—Medium Sprün KW).

Several smoke mortar batteries and one decontamination battery took part in the campaign in Poland, although the latter was not required for active duty.

1940.—Several smoke mortar batteries were used in the spring offensive in France, where they proved of particular value during river crossings. The decontamination batteries were never used for the purpose for which they were trained. Mortar batteries were also sent to Norway.

The Nebelwerfer 40, a breech loaded mortar with a range of 6,500 metres, was introduced in this year. On account of its weight and angle of fire it was never popular with the troops and in a short time it was withdrawn.

Towards the end of this year, a special Smoke-regiment staff was formed, so that smoke-mortar batteries and decontamination batteries could be grouped together in one regiment, for the sake of easier and more concerted control. For this purpose two smoke-mortar batteries and one decontamination battery were linked to form the smoke regiment.

In the autumn, an instructional battery was formed in Poland to train smoke-mortar and decontamination batteries in the use of the heavy projector 40 (Wurfgerät 40). This appliance consisted of a wooden frame which could be anchored firmly to the ground and adjusted at will to shoot at any angle from 20° to 45°. The ammunition, a 28-cm. H.E. or a 32-cm. incendiary projectile, was transported in packing cases which could be bolted firmly on to the wooden frames, to serve as a species of mortar barrel from which the projectile could be fired. The range for H.E. was 2,200 metres and for incendiary 1,950 metres. Each frame accommodated four projectiles, which could be fired electrically in rapid succession (6–8 seconds). Each battery had 48 such frames and could thus fire 192 projectiles in 8 seconds. The scatter was considerable and the weapon was therefore only suitable for area shoots as an assault weapon (Schwerpunktwaffe).

1941.—At the beginning of this year it was possible, with the formation of the special regimental staffs, to organize five regiments of three batteries by utilizing the nine smoke-mortar batteries, five decontamination batteries and one Wurfgerät instructional battery.

Smoke-mortar batteries were used in the spring campaign in the Balkans with success and great use was made of these units during the campaign in Russia in mid-summer, particularly in the crossing of the Bug and in the attack on Brest Litowsk in which nine batteries of Wurfgerät were employed.

In August, 1941, the five decontamination batteries were finally equipped with the heavy Wurfgerät 40 and in September their decontamination equipment was handed over to reserve troops. Towards the end of 1941 the wooden frame of the Wurfgerät was replaced by an iron frame with an improved anchoring device. This apparatus was known as the Heavy Projector 41 (Schweres Wurfgerät 41).

To make up for the loss of the decontamination batteries, three road decontamination batteries were formed in the autumn of 1941. They were equipped with motor vehicles on which could be fitted an improvised bleach spreader, but which were used essentially for the transport of the bleach in drums. The chief purpose of the battery was to deal with decontamination of transport immediately behind the front line.

1942.—At the beginning of this year the Smoke Arm comprised :—

Five Nebelwerfer regiments, equipped with 15-cm. Nebelwerfer.
Two Heavy Werfer regiments, equipped with Wurfgerät or their successors (*see* comment below).
One Smoke Instruction regiment.
One Mountain Werfer battery.
Three road decontamination batteries.

(*Comment*.—Hirsch does not make any previous mention of the equipment of Smoke Troops with 15-cm. Nebelwerfer 41, *i.e.*, 6-barrel 15-cm. rocket projectors. It is believed that they were used on the Russian front in 1941 and it is possible that some of the nine batteries of Wurfgerät which he says were employed against Brest Litowsk in fact included some Nebelwerfer 41 batteries, particularly since he says only two Heavy Werfer regiments existed in 1942.)

The regiments were usually composed of three batteries but one Heavy Werfer regiment had only two batteries. For firing the 28/32-cm. ammunition an apparatus (the 28/32-cm. Nebelwerfer 41) for firing six rounds was introduced in place of the Wurfgerät 41. In addition a 32-cm. Nebelwerfer with a range of 5,600 metres was developed ultimately to replace the 28/32 Werfer. (Possibly by this Hirsch meant the 30-cm. Nebelwerfer 42 but the range is rather overstated.)

For armoured pioneer companies another device (the 28/32 Schweres Wurfrahman 40) was developed, comprising a frame permanently fitted to the side of M.T., adjustable for angle of fire and capable of firing electrically from their cases six 28 cm. H.E. or 32-cm. incendiary projectiles. In this year also the 15-cm. Panzer Werfer 42 was developed. This appliance consisted of ten 15-cm. tubes on a pivoted base, which could be fitted to armoured vehicles, providing a remarkable rate of fire.

Towards the end of 1942 a brigade staff was organized for the Smoke Arm and used in the larger battles.

1943–44.—At the beginning of 1943, on account of the scarcity of Werfer transport and high demands for Werfer batteries, all the special vehicles for decontamination and bulk contamination were changed over to Werfer vehicles. In this year Panzer Werfer troops were organized, each troop having four to six A.F.Vs. equipped with the 10-tube projector. Organization of the brigade staff was completed and Werfer brigades formed, each consisting of one light and one heavy regiment.

By the spring of 1944 four Werfer brigades existed and the numbers increased to eight during the year with a ninth brigade in formation. The equipment of these brigades was 15-cm. Nebelwerfer 41 for the light batteries and 21-cm. and 30-cm. Nebelwerfer 42 for the heavy batteries. The road decontamination batteries were disbanded and the personnel transferred to Werfer batteries. The Mountain Werfer battery was still maintained.

1945.—Three further brigades should have been formed during this year but at the cessation of hostilities only the staff had been formed. Consideration was being given to the possibility of equipping the Werfer regiments with 35-cm. projectors and later with even larger calibres but improvements in the design of projector and in the ballistics of the ammunition were considered necessary to effect a reduction in scatter of the rounds.

(*Comment.*—It will be noted from the foregoing that between 1937 and 1945 the Smoke Arm developed from 3 batteries to 9 brigades of about 54 batteries. This growth is summarized in the following table :—

	1935	1936	1937	1938	1939	1940	1941	1942	1943	1944	1945
Nebelabteilung (Smoke batteries)	1	2	3	5	6	7	(10)	(20)	(21)	(48)	(54)
Decontamination batteries..					5	5	(5)				
Road decontam. batteries ..							3	3	3		
Werfer Regt.							5	7	(7)	(16)	(18)
Brigades									3	8	9
Mountain battery								1	1	1	1
Werfer Replacement Regt...								1	1	3	3

It will be observed, however, that the character of the Arm also changed considerably, beginning as a small force combining the function of smoke screening, decontamination and contamination units, passing through a phase when smoke screening and assault duties and decontamination/contamination duties were catered for by separate units, and emerging finally as a series of units of considerable strength, specifically equipped for delivering massive attacks on area targets with H.E. or incendiary weapons. Despite this marked change in their actual role, it is clear that, had gas warfare been started, the organization and arming of these units was such that they could equally well have delivered massive attacks with gas. In fact, however, the quantity of gas ammunition suitable for use with their weapons was quite inadequate at the time when hostilities ceased. Papers in Wa.Prüf. development files show that there was an OKH requirement for gas charged 30 cm. rockets, but the designs had not been completed in 1944 and none had been charged.)

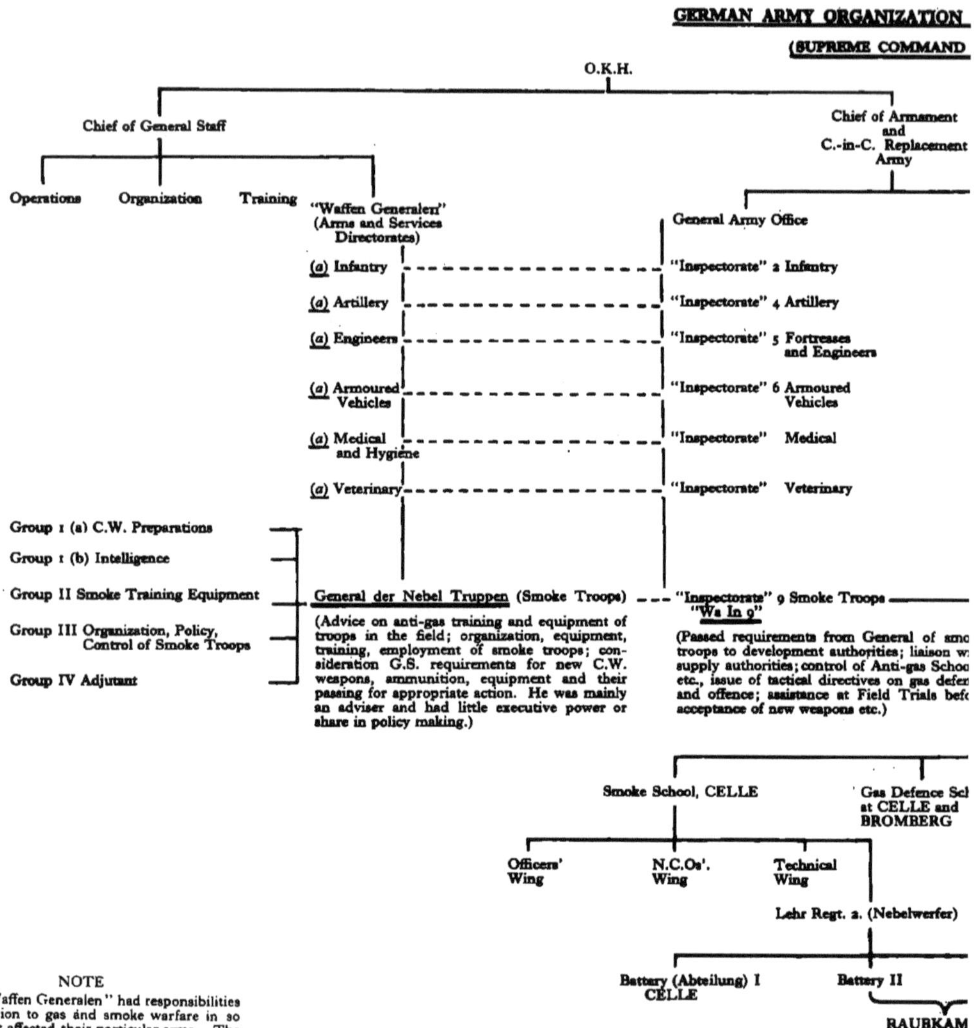

NOTE

(a) All "Waffen Generalen" had responsibilities in relation to gas and smoke warfare in so far as it affected their particular arms. The same applies to the parallel "Inspectorates".

To face page 326

FOR GAS AND SMOKE WARFARE
OF THE GERMAN ARMY)

APPENDIX XXVI
(See page 243)

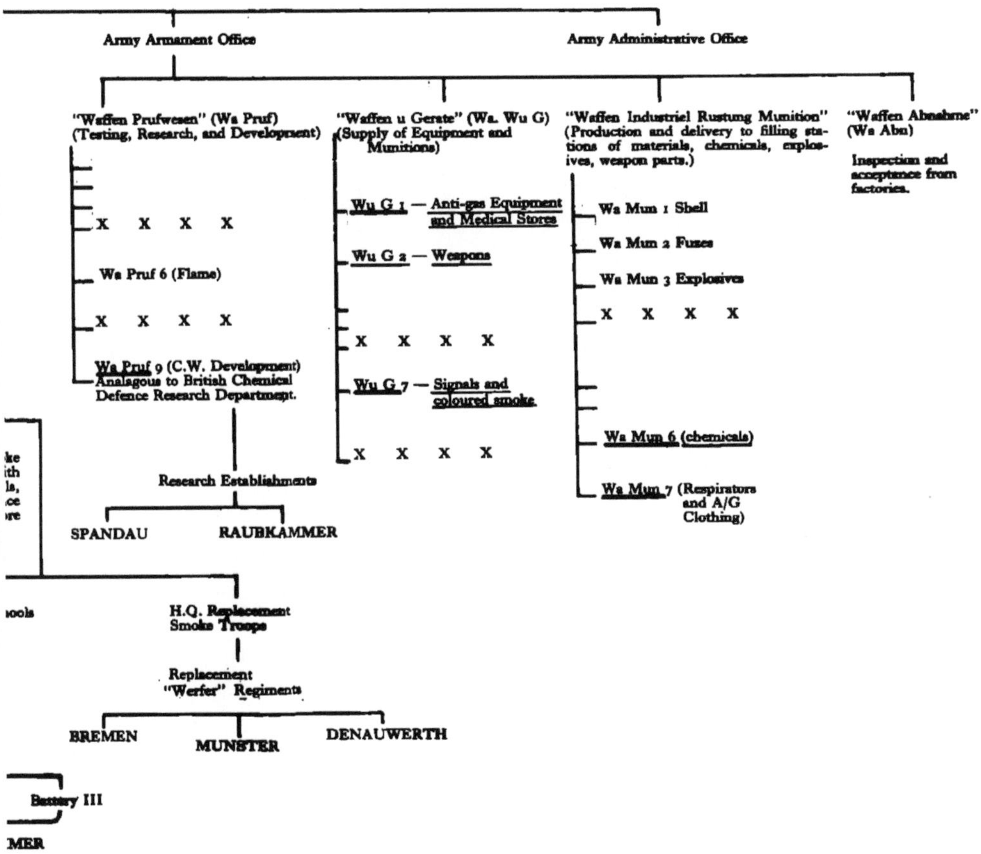

BIBLIOGRAPHY

(For the Second World War)

Note 1.—Pamphlets, etc., which are not wholly concerned with gas warfare are obvious from the title.

Note 2.—Manuals current at the end of the war are starred.

Training

*Gas Training, 1942.

*Advanced Gas Training, 1943.

*Regulations for P.A.D. Organization, 1944.

Protection against Gas and Air Raids, Pamphlet No. 1.—Protection against Gas in the Field, 1939.

Protection against Gas and Air Raids, Pamphlet No. 2.—Respirators, 1939.

Protection against Gas and Air Raids, Pamphlet No. 3.—Passive Air Defence, 1939.

Specimen Standing Orders for Protection against Gas and Air Raids in Billets, Camps and Bivouacs for all H.Q. and Units (1939).

Defence against Gas, 1935.

Tactical Notes on Defence against Gas, 1934.

Defence against Gas, 1927.

Provisional Memorandum, The Employment of Gas (1938).

*Royal Army Service Corps Training Pamphlet No. 7. 1942 :—

Part III.—Distribution of Blankets, Anti-gas Equipment, and Reserve Clothing in advance of Railhead.

*Royal Army Service Corps Training Pamphlet No. 19. 1942 :—

Protection of Army Food Supplies against Gas.

*Military Training Pamphlet No. 32.—The Tactical and Technical Employment of Chemical Weapons :—

Part I.—Chemical Warfare Units—Organization, Employment and Training (1940).

Part II.—The Projector (1940—revised edition, 1941).

Part III.—The Cylinder (1940).

Part IV.—The Bulk Contamination Vehicle (1940).

Part V.—The Chemical Mine (1940).

Part VI.—Bombs, Ground, 6-lb. (1940).

Artillery Training Volume II. Gunnery, 1934. Supplement No. 2.—1940.—Programme Shoots.—Employment of Base Ejection Smoke and Gas Shell, etc.

*Artillery Training Volume III. Pamphlet No. 5.—The Employment of Base Ejection Smoke and Chemical Shell (1943).

*Military Training Pamphlet No. 5.—The Use of Gas in the Field, 1943.

Military Training Pamphlet No. 23.—Part V. The Use of Gas in the Field, 1940.

*Military Training Pamphlet No. 66.—Anti-gas Measures. (Railways and Ports) 1944.

Intelligence

*New Notes on German Chemical Warfare (M.I. 10), May, 1943.

New Notes on the German Army. No. 2.—Chemical Warfare and Smoke, July, 1942.

*Notes on Italian Chemical Warfare (M.I. 10), July, 1941.

*Notes on Japanese Chemical Warfare (M.I. 10), June, 1942.

Medical

*Medical Manual of Chemical Warfare (1940), (1941), (1943).
*Purification of Water Supplies in the Field.
Water Supplies in the Field, 1941. (Published by Director of Hygiene, War Office.)

Ordnance, Transportation, etc.

*R.A.O.S., Part 8. Pamphlet No. 26.—Respirators—Instructions for Examination, Repair, Sentencing, Replacement, etc. (1941), (1942).
*R.A.O.S., Part 8. Pamphlet No. 26.—Sub-pamphlet No. 1.—Respirators, Anti-Gas, Horse and Mule, M.K. 1/L (1941).
*R.A.O.S., Part 8. Pamphlet No. 26.—Sub-pamphlet No. 2.—Respirators, Anti-Gas, Light (1943).
*R.A.O.S., Part 8. Pamphlet No. 29.—Instructions for the Examination, Repair and Sentencing of Anti-Gas Clothing (1942).
Decontamination of Clothing and Equipment Procedure (Issued by D.O.S., 1941—54/Misc./4659).
*R.A.O.S., Part 8. Pamphlet No. 31.—Instructions for the Decontamination of Clothing and Equipment.
*R.A.O.S., Part 8. Pamphlet No. 32.—Instructions for the Operation of a Mobile Laundry and Bath Unit (Type A), R.A.O.C. (1942).
*R.A.O.S., Part 8. Pamphlet No. 33.—Instructions for the Operation of a Base Laundry, R.A.O.C. (1942).
*R.A.O.S., Part 8. Pamphlet No. 35.—Instructions for the Operation of a Mobile Laundry and Bath Unit (Type B), R.A.O.C. (1942).
*R.A.O.S., Part 6. Supply of Ammunition in the Field (1944).
*D.W.S. Notes on Ammunition (Chemical) No. 7A (1943) (superseded No. 7 and preceded R.A.O.S., Part 7. Pamphlet No. 7 (1946)).
D.W.S. Notes on Ammunition. Issue No. 20. The Storage of Ammunition and Explosives in the Field (1942).
*B.510—Regulations for the Storage and Transport, Maintenance and Disposal of Chemical Weapons (1941).
*Notes on the Storage and Transport of C.W. Weapons and Chemical Defence Preparation (1944). (Published by the Explosives Storage and Transport Committee.)
*Special Regulations for the Conveyance of Government Explosives in the Present Emergency (1941), (1943). (Published by the Explosives Storage and Transport Committee.)
*Notes on the Transport of Explosives by Land in Great Britain (1941). (Published by the Explosives Storage and Transport Committee.)
*Instructions as to the Conveyance of Explosive, Inflammable Liquid, Corrosive and Poisonous Chemicals, Compressed or Liquefied Gases and other Dangerous Goods by Rail, Road or Water on behalf of H.M.'s Government (1939), (1944). (Published by Railway Executive Committee.)
*Instructions to Surveyors and Notice to Ship Owners, Ship Masters and Skippers on the Conditions of Stowage of Explosives and of Dangerous and certain other Goods being Shipped. (Circular 1768, Ministry of War Transport.)

Ordnance, Transportation, etc.—continued.

*T.152—Conveyance of Government Explosives in Freight Ships during the Present Emergency (1941 and revised annually to 1945). (Ministry of War Transport.)

*Notes on the care and preservation of Ammunition and Explosives in the Field. Part II. (G.S. Publication.)

Meteorology

Notes on Meteorology, 1940.

*S.D. 216.—The Meteorology of Gas and Smoke. (Air Ministry.)

*M.O. 247 (A.P. 850).—A Short Course in Elementary Meteorology by W. A. Pick, B.Sc., F.C.P., F.Imt.P. (Air Ministry.)

Miscellaneous and Technical

*The Chemical Warfare Pocket Book (1940), (1942).

*Field Service Pocket Book, Pamphlet No. 10.—Gas.

*Chemical Warfare and Animals.

*Animal Management—Symptoms and Treatment—R.A.V.C. (1945).

*B.498.—Scale of Anti-Gas Equipment and Clothing for all Units (1941).

*United States War Department, Pamphlet No 3.—1. Comparison of U.S. and British Chemical Warfare Offensive Equipment (1944).

*United States War Department, Pamphlet No. 3.—2. Comparison of U.S. and British Chemical Warfare Protective Equipment (1944).

*United States War Department Field Manual, FM 101–10.—Staff Officers Field Manual, Organization, Technical and Logistical Data (includes data on the Chemical Warfare Service).

*1 Canadian Chemical Warfare Laboratory—Report No. 44, German Chemical Ammunition (1945).

*Project Co-ordination Staff, Edgewood Arsenal, Md.—Report No. 9.—Résumé of recent knowledge on the Technical aspects of Chemical Warfare (1945). (Republished as "Technical Aspects of Chemical Warfare in the Field," Parts I and II, 1946.)

*Gas Training (India), 1943.

*Advanced Gas Training (India) 1944.

*Handbook of Chemical Weapon (India) (1st and 2nd Editions).

**Air Ministry*

A.P. 1510.	R.A.F. Manual, Defence Against Gas, Volumes I and II.
B.C. 46.	The Use of Gas from the Air.
S.D. 153.	The Use of Gas in Air Warfare.
S.D. 357.	Manual of Chemical Warfare, Volumes I and II.
C.D. 206.	Notes on the Treatment and Handling of Unexploded Enemy Bombs.
S.D. 369.	Bomb, A/C L.C., 30-lb., Marks I and II.
S.D. 370.	Bomb, A/C L.C., 250-lb., Mark II.
S.D. 371.	Bomb, A/C L.C., 65-lb., Marks I and II.
S.D. 478.	Installation S.C. 400-lb., Type S.L., Mark I.
S.D. 366.	Installation S.C., 250-lb., Type S.G., Mark IV.
S.D. 367.	Installation S.C., 500-lb., Type S.G., Mark III.
C.D. 276.	Cluster Projectiles.

SECRET

**Ministry of Home Security*

A.R.P. Handbook No. 1.	Personal Protection against Gas.
A.R.P. Handbook No. 2.	First Aid and Nursing for Gas Casualties.
A.R.P. Handbook No. 3.	Medical Treatment of Gas Casualties.
A.R.P. Handbook No. 4.	Decontamination of Materials—Decontamination of Streets, Buildings and their Contents, Vehicles and Plant.
A.R.P. Handbook No. 4a.	Decontamination of Clothing and Equipment.
A.R.P. Handbook No. 5.	Structural Defence.
A.R.P. Handbook No. 5a.	Bomb Resisting Shelters.
A.R.P. Handbook No. 6.	Air Raid Precautions in Factories and Business Premises.
A.R.P. Handbook No. 7.	Anti-Gas Precautions for Merchant Shipping.
A.R.P. Handbook No. 8.	The Duties of Air Raid Wardens.
A.R.P. Handbook No. 9.	Incendiary Bombs and Fire Precautions.
A.R.P. Handbook No. 10.	Training and Work of First Aid Parties.
A.R.P. Handbook No. 12.	Air Raid Precautions for Animals.
Memoranda. No. 3.	Organization of Decontamination Services.
Memoranda. No. 5.	Anti-Gas Training.
Memoranda. No. 11.	Gas Detection and Identification Service.
Memoranda. No. 13.	Inspection and Repair of Respirators and Oilskin Clothing.
Pamphlet.	Anti-Gas Protection of Babies and Young Children.
Pamphlet.	Atlas of Gas Poisoning.
Pamphlet.	Protection of Foodstuffs against Poison Gas.
Pamphlet.	The Detection and Identification of War Gases.
Pamphlet.	S.D. Methods for Detection of Mustard Gas Vapour.

A.R.P. Training Manual, No. 3.—Rescue Service Manual.
Training Pamphlet No. 4.—Notes on Gas Tests and Exercises.

INDEX

(C29058) N**

Page

(C29058) N** 2

(C29058) 375 7/51

www.ingramcontent.com/pod-product-compliance
Ingram Content Group UK Ltd.
Pitfield, Milton Keynes, MK11 3LW, UK
UKHW051118290926
15471UKWH00010B/1748

9 781783 313655